Mach 5 and Beyond

Hypersonic Warfare, Missile Technology and Global Strategic Implications

Richard Skiba

After Midnight Publishing

This publication is designed to provide accurate and authoritative information in regard to the subject matter covered. While the publisher and author have used their best efforts in preparing this book, they make no representations or warranties with respect to the accuracy or completeness of the contents of this book and specifically disclaim any implied warranties of merchantability or fitness for a particular purpose. No warranty may be created or extended by sales representatives or written sales materials. The advice and strategies contained herein may not be suitable for your situation. You should consult with a professional when appropriate. Neither the publisher nor the author shall be liable for any loss of profit or any other commercial damages, including but not limited to special, incidental, consequential, personal, or other damages.

Skiba, Richard (author)

Mach 5 and Beyond: Hypersonic Warfare, Missile Technology and Global Strategic Implications

ISBN 978-1-7637173-7-4 (Paperback) 978-1-7637173-8-1 (eBook) 978-1-7637173-9-8 (Hardcover)

Non-fiction

Contents

INTRODUCTION

The term "hypersonic missile" does not have a universally accepted definition, but it generally refers to missiles capable of traveling at speeds greater than Mach 5—five times the speed of sound—while maintaining manoeuvrability throughout their flight. Unlike ballistic missiles, which follow predictable, parabolic trajectories, hypersonic missiles can alter their flight path during travel, making them difficult to detect, track, and intercept. These missiles operate within Earth's atmosphere for most of their flight, where they are capable of performing sharp, evasive manoeuvres that make them a challenge for existing defence systems [1].

Several countries are heavily investing in the development of hypersonic missile technology. Russia and China are reported to have already deployed hypersonic missiles, positioning themselves at the forefront of this military technology. The United States has multiple ongoing hypersonic weapons programs and continues to conduct tests aimed at bringing these systems into active deployment. The United Kingdom does not currently possess hypersonic missile capabilities, but it has announced plans to develop future hypersonic concepts and technologies. In 2022, the UK, US, and Australia committed to joint development of hypersonic and counter-hypersonic capabilities through the AUKUS security partnership, underscoring the importance of hypersonic technology in modern defence strategies [1].

The missions for which hypersonic missiles may be used depend on various factors, including the missile's range, speed, warhead type, and launch platform. Potential applications include the ability to strike high-value, time-sensitive targets quickly and with precision, such as mobile military units or critical infrastructure. Hypersonic missiles may also be employed for long-range precision strikes, enabling militaries to attack distant targets rapidly while minimizing the risk of interception. Another possible role for hypersonic missiles is in nuclear deterrence, although the effectiveness of hypersonic weapons in this context remains debated among experts. Some believe that their extreme speed and manoeuvrability make them ideal for delivering nuclear payloads, while others question whether they add significant value over existing nuclear delivery systems, such as ballistic missiles [1].

One of the most important implications of hypersonic missiles is their potential impact on missile defence systems. Current missile defence systems, particularly those designed to counter intercontinental ballistic missiles (ICBMs), are not optimized to defend against the unique flight characteristics of hypersonic weapons. Hypersonic missiles fly at lower altitudes than ballistic missiles, which makes them harder to detect with long-range radar systems. Their manoeuvrability allows them to change trajectory mid-flight, further complicating interception. This combination of speed, altitude, and manoeuvrability poses a significant challenge for missile defence, which typically relies on early detection and prediction of a missile's trajectory to intercept it. Defence systems would need to be redesigned or significantly upgraded to counter hypersonic threats, likely requiring considerable investment in research and development [1].

There are two main types of hypersonic missiles: Hypersonic Glide Vehicles (HGVs) and Hypersonic Cruise Missiles (HCMs). Each type operates differently but shares the same underlying characteristics of high speed and manoeuvrability. HGVs are typically mounted onto rocket boosters, which accelerate them to hypersonic speeds. Once the HGV reaches its desired altitude—often between 30 to 80 kilometres—it detaches from the booster and glides unpowered toward its target at high speed. These glide vehicles can reach speeds of Mach 20 or more during their descent, making them nearly impossible to intercept with current defence technologies [1].

On the other hand, HCMs are powered throughout their flight by ramjet or scramjet engines, which enable them to maintain hypersonic speeds at altitudes ranging between 20 and 40 kilometres. The propulsion systems in HCMs allow for sustained flight at high speeds, making them effective for long-range missions and difficult to detect in time for interception [1].

Russia and China have reportedly deployed hypersonic missiles that are capable of carrying either conventional or nuclear warheads, highlighting the potential for these weapons to be used in both tactical and strategic contexts. The United States is testing multiple hypersonic technologies as part of its ongoing effort to catch up in the global hypersonic arms race. Meanwhile, the AUKUS agreement between the UK, US, and Australia reflects a commitment to developing not only hypersonic capabilities but also counter-hypersonic technologies, which are seen as essential for maintaining military balance in the future [1].

The development of hypersonic missiles is a highly complex and expensive endeavour. The technical challenges involved in creating missiles capable of sustained hypersonic flight include overcoming issues related to heat management, manoeuvrability, and the integration of advanced propulsion systems. These difficulties contribute to the high development and manufacturing costs associated with hypersonic missiles. Despite these challenges, the strategic advantages offered by these weapons—such as the ability to evade traditional missile defence systems and strike targets rapidly—make them a priority for military powers around the world [1].

The combination of speed, manoeuvrability, and altitude that characterizes hypersonic missiles presents new challenges for existing missile defence systems. While missile defence technologies are continually evolving, current systems are generally not equipped to deal with the unique threats posed by hypersonic weapons. Hypersonic missiles offer reduced detection time, giving the target less time to respond and launch countermeasures. Defence systems will need to undergo significant upgrades to handle hypersonic threats, likely involving new sensors, faster interceptors, and advanced tracking capabilities [1].

There is ongoing debate among defence analysts about the broader implications of hypersonic missiles for global peace and stability. Some argue that the introduction of these weapons could escalate tensions and increase the risk of conflict, particularly between nuclear-armed states. The speed and unpredictability of hypersonic weapons might pressure countries to act more aggressively or pre-emptively, for fear of being unable to defend against a hypersonic strike. On the other hand, some experts believe that hypersonic missiles will not fundamentally change the strategic balance between major nuclear powers, given that traditional nuclear deterrents like ballistic missiles remain highly effective [1].

Arms control, export controls, and other measures could help mitigate the risks posed by hypersonic weapons, but these approaches face significant challenges. As hypersonic technology continues to evolve, it will be essential for global powers to engage in discussions about regulation and arms control to ensure that the development and deployment of these advanced weapons do not lead to unintended consequences for global stability. The future of hypersonic technology will depend not only on overcoming technical barriers but also on finding ways to manage the geopolitical risks associated with its deployment [1].

This book on hypersonic weapons is suited for a diverseaudience, particularly those interested in the intersection of modern militarytechnology, defence strategies, and global security. It offers valuableinsights for military professionals, defence analysts, and policymakers,discussing the military applications and strategic advantages of hypersonicweapons in global conflicts. Engineers and researchers in aerospace and defencesectors can benefit from the technical chapters on propulsion, materials, andtesting. Additionally, students and academics studying military technology orinternational relations may find it a useful resource for understanding theimplications of these emerging technologies. General readers fascinated by thefuture of warfare will also appreciate its comprehensive overview of howhypersonic weapons may reshape conflict dynamics.

PART I - The Basics of Hypersonic Technology

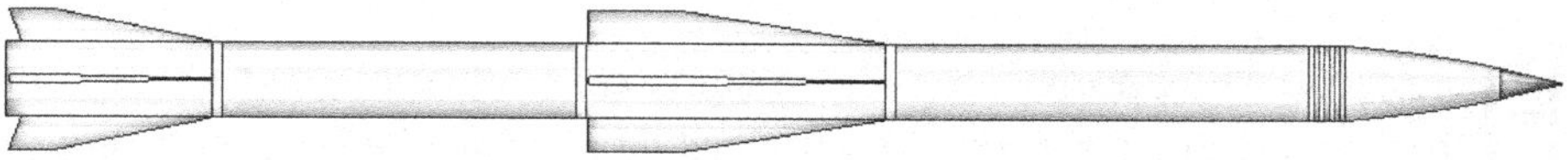

What Are Hypersonic Weapons?

Definition and Characteristics

Hypersonic weapons represent a significant advancement in military technology, characterized by their ability to travel at speeds exceeding Mach 5, which is five times the speed of sound. These weapons can be broadly categorized into two types: hypersonic glide vehicles (HGVs) and hypersonic cruise missiles (HCMs). HGVs are typically launched from a rocket and glide at hypersonic speeds, while HCMs utilize air-breathing engines to sustain their speed during flight. Both types are designed to evade traditional missile defence systems due to their speed, manoeuvrability, and low-altitude flight profiles, which complicate detection and interception efforts [2, 3].

Mach 5 refers to a speed that is five times the speed of sound. The speed of sound, or Mach 1, varies slightly depending on factors such as altitude and air temperature, but at sea level, it's approximately 767 miles per hour (mph) or 1,235 kilometres per hour (km/h).

Therefore, Mach 5 is:

- 3,835 miles per hour (mph) in imperial units.
- 6,174 kilometres per hour (km/h) in metric units.

Examples to put Mach 5 into perspective:

- Commercial Aircraft:
  - A typical commercial airliner like a Boeing 747 flies at around 570

mph (917 km/h), which is only about 0.75 Mach. Mach 5 would be more than 6 times faster than a 747's cruising speed.

- Concorde (Supersonic Airliner):
    - The Concorde, a supersonic passenger plane that flew from 1976 to 2003, reached speeds of 1,354 mph (2,180 km/h), or about Mach 2. Mach 5 would be more than twice the speed of the Concorde.
- New York to London Flight Time:
    - A flight from New York to London (about 3,460 miles, or 5,570 kilometres) takes around 7 hours on a commercial flight. At Mach 5, the trip could theoretically be done in under 1 hour.
- Bullet Speed:
    - A typical handgun bullet travels at about 1,200 mph (1,930 km/h), or Mach 1.6. A Mach 5 object would be traveling more than 3 times the speed of a bullet.

At Mach 5 speeds, a hypersonic vehicle or weapon covers over 1 mile (1.6 km) every second, highlighting the immense velocity and challenge of intercepting or defending against such fast-moving systems.

The strategic implications of hypersonic weapons are profound. Their rapid flight capabilities and ability to perform evasive manoeuvres make them a formidable challenge for existing air defence systems, which are often not equipped to handle such fast-moving targets. This has led to concerns regarding national security and the potential for destabilization in global military balances, as countries like the United States, Russia, and China race to develop these technologies [4-6]. The unique characteristics of hypersonic weapons, including their unpredictable trajectories and high speeds, significantly reduce the response time for adversaries, potentially encouraging pre-emptive strikes and altering traditional deterrence strategies [7, 8].

The development of hypersonic weapons is not merely a technological challenge but also an economic one, as nations invest heavily in research and development to maintain strategic advantages. The arms race surrounding hypersonic technology reflects broader geopolitical tensions and the desire for military superiority. This competition is likely to drive further innovations in both offensive and defensive capabilities, as nations seek to counter the threats posed by hypersonic systems [5, 9]. The implications extend beyond immediate military applications, as advancements in hypersonic technology could influence future warfare strategies and international relations [2, 10].

Hypersonic weapons are a transformative element in modern military arsenals, offering unprecedented speed and manoeuvrability that challenge existing defence frameworks. Their development is indicative of a broader trend in military technology, where speed and precision are paramount, and the implications for global security and military strategy are profound and far-reaching [11, 12].

Hypersonic weapons are primarily categorized into two types: Hypersonic Glide Vehicles (HGVs) and Hypersonic Cruise Missiles (HCMs). Each type exhibits distinct operational characteristics and capabilities that enhance their effectiveness in modern warfare.

Hypersonic Glide Vehicles (HGVs) are launched into the upper atmosphere using a rocket and then glide towards their targets at hypersonic speeds, typically exceeding Mach 5. Unlike traditional ballistic missiles, which follow a predictable trajectory, HGVs possess the ability to manoeuvre during flight. This manoeuvrability is a critical feature that complicates interception efforts by missile defence systems. HGVs are characterized by their high speed, precision, and strong manoeuvrability, which allow them to operate in near-space environments with minimal power requirements during the glide phase [13-15]. The trajectory design and guidance methods for HGVs are essential areas of research, focusing on optimizing their flight paths to enhance penetration capabilities against advanced defence systems [16-18].

In contrast, Hypersonic Cruise Missiles (HCMs) utilize advanced propulsion systems, such as scramjets, to maintain sustained hypersonic flight. These missiles can operate at lower altitudes compared to HGVs, which further complicates detection and interception by defence systems. HCMs are designed for rapid response and can strike targets with high precision, making them formidable components of modern military arsenals [19, 20]. The integration of sophisticated guidance laws and control systems is crucial for HCMs, as they must navigate complex environments while maintaining hypersonic speeds [21, 22]. The development of these systems involves addressing challenges related to aerodynamics, thermal management, and trajectory optimization to ensure effective performance during their operational phases [23-25].

Both HGVs and HCMs represent a shift in military strategy, emphasizing speed, manoeuvrability, and the ability to evade traditional defence mechanisms. The ongoing research and development in this field aim to enhance the capabilities of these hypersonic weapons, ensuring they remain effective against evolving threats in the global security landscape [11, 26, 27].

Hypersonic weapons are characterized by several key features that enhance their effectiveness on the battlefield. These characteristics include speed, manoeuvrability, range, and survivability, each contributing to their strategic advantages over traditional missile systems.

Speed is one of the most defining characteristics of hypersonic weapons, as they travel at speeds exceeding Mach 5. This high velocity drastically reduces the reaction time available for defence systems to intercept them, making traditional missile defence strategies less effective. For instance, Cernat notes that hypersonic missiles can achieve speeds over five times the speed of sound, complicating interception efforts by conventional systems [2]. Additionally, Chekov emphasizes that the combination of speed and manoeuvrability makes these weapons particularly destabilizing in military contexts [28].

Manoeuvrability is another critical feature that distinguishes hypersonic weapons from conventional ballistic missiles. Unlike their traditional counterparts, hypersonic glide vehicles (HGVs) can change direction mid-flight, allowing them to evade interceptors and complicate defence strategies. Xue et al. [13] highlight that hypersonic vehicles possess operational advantages such as efficient penetration and wide-range manoeuvrability, which are essential for avoiding detection and interception. This capability is further supported by the findings of Fan et al. [20], who describe HGVs as rapid strike weapons capable of gliding long distances while maintaining high speeds.

The range of hypersonic weapons is also noteworthy, as they can cover vast distances quickly, making them suitable for long-range precision strikes. According to Chen et al., hypersonic weapons are designed for long-range fast strike capabilities, which enhances their operational effectiveness [29]. The ability to strike targets from significant distances with high precision is a game-changer in modern warfare, as it allows for rapid response to threats and the ability to engage targets that are far beyond the reach of traditional systems.

Finally, the survivability of hypersonic weapons is enhanced by their speed and manoeuvrability, which pose significant challenges to current missile defence technologies. As noted by McFarland [30], the development of hypersonic weapons by major powers like the US, China, and Russia aims to create systems that can evade existing defence mechanisms, thereby increasing their effectiveness in combat scenarios. The combination of high speed and the ability to manoeuvre makes these weapons particularly difficult to track and intercept, as highlighted by Wright [31], who discusses the implications of hypersonic missile proliferation and the challenges they pose to strategic stability.

Hypersonic weapons are currently being developed and deployed by several countries, notably the United States, Russia, China, and India. These nations are engaged in a significant arms race to enhance their military capabilities through hypersonic technology, which offers advantages such as speed, manoeuvrability, and the ability to evade traditional missile defence systems [2, 4, 5].

The United States has been actively pursuing hypersonic weapon systems, with multiple programs underway aimed at developing both hypersonic cruise missiles and hypersonic glide vehicles. The U.S. Department of Defense has emphasized the strategic importance of these weapons, asserting that they will provide revolutionary capabilities that challenge existing defence frameworks [4, 32]. Russia has also made substantial advancements, with operational hypersonic systems such as the Kinzhal air-launched ballistic missile and the Zircon cruise missile already being deployed [33]. These systems exemplify the potential of hypersonic technology to alter the balance of power in military engagements.

China is similarly advancing its hypersonic capabilities, reportedly developing a range of hypersonic glide vehicles and cruise missiles designed to enhance its strategic deterrence and strike capabilities [2, 5]. The Chinese military's focus on hypersonic technology is part of a broader strategy to counteract U.S. military dominance in the Asia-Pacific region [4]. Additionally, India has entered the hypersonic arms race, with ongoing projects aimed at developing indigenous hypersonic missile systems, reflecting its strategic priorities in regional security [34].

The proliferation of hypersonic weapons is not limited to these major powers. Emerging trends suggest that other nations, particularly in Europe and Asia, may soon pursue similar capabilities. The technological advancements and the perceived necessity for hypersonic systems could lead to a broader diffusion of this technology, potentially destabilizing global security dynamics [31]. As hypersonic weapons become more accessible, the implications for arms control and strategic stability will be profound, necessitating a re-evaluation of existing frameworks to address the challenges posed by these advanced systems [35, 36].

The following provides some examples of hypersonic missi es developed or in development by major military powers:

1. Russia: Avangard

- **Type**: Hypersonic Glide Vehicle (HGV)
- **Speed**: Estimated to exceed Mach 20 (over 15,000 mph or 24,000 km/h)
- **Capabilities**: Launched atop an intercontinental ballistic missile (ICBM), the Avangard can glide and manoeuvre at hypersonic speeds, making it difficult to intercept. It is capable of carrying both conventional and nuclear warheads.
- **Status**: Reportedly deployed in Russia since 2019.

Figure 1: The UR-100UTTKh ICBM launched from the Dombarovsky Air Base carrying the Avangard. Mil.ru, CC BY 4.0, via Wikimedia Commons.

The Avangard hypersonic glide vehicle (HGV) developed by Russia represents a significant advancement in military technology, characterized by its remarkable speed, manoeuvrability, and payload versatility. The

Avangard is reported to exceed speeds of Mach 20, translating to over 15,000 mph (24,000 km/h), which positions it among the fastest military vehicles currently in existence [13, 15, 37]. This extraordinary speed is achieved through its operational design, which involves launching atop an intercontinental ballistic missile (ICBM) before gliding unpowered through the atmosphere, utilizing aerodynamic lift to maintain its trajectory [14, 16].

One of the defining features of the Avangard is its ability to manoeuvre during flight, which significantly complicates interception efforts by missile defence systems. The vehicle's design allows it to navigate through a series of flight phases, including an initial powered ascent followed by a gliding phase where it can adjust its trajectory in response to threats or obstacles [17, 38]. This capability is crucial for evading interception, as traditional missile defence systems are typically optimized for predictable ballistic trajectories rather than the erratic paths that HGVs can take [39, 40]. The Avangard can carry both conventional and nuclear warheads, enhancing its strategic deterrent role within Russia's military arsenal [15, 41].

Since its reported deployment in 2019, the Avangard has been a focal point of international military discourse, highlighting the growing importance of hypersonic technologies in modern warfare. The vehicle's operational status has been confirmed by various military assessments, indicating that it is fully integrated into Russia's strategic capabilities [24, 42]. The implications of such technology extend beyond mere speed; they encompass a shift in defence strategies globally, as nations seek to develop countermeasures against hypersonic threats [43, 44].

2. China: DF-ZF (Dong Feng - Zhengfeng)

- **Type**: Hypersonic Glide Vehicle (HGV)
- **Speed**: Estimated to reach speeds between Mach 5 and Mach 10 (3,800 to 7,600 mph or 6,174 to 12,348 km/h)
- **Capabilities**: The DF-ZF can be mounted on various Chinese ballistic missiles, including the DF-17, which is believed to have an operational range of 1,000 to 1,500 miles. It is designed for high-speed, unpredictable manoeuvring to avoid interception.
- **Status**: Reportedly operational.

Figure 2: The Dongfeng-17 missile launch vehicle exhibited at the Beijing Exhibition Center's "Striving for a New Era" thematic achievement exhibition. Yiyuanju, CC BY-SA 4.0, via Wikimedia Commons.

The DF-ZF (Dong Feng - Zhengfeng) is a hypersonic glide vehicle (HGV) developed by China, designed to enhance its strategic military capabilities. This vehicle is reported to achieve speeds estimated between Mach 5 and Mach 10, translating to approximately 3,800 to 7,600 mph (6,174 to 12,348 km/h) [45]. The operational speed range positions the DF-ZF among the fastest military technologies currently in development, allowing it to evade traditional missile defence systems effectively.

The DF-ZF is engineered to be versatile, capable of being mounted on various Chinese ballistic missiles, including the DF-17. The DF-17 itself is believed to have an operational range of 1,000 to 1,500 miles, which complements the DF-ZF's hypersonic capabilities by enabling it to strike targets at considerable distances with high precision [46]. The design of the DF-ZF emphasizes high-speed, unpredictable manoeuvring, which is crucial for avoiding interception by enemy defence systems. This capability is particularly significant in modern warfare, where rapid response and evasion tactics are essential for maintaining strategic advantages [47].

As of recent reports, the DF-ZF is considered operational, indicating that it has progressed beyond the testing phase and is now part of China's active military arsenal [48]. The operational status of the DF-ZF reflects China's commitment to advancing its hypersonic technology, which is a focal point in the current global arms race. The development of such advanced weaponry not only enhances China's deterrence capabilities but also raises concerns among other nations regarding regional and global security dynamics [49].

3. United States: AGM-183A Air-Launched Rapid Response Weapon (ARRW)

- **Type**: Hypersonic Glide Vehicle (HGV)
- **Speed**: Expected to reach speeds of Mach 5 to Mach 8 (3,835 to 6,100 mph or 6,174 to 9,656 km/h)
- **Capabilities**: Launched from an aircraft, the ARRW is designed to engage high-value, time-sensitive targets with precision. Its ability to travel at hypersonic speeds makes it difficult to detect and intercept.
- **Status**: Under development, with tests continuing as of 2023.

Figure 3: B-52 out of EDW carries ARRW IMV asset for its first captive carry flight over Edwards Air Force Base. U.S. Air Force / Christopher Okula - Edwards Air Force Base, Public Domain, via Wikimedia Commons.

The AGM-183A Air-Launched Rapid Response Weapon (ARRW) represents a significant advancement in the United States' hypersonic weapon capabilities. As a hypersonic glide vehicle (HGV), the ARRW is designed to achieve speeds ranging from Mach 5 to Mach 8, translating to approximately 3,835 to 6,100 mph (6,174 to 9,656 km/h) [4, 50]. This remarkable speed is a defining characteristic of hypersonic weapons, which are classified as vehicles traveling faster than five times the speed of sound. The development of such weapons has become a focal point in the strategic military landscape, as they pose substantial challenges to existing defence systems due to their high velocity and manoeuvrability, making them difficult to detect and intercept [2, 5].

The capabilities of the ARRW are particularly noteworthy. It is designed to be launched from aircraft, allowing for rapid deployment against high-value, time-sensitive targets with precision [51]. The hypersonic speed of the ARRW not only enhances its strike capability but also complicates the adversary's defence strategies, as traditional missile defence systems are often ill-equipped to counter such fast-moving threats [2, 7]. The ongoing development and testing of the ARRW, as of 2023, underscore the United States' commitment to maintaining a technological edge in hypersonic warfare [52].

The status of the ARRW is currently under development, with testing phases actively ongoing. This reflects a broader trend among major powers, including the United States, Russia, and China, who are engaged in a competitive race to develop hypersonic weapons [4, 35]. The strategic implications of these developments are profound, as hypersonic weapons like the ARRW could potentially alter the dynamics of military engagements and deterrence strategies globally [5, 53]. The emphasis on hypersonic technology is not merely a response to existing threats but also a proactive measure to ensure that the United States remains at the forefront of military innovation [54].

4. Russia: 3M22 Zircon (Tsirkon)

- **Type**: Hypersonic Cruise Missile (HCM)
- **Speed**: Capable of flying at speeds of up to **Mach 9** (approximately 6,900 mph or 11,000 km/h)
- **Capabilities**: Designed primarily for anti-ship operations, the Zircon can be launched from surface ships or submarines. It can travel at low altitudes, making it harder to detect and intercept.

- **Status**: In the final stages of testing and expected to enter service soon.

The 3M22 Zircon (Tsirkon) is a hypersonic cruise missile developed by Russia, designed primarily for anti-ship operations. It is capable of reaching speeds up to Mach 9, which translates to approximately 6,900 mph or 11,000 km/h, making it one of the fastest missiles in existence [55]. The missile's design allows it to be launched from both surface ships and submarines, enhancing its versatility in naval warfare. Its ability to fly at low altitudes significantly complicates detection and interception efforts by enemy defence systems, thereby increasing its operational effectiveness [56].

The Zircon's development is currently in the final stages of testing, with expectations for it to enter service imminently. This status reflects a broader trend in military technology where nations are increasingly investing in hypersonic capabilities to maintain strategic advantages [57]. The missile's advanced features, including its speed and low-altitude flight profile, are indicative of a significant leap in missile technology, which poses new challenges for existing missile defence systems [56].

Moreover, the testing of the Zircon involves sophisticated tracking and telemetry systems that are crucial for evaluating missile performance during flight tests. These systems utilize a combination of electro-optical tracking stations and radar to ensure accurate data collection and analysis, which is essential for refining missile capabilities [58]. The integration of advanced guidance systems further enhances the missile's precision and effectiveness in combat scenarios [59].

5. India: BrahMos-II

- **Type**: Hypersonic Cruise Missile (HCM)
- **Speed**: Expected to reach speeds of **Mach 7** (around 5,370 mph or 8,650 km/h)
- **Capabilities**: A joint venture between India and Russia, the BrahMos-II missile is designed for anti-ship and land-attack missions. It is the successor to the highly successful BrahMos-I, which is a supersonic cruise missile.
- **Status**: Under development.

The BrahMos-II missile, a hypersonic cruise missile (HCM) under development as a joint venture between India and Russia, is anticipated to achieve speeds of up to Mach 7, approximately 5,370 mph or 8,650 km/h. This missile is designed for both anti-ship and land-attack missions, serving as a successor to the BrahMos-I, which is recognized for its supersonic capabilities. The development of hypersonic missiles like BrahMos-II is significant due to their potential to evade existing missile defence systems, enhancing strike efficiency and operational effectiveness in modern warfare scenarios.

Hypersonic weapons, including cruise missiles, are characterized by their high speed and manoeuvrability, which complicates interception efforts by traditional defence systems. The BrahMos-II, leveraging advanced scramjet technology, is expected to operate effectively at hypersonic speeds, thus presenting a formidable challenge to adversaries' air defence capabilities [32]. The missile's design incorporates features that allow it to maintain high velocities while executing complex flight paths, which is crucial for penetrating enemy defences [60].

The strategic implications of BrahMos-II's development are profound, particularly in the context of regional security dynamics. As nations like India enhance their hypersonic capabilities, the balance of power in South Asia may shift, prompting responses from neighbouring countries, particularly China and Pakistan [61]. The BrahMos-II's anticipated capabilities could serve as a deterrent against potential aggressors, reinforcing India's military posture in a rapidly evolving geopolitical landscape.

Furthermore, the BrahMos-II is part of a broader trend in military technology where hypersonic systems are increasingly sought after for their ability to deliver payloads with minimal warning and high precision. The integration of advanced navigation systems and the ability to execute coordinated attacks further enhance its operational utility [24]. The ongoing development of BrahMos-II reflects not only technological advancements but also strategic considerations that underscore the importance of maintaining a competitive edge in missile technology.

6. France: ASN4G

- **Type**: Hypersonic Cruise Missile (HCM)
- **Speed**: Estimated to exceed **Mach 5**
- **Capabilities**: This missile is expected to replace the current air-launched nuclear missile in France's arsenal. It will be capable of high-speed strikes and nuclear deterrence.
- **Status**: In development.

The ASN4G, a hypersonic cruise missile (HCM) currently under development in France, is designed to exceed speeds of Mach 5, thereby significantly enhancing France's military capabilities. This missile is intended to replace the existing air-launched nuclear missile in France's arsenal, thereby reinforcing its nuclear deterrence strategy. The development of hypersonic missiles like the ASN4G is a response to evolving global security dynamics, particularly in light of recent geopolitical tensions that have highlighted the need for advanced military technologies.

Hypersonic missiles, including the ASN4G, are characterized by their ability to fly at lower altitudes than traditional ballistic missiles, which complicates detection and interception by existing missile defence systems. This feature is crucial for maintaining a strategic advantage, as hypersonic missiles can evade early-warning radars and strike targets with minimal warning [3, 62]. The ASN4G's capabilities align with the broader trend of military modernization among major powers, where hypersonic technology is increasingly seen as essential for effective deterrence and rapid response [31, 63].

The strategic implications of the ASN4G extend beyond mere speed; its development reflects a shift in France's nuclear posture. The missile is expected to play a critical role in ensuring assured retaliation, a key component of nuclear deterrence. The integration of hypersonic capabilities into France's nuclear arsenal could alter the strategic calculus of potential adversaries, as the speed and manoeuvrability of such missiles may render existing missile defences obsolete [7]. Furthermore, the ASN4G's development is indicative of a broader trend in Europe, where several nations are exploring or advancing their own hypersonic capabilities in response to perceived threats [31].

As the ASN4G progresses through its development phases, it is essential to consider the implications for international arms control and strategic stability. The proliferation of hypersonic technologies poses challenges to existing frameworks, such as the Missile Technology Control Regime (MTCR), which seeks to regulate missile technology transfers [32]. The potential for hypersonic missiles to destabilize regional security dynamics cannot be overlooked, as their deployment may provoke arms races and increase the risk of miscalculation during crises [36].

These hypersonic missiles represent a significant leap in military technology, combining speed, manoeuvrability, and range, making them formidable tools in modern warfare and driving a global hypersonic arms race.

Manoeuvrability

Hypersonic weapons, particularly Hypersonic Glide Vehicles (HGVs), stand out from traditional ballistic missiles due to their ability to manoeuvre unpredictably during flight. Unlike ballistic missiles, which follow a largely fixed, predictable trajectory once launched, hypersonic glide vehicles are designed to make continuous adjustments to their flight path. This dynamic ability to alter course in mid-flight creates a formidable challenge for current missile defence systems, which rely on tracking and predicting an incoming missile's trajectory to intercept it. Hypersonic weapons break this predictability, giving them a tactical advantage that increases their effectiveness in penetrating even the most advanced defence networks.

In the glide phase, HGVs are launched into the upper atmosphere by rockets, much like ballistic missiles. However, after reaching a certain altitude, the vehicle disengages from its booster and begins to glide toward its target at hypersonic speeds. During this phase, the weapon can make constant small adjustments to its trajectory, altering its direction and altitude to evade detection and tracking. This manoeuvrability is a key feature of HGVs and is one of the main reasons why these weapons are so difficult to intercept. Traditional missile defence systems are designed to predict an incoming missile's path, but the ability of HGVs to continually change their direction makes such predictions highly unreliable. This also complicates the timing and targeting of interceptors, as HGVs can appear to be heading in one direction before quickly veering off-course.

In the terminal phase of their flight, just before reaching the target, both Hypersonic Cruise Missiles (HCMs) and some HGVs can perform additional evasive manoeuvres. The terminal phase is crucial because it is the point at which missile defence systems would typically attempt to intercept the weapon. However, hypersonic weapons' ability to alter their path in this phase makes interception extremely difficult. For HCMs, which are powered by scramjet engines throughout their flight, manoeuvrability is retained due to the air-breathing nature of their propulsion. As a result, HCMs can fly at lower altitudes and execute complex, evasive manoeuvres as they approach their target.

In this final stage, HGVs can also increase their manoeuvrability by adjusting their angle, speed, and flight path. This capability is especially useful in evading last-minute attempts by defence systems to track and destroy the weapon. Even if a missile defence system manages to lock onto the weapon, the unpredictable movement in the terminal phase increases the likelihood that the interceptors will miss their target. This manoeuvrability in the final phase not only allows hypersonic weapons to evade interception but also increases the precision of their strikes, enabling them to hit high-value targets with greater accuracy.

Distinguishing Hypersonic Weapon Systems

While speed is often emphasized as a defining feature of hypersonic missiles, it is not the only factor that distinguishes these weapons from other missile systems. Many ballistic missiles, which predate hypersonic missile technology by nearly a century, also achieve speeds far in excess of Mach 5. For instance, the V-2, the ancestor of modern ballistic missiles, could reach speeds exceeding Mach 5 during both ascent and re-entry in the 1940s. However, no one would classify the V-2 as a hypersonic missile because it lacks the other critical characteristics that define modern hypersonic systems [64].

Modern intercontinental ballistic missiles (ICBMs) often achieve speeds exceeding Mach 20 during certain phases of their flight, particularly during ascent and re-entry, but they are not typically referred to as hypersonic weapons either. This is because, despite their extreme speeds, ballistic missiles follow predictable trajectories and lack the manoeuvrability that characterizes hypersonic weapons. Therefore, the label of "hypersonic missile" applies not just based on speed, but also on the weapon's ability to manoeuvre during flight while traveling at hypersonic speeds through the atmosphere [64].

Hypersonic weapons are commonly defined by a combination of two key features: speed exceeding Mach 5, and the ability to manoeuvre both vertically and horizontally while traveling at these speeds. This manoeuvrability is what sets hypersonic missiles apart from traditional ballistic missiles, which follow parabolic trajectories and are much easier for missile defence systems to predict and intercept. While public discussions, media reports, and even definitions found on platforms like Wikipedia often emphasize speed alone, experts in the field recognize that manoeuvrability is just as crucial in defining hypersonic weapons. As the US Missile Defense Advocacy Alliance points out, hypersonic weapons are unique not just because they travel faster than Mach 5, but because they can manoeuvre throughout their entire flight path, making them much harder to intercept [64].

The manoeuvrability of hypersonic weapons presents significant challenges to missile defence systems, which are designed to track and intercept missiles based on predictable flight paths. Ballistic missiles, for example, can be intercepted because their arc-like trajectories are relatively straightforward to calculate once they are detected. Hypersonic missiles, however, can adjust their altitude, speed, and direction mid-flight, evading traditional interception methods. The ability to alter their flight path makes hypersonic weapons a game-changer in modern military strategy, as they can potentially bypass even the most advanced missile defence systems [64].

Types of Hypersonic Weapons (Glide Vehicles, Cruise Missiles)

Hypersonic weapons are generally divided into two main categories: hypersonic cruise missiles (HCMs) and hypersonic glide vehicles (HGVs). Both types of missiles travel at hypersonic speeds, but they differ in terms of their propulsion systems and flight characteristics.

However it should be noted that hypersonic missiles can also be categorised as aero-ballistic missiles, hypersonic glide vehicles, and hypersonic cruise missiles to account for older technologies. Each type employs distinct mechanisms to achieve and sustain hypersonic speeds, which are generally defined as speeds exceeding Mach 5.

Aero-ballistic missiles, such as Russia's Kinzhal, are launched from aircraft and utilize rocket propulsion to reach hypersonic speeds before following a ballistic trajectory. This system allows for rapid deployment and high-speed strikes, making interception challenging due to its unpowered flight phase after the initial boost. The Kinzhal missile has been reported to reach speeds of Mach 10 and perform evasive manoeuvres, complicating defence strategies against it [65, 66]. The development of such systems dates back to the 1980s, highlighting a long-standing interest in hypersonic technology [67].

Figure 4: Training mockup 9-S-7760-EDM of Kh-47M2 Kinzhal hypersonic aero-ballistic air-to-surface missile mounted on Russian Air Force MiG-31K, blue 36, registration RF-92462 on static display at military-technical forum Army-2022, Kubinka Air Base, Moscow region, Russian Federation. Boevaya mashina, CC BY-SA 3.0, via Wikimedia Commons.

Hypersonic Cruise Missiles (HCMs) maintain a constant hypersonic speed and typically travel at a fixed altitude throughout their flight. These missiles are powered continuously by engines—often scramjet or ramjet engines—that enable them to sustain high speeds across long distances. HCMs are particularly effective for long-range precision strikes, as their propulsion system allows them to fly at consistent speeds and altitudes, making them ideal for targeting distant threats.

In contrast, Hypersonic Glide Vehicles (HGVs) are typically launched by ballistic missiles in what is known as a boost-glide system. After the initial rocket launch propels the vehicle to high altitudes, the HGV detaches from the booster and glides back through the atmosphere toward its target at hypersonic speeds. During this glide phase, the vehicle is unpowered but can still manoeuvre, using its speed and altitude to evade detection and interception. HGVs are particularly well-suited for evading missile defence systems because they can make complex, unpredictable manoeuvres during the descent phase of their flight.

When comparing hypersonic missiles to ballistic missiles, several factors come into play, including speed, range, and the time it takes to reach a target. Ballistic missiles, which follow a parabolic trajectory, generally reach their targets more quickly than hypersonic missiles at distances beyond 600 to 800 kilometres, especially during the ascent phase when they reach extreme speeds. However, Hypersonic Cruise Missiles (HCMs) traveling at a constant speed of Mach 5 tend to be faster than ballistic missiles at ranges beyond 1,900 to 2,500 kilometres, depending on the altitude at which they are flying.

Because HGVs are launched by ballistic missiles, their time to target depends largely on the trajectory of the booster rocket and the manoeuvres performed during the glide phase. The ability to manoeuvre, combined with their speed, makes hypersonic missiles highly versatile and difficult to defend against, but the challenges of maintaining precision, survivability, and managing heat loads make their development technically demanding.

Aero-Ballistic Missiles

Aero-ballistic missiles represent a significant evolution in missile technology, combining features of both traditional ballistic missiles and air-launched systems. These missiles are characterized by their launch from aircraft, allowing them to utilize high altitudes and speeds at launch, which enhances their operational effectiveness. The Kinzhal missile, specifically the Kh-47M2 variant developed by Russia, exemplifies this technology and has garnered attention for its unique capabilities.

Aside from Russia's Kinzhal missile, several other aero-ballistic missiles have been developed or are under development by various countries. These missiles are characterized by their air-launch capability and their ability to travel along a semi-ballistic trajectory, combining elements of both ballistic and cruise missiles. Here are notable examples of aero-ballistic missiles:

1. AGM-183A ARRW (Air-Launched Rapid Response Weapon) – United States

- **Country**: United States
- **Overview**: The AGM-183A ARRW is a U.S. Air Force program designed to deliver a hypersonic missile launched from an aircraft. The missile is a boost-glide system, meaning that after being launched from an aircraft, it uses a rocket booster to accelerate to hypersonic speeds before releasing a glide vehicle that travels at very high speeds toward its target.

- **Speed**: Reported speeds of Mach 5 to Mach 20.
- **Range**: Estimated range is around 1,000 miles (1,600 kilometres).
- **Purpose**: The ARRW is designed for long-range strike missions, targeting high-value enemy assets. It is meant to provide the U.S. with a hypersonic strike capability that can bypass current air defence systems.

2. Iskander-M (SS-26 Stone) – Russia

- **Country**: Russia
- **Overview**: The Iskander-M is a short-range tactical missile system developed by Russia. Although primarily a ground-launched ballistic missile, some variations and modifications have been tested as air-launched systems. The missile has a semi-ballistic flight path and can perform high-speed manoeuvres during flight, making it difficult to intercept.
- **Speed**: Capable of traveling at speeds between Mach 6 to Mach 7.
- **Range**: Up to 500 kilometres (310 miles).
- **Purpose**: The Iskander-M is designed to destroy missile defence systems, command centres, and other high-value targets. Its versatility and ability to perform evasive manoeuvres make it a formidable weapon in regional conflicts.

Figure 5: A launcher 9P78-1 of Russian missile system 9K720 Iskander in foreground and a transloader 9T250 in the background shown at Alabino training ground during military technical forum ARMY-2016. Boevaya mashina, CC BY-SA 4.0, via Wikimedia Commons.

Figure 6: Iskander/SS-26 Stone launch cruise missile. Ermaleksandr, Public Domain, via Flickr.

3. ASMP-A (Air-Sol Moyenne Portée Amélioré) – France

- **Country**: France
- **Overview**: The ASMP-A is a nuclear-capable aero-ballistic missile designed by France as part of its nuclear deterrence strategy. It is launched from fighter jets, including the Rafale, and follows a semi-ballistic trajectory. The ASMP-A is the primary nuclear delivery system for the French Air Force.
- **Speed**: Approximately Mach 3.
- **Range**: Around 500 kilometres (310 miles).
- **Purpose**: The ASMP-A is designed to carry a nuclear warhead and is part of France's strategic nuclear deterrent. It provides the French military with a fast, precise nuclear delivery system.

Figure 7: ASMP-A. David Monniaux, CC BY-SA 3.0, via Wikimedia Commons.

4. Kh-15 (AS-16 'Kickback') – Russia

- **Country**: Russia (formerly Soviet Union)
- **Overview**: The Kh-15, also known by its NATO designation AS-16 "Kickback", is an air-launched nuclear-capable ballistic missile. It was developed during the Cold War and is launched from Tu-22M, Tu-95, or Tu-160 strategic bombers. After launch, it climbs to high altitudes before diving down to its target at high speeds.
- **Speed**: Capable of reaching speeds of up to Mach 5.
- **Range**: Between 150 and 300 kilometres (93 to 186 miles).

- **Purpose**: Designed for tactical nuclear strikes, the Kh-15 was developed as an air-launched system capable of penetrating enemy defences and delivering nuclear payloads against hardened targets.

5. Kh-22 (AS-4 'Kitchen') – Russia

- **Country**: Russia (formerly Soviet Union)
- **Overview**: The Kh-22 is a long-range, air-launched missile that was originally designed to destroy aircraft carriers and large naval vessels. It is launched from bombers like the Tu-22M and follows a high-speed, semi-ballistic flight path. A modernized version, the Kh-32, has improved speed and range.
- **Speed**: Mach 4 to Mach 5.
- **Range**: Up to 600 kilometres (370 miles).
- **Purpose**: The Kh-22 is primarily an anti-ship missile, capable of carrying either conventional or nuclear warheads. It was developed to be used against high-value maritime targets such as aircraft carriers and large surface ships.

6. Ra'ad-II (Hatf-VIII) – Pakistan

- **Country**: Pakistan
- **Overview**: The Ra'ad-II is an air-launched cruise missile developed by Pakistan, based on an earlier version, the Ra'ad. It has a semi-ballistic flight path and is designed for nuclear or conventional payloads. It can be launched from aircraft such as the JF-17 Thunder or the Mirage III.
- **Speed**: **Supersonic** but not hypersonic, with some reports suggesting speeds of Mach 2 to Mach 3.
- **Range**: Approximately 600 kilometres (370 miles).
- **Purpose**: Ra'ad-II is part of Pakistan's nuclear deterrent and can be used for precision strikes against strategic targets. It is designed to be able to evade detection and missile defences.

7. Kh-47M2 Kinzhal – Russia

- **Country**: Russia

- **Overview**: While already mentioned, the Kh-47M2 Kinzhal deserves a special mention as one of the most advanced aero-ballistic missiles in the world. The Kinzhal is an air-launched hypersonic missile capable of striking ground and naval targets.
- **Speed**: Up to Mach 10.
- **Range**: Up to 2,000 kilometres (1,243 miles).
- **Purpose**: The Kinzhal is designed for anti-ship and land-attack missions, and its speed and manoeuvrability make it difficult to intercept. It can carry either a conventional or nuclear warhead.

An aero-ballistic missile is defined by its air-launch capability and its subsequent ballistic trajectory. Unlike conventional ballistic missiles that are launched from ground-based platforms, aero-ballistic missiles are deployed from aircraft, which allows them to achieve greater initial velocity and altitude. This air-launch method provides a tactical advantage, as it enables the missile to evade ground-based detection systems during its ascent phase [65]. Once launched, these missiles follow a high-speed, arcing trajectory influenced by gravity and momentum, similar to traditional ballistic missiles, but with enhanced manoeuvrability that complicates interception efforts [65, 67].

The Kinzhal missile is a prime example of an aero-ballistic missile, capable of reaching speeds up to Mach 10 (approximately 12,250 km/h) and performing evasive manoeuvres throughout its flight [65]. This combination of speed and agility makes it particularly challenging for existing missile defence systems to track and intercept, as these systems are typically designed for slower, more predictable targets [65]. Moreover, the Kinzhal's operational range of around 2,000 kilometres allows it to strike targets far beyond the reach of its launch platform, enhancing its strategic utility [65].

The Kinzhal missile's design incorporates several advanced features that contribute to its effectiveness as a weapon system. Its hypersonic speed significantly reduces the time available for interception, thereby increasing the likelihood of successful strikes against high-value targets such as naval vessels and missile defence installations [65, 67]. Additionally, the missile's ability to manoeuvre in-flight complicates the targeting process for missile defence systems, which may struggle to adapt to its unpredictable trajectory [65, 67].

The Kinzhal is launched from modified MiG-31K fighter jets, which not only provide a high-speed and high-altitude launch platform but also extend the missile's operational range [65]. This air-launch capability allows the Kinzhal to engage targets from a safe distance, minimizing the risk to the aircraft carrying it. Furthermore, the missile can be equipped with either conventional or nuclear warheads, offering flexibility in its deployment for various military scenarios [65].

The introduction of the Kinzhal missile has significant implications for military strategy and deterrence. It fits within Russia's anti-access/area denial (A2/AD) strategy, which aims to restrict adversaries' movements within specific regions [65, 67]. The Kinzhal's ability to target naval fleets and critical military infrastructure poses a direct challenge to NATO and U.S. forces operating near Russian territories, thereby enhancing Russia's strategic posture in regional conflicts [65, 67].

Moreover, the Kinzhal's potential nuclear capability raises concerns regarding escalation in conflicts, as its speed and manoeuvrability make it a formidable delivery system for nuclear warheads [65, 67]. This capability contributes to Russia's nuclear deterrence strategy, complicating adversaries' response calculations and increasing the ambiguity surrounding the missile's intended use [65, 67].

The Kinzhal missile presents significant challenges for contemporary missile defence systems. Its hypersonic speed and ability to manoeuvre during flight make it difficult to intercept, as existing defence technologies are primarily designed to counter slower, less agile threats [65, 67]. Additionally, the Kinzhal's lower flight path compared to traditional ballistic missiles reduces early detection opportunities, compressing the response time for defensive measures [65, 67]. The versatility of the Kinzhal, being deployable from various aircraft platforms, further complicates the prediction of potential launch sites and attack vectors [65, 67].

One of the primary technologies enabling aero-ballistic missiles is rocket propulsion systems. Unlike cruise missiles, which typically rely on jet engines or scramjets, aero-ballistic missiles use rocket engines to accelerate rapidly after being launched from an aircraft. Most of these missiles employ solid-fuel rockets, which are known for being compact, reliable, and capable of providing high thrust. The benefit of solid-fuel rockets is their simplicity and long-term storability. While liquid-fuel rockets are sometimes used in older designs, they are more complex and less common in modern systems. For example, Russia's Kh-47M2 Kinzhal uses a solid-fuel rocket booster to achieve speeds of up to Mach 10, and the U.S.'s AGM-183A ARRW uses rocket propulsion to reach hypersonic speeds.

Another key component of aero-ballistic missiles is the air launch platform. These missiles are typically launched from high-speed aircraft, providing them with an initial advantage in terms of altitude and velocity. Launching from an aircraft allows the missile to bypass certain air defence systems and increase its range. For instance, the Kinzhal is launched from aircraft such as the MiG-31K or Tu-22M bombers, which act as first-stage boosters, allowing the missile to gain a significant operational edge.

Aero-ballistic missiles follow a semi-ballistic flight path, where they ascend to high altitudes and then follow a curved trajectory toward the target. This trajectory is influenced by gravity, but unlike traditional ballistic missiles, aero-ballistic missiles can perform in-flight manoeuvres. Their manoeuvrability is achieved through aerodynamic control surfaces or thrust vectoring, making them harder to intercept than conventional ballistic missiles. Moreover, missiles that reach very high altitudes must cope with intense forces upon re-entry into denser atmospheric layers, requiring re-entry technology that can withstand the extreme heat and mechanical stresses.

In terms of guidance and navigation systems, aero-ballistic missiles use highly sophisticated systems to ensure they hit their target with accuracy. Inertial navigation systems (INS) are commonly used, relying on internal sensors such as accelerometers and gyroscopes to measure the missile's movement and position. This system remains effective even when GPS or satellite-based navigation is unavailable or compromised. Many modern missiles also incorporate satellite navigation systems like GPS or GLONASS for improved accuracy, and some use terminal guidance systems such as infrared (IR) seekers, radar homing, or electro-optical sensors for precise targeting in the final stage of flight.

Manoeuvrability is a hallmark of aero-ballistic missiles, making them more elusive than traditional ballistic systems. These missiles often employ aerodynamic control surfaces, like fins or canards, to alter their trajectory during flight. Thrust vectoring technology, where the missile redirects its exhaust gases to change direction, allows for rapid course corrections at high speeds, increasing unpredictability and evasion capabilities against enemy missile defences.

At the high speeds aero-ballistic missiles travel, they experience extreme aerodynamic heating due to friction with the atmosphere, especially during re-entry. To protect their structure and internal components, these missiles are equipped with thermal protection systems such as ablative materials, which vaporize slowly to dissipate heat, and ceramic composites, which provide both heat resistance and structural integrity.

Aero-ballistic missiles can carry various types of warheads, depending on the mission profile. Conventional warheads are used for high-precision strikes against military infrastructure, air defence systems, or naval targets. Some, like the Kh-47M2 Kinzhal and Kh-15, are designed to carry nuclear warheads, making them critical components of a nation's nuclear deterrence strategy. Additionally, penetrator warheads can be deployed to destroy fortified targets like underground bunkers or missile silos.

Stealth technology also plays a role in some aero-ballistic missile designs. Radar-absorbing materials (RAM) are used to reduce the missile's radar cross-section, making it more difficult to detect by enemy defences. Combined with their ability to fly at low altitudes, these missiles can avoid detection by radar systems until it's too late for effective interception.

Some aero-ballistic missiles are equipped with electronic countermeasures (ECM) to confuse or jam enemy radar and missile defence systems. Active jamming systems disrupt radar signals, while decoys can mislead defence systems, diverting interceptors away from the actual threat.

In terms of their strategic and tactical importance, aero-ballistic missiles like Russia’s Kinzhal or France’s ASMP-A offer unique advantages. Their high-speed precision strikes allow them to engage high-value targets before enemy defences can react, making them effective against aircraft carriers, radar installations, and command centres. For countries with nuclear arsenals, these missiles provide a flexible, hard-to-intercept nuclear delivery platform. Additionally, they are a key component of Anti-Access/Area Denial (A2/AD) strategies, helping countries control regional airspace and deny access to foreign forces.

Hypersonic Cruise Missiles (HCMs)

Hypersonic cruise missiles utilize air-breathing engines, specifically scramjets, to maintain hypersonic speeds after being boosted by rockets. This design allows for a more efficient propulsion system, enabling these missiles to be launched from a wider variety of platforms and at potentially lower costs compared to HGVs. The U.S. and China are actively developing hypersonic cruise missiles, with the U.S. reportedly conducting successful test flights of scramjet-powered systems [60]. The ability of these missiles to sustain hypersonic speeds through scramjet technology represents a significant leap in missile design and operational capability [68].

Unlike ballistic missiles, which follow a high-arcing trajectory outside the Earth's atmosphere, hypersonic cruise missiles remain within the atmosphere, maintaining a high level of manoeuvrability and sustained speed over the entire duration of their flight. Their combination of speed, manoeuvrability, and ability to fly at lower altitudes makes them a game-changing weapon in modern warfare. HCMs are designed to travel at speeds exceeding Mach 5, which allows them to cover extensive distances rapidly. This capability drastically reduces the reaction time available for adversaries to detect, track, and respond to these threats, thereby enhancing their effectiveness in combat scenarios [32]. Traditional cruise missiles, such as the U.S. Tomahawk, operate at subsonic or supersonic speeds, making them comparatively easier to intercept [32]. The ability of HCMs to maintain hypersonic speeds over long distances is crucial for their role in modern warfare, as it complicates defensive measures against them [32].

Unlike ballistic missiles that rely on rocket propulsion, HCMs utilize air-breathing engines, such as ramjets and scramjets. These engines are capable of sustaining high speeds by utilizing atmospheric oxygen for combustion, which eliminates the need for onboard oxidizers that are essential for rocket propulsion [69, 70]. The scramjet engine, in particular, is vital for achieving and maintaining hypersonic velocities, as it compresses incoming air, mixes it with fuel, and ignites the mixture to produce thrust at supersonic speeds [70, 71]. This technology not only enhances speed but also increases the operational range of hypersonic missiles [72].

HCMs are engineered for high manoeuvrability, allowing them to evade existing missile defence systems. Unlike traditional ballistic missiles that follow predictable trajectories, HCMs can perform sharp manoeuvres during flight, making their paths difficult to predict [32]. This capability poses significant challenges for missile defence systems, which are typically not designed to intercept such fast-moving and erratic threats [32]. The integration of advanced control strategies, such as blended control technology, further enhances the manoeuvrability of these missiles [69].

Typically, HCMs operate at low altitudes (between 20 to 40 kilometres), which complicates detection by ground-based radar systems that are more effective at tracking high-altitude ballistic missiles [32]. This low-altitude flight profile not only reduces the chances of detection by space-based sensors but also provides a stealth advantage during their approach to targets [32]. The ability to fly at these altitudes while maintaining hypersonic speeds is a critical feature that enhances their survivability and effectiveness in combat operations [32].

HCMs are designed for long-range precision strikes, capable of penetrating deep into enemy territory to target high-value assets such as military bases and command centres [32]. Their combination of high speed and manoeuvrability allows them to strike strategic targets with conventional warheads within minutes, making them a key component of prompt global strike capabilities [32]. This precision and speed are critical in modern military strategies, where rapid response times can significantly influence the outcome of conflicts [32].

The key technological components of Hypersonic Cruise Missiles include:

Air-Breathing Scramjet Engines: One of the most critical technological advancements that allow hypersonic cruise missiles to maintain speeds of Mach 5 or higher is the scramjet engine. Scramjets (supersonic combustion ramjets) differ from traditional jet engines in their operation. Traditional jet engines work by slowing down the incoming air to subsonic speeds before combustion, but scramjets compress and ignite the air at supersonic speeds. This is a key distinction that allows scramjets to operate effectively in the hypersonic regime.

In a scramjet engine, the air enters the engine at extremely high velocities, where it is compressed by the shape of the inlet, mixed with fuel, and then ignited to produce thrust. This high-speed combustion process allows the missile to fly at sustained hypersonic speeds without needing large amounts of oxidizer onboard, making it lighter and more fuel-efficient compared to traditional rocket-propelled systems. The scramjet's ability to utilize the oxygen from the atmosphere rather than carrying an oxidizer is what makes hypersonic cruise missiles more efficient for long-range flight at high speeds.

However, the operation of scramjets at such extreme velocities presents significant engineering challenges. Stable combustion at hypersonic speeds is difficult to achieve, and the materials used must withstand the extreme heat and pressure generated during flight. This requires precise engineering and the use of advanced materials capable of withstanding these harsh conditions, ensuring that the engine performs reliably throughout the missile's mission.

Thermal Protection Systems: At hypersonic speeds, the missile encounters extreme aerodynamic friction, which generates intense heat. The temperatures can soar to between 2,000°C and 3,000°C (about 3,632°F to 5,432°F), which poses a significant threat to the structural integrity of the missile, as well as to its onboard systems. Without adequate thermal protection, the missile's airframe, engine components, and sensitive electronics would degrade or fail due to this immense heat.

To manage these high temperatures, hypersonic cruise missiles are typically constructed with advanced heat-resistant materials. These include ceramic composites, which can endure extreme heat without melting, and ablative coatings, which protect the missile by gradually vaporizing and dissipating heat away from critical components. In some cases, thermal shielding is also used to insulate particularly sensitive electronics and payloads from the heat generated during flight.

Effective thermal management is one of the major technological challenges in developing hypersonic cruise missiles. The missile's materials and design must ensure that it can endure prolonged exposure to high temperatures, especially during long-range missions or while performing complex manoeuvres, without compromising its performance or accuracy.

Guidance and Navigation Systems: Maintaining accuracy at hypersonic speeds requires highly advanced guidance and navigation systems. Traditional systems like GPS can be used for general guidance, but they face limitations at hypersonic speeds due to the plasma cloud that forms around the missile. This plasma, created by the intense heat and friction, can disrupt GPS signals, making it harder for the missile to maintain a precise course.

As a result, inertial navigation systems (INS) are essential for hypersonic cruise missiles. INS relies on internal sensors, such as accelerometers and gyroscopes, to calculate the missile's position, orientation, and velocity. These systems do not depend on external signals, which means they can continue functioning even if the plasma cloud or jamming efforts from enemy defences interfere with GPS signals. The INS ensures that the missile stays on course, even at extremely high speeds.

In addition to INS, AI-based predictive algorithms are being developed to further enhance the missile's guidance systems. These algorithms optimize the missile's flight path, helping it choose the most efficient and evasive route to the target. By constantly recalculating the optimal trajectory, these AI systems allow the missile to adjust its course in real-time to avoid threats and countermeasures, improving its chances of successfully reaching the target.

Stealth Technology: While stealth technology is not as prominent in hypersonic cruise missiles as it is in aircraft like stealth bombers, it still plays an important role in reducing the missile's detectability. Given the high speeds at which hypersonic missiles travel, the window of time for detection and interception is already small. However, incorporating stealth technology further reduces the chance that enemy radar systems will detect the missile early enough to mount a successful defence.

Radar-absorbing materials (RAM) can be applied to the missile's exterior to reduce its radar cross-section, making it harder for radar systems to detect. In addition to these materials, hypersonic cruise missiles are often designed with low-profile aerodynamic shapes that minimize the surface area that radar waves can reflect off of. This combination of RAM and aerodynamic design helps to decrease the missile's visibility to radar systems.

Moreover, hypersonic cruise missiles typically fly at low altitudes within the atmosphere, often below the radar horizon of ground-based detection systems. This low-altitude flight, coupled with high speed and stealth features, makes these missiles incredibly difficult to detect and intercept, further complicating enemy defence efforts.

The technological components of hypersonic cruise missiles, from their scramjet engines to thermal protection, guidance systems, and stealth features, are critical to their effectiveness. Each of these technologies addresses specific challenges—such as maintaining hypersonic speeds, managing heat, ensuring precision, and avoiding detection—all of which enable these missiles to perform as game-changing assets in modern military operations.

The strategic and tactical implications of hypersonic cruise missiles (HCMs) are profound, particularly in the context of deterrence and defence. HCMs serve dual roles as both deterrents and first-strike capabilities due to their unique operational characteristics. Their ability to strike high-value targets rapidly and with precision allows states to threaten critical enemy assets, including military command centres and missile silos, thereby enhancing their strategic posture [5]. The development of HCMs is seen as a game-changer in modern warfare, as they can potentially dominate the early stages of conflict and provide a conventional alternative to nuclear weapons in prompt global strike strategies [5]. This capability is particularly relevant in scenarios where rapid response is essential, allowing states to maintain a credible deterrent without resorting to nuclear options.

However, the emergence of HCMs also presents significant defensive challenges. Current missile defence systems are inadequately equipped to counter the speed, manoeuvrability, and low-altitude flight profiles of HCMs [73]. Existing surface-to-air missile systems and radar networks will require substantial upgrades to effectively detect and intercept these advanced threats before they reach their targets [26]. The ability of HCMs to evade traditional defences complicates military planning, especially in naval warfare and within highly defended airspaces, where the rapid engagement of targets is critical [5]. The challenges posed by HCMs necessitate a re-evaluation of current defence strategies and the development of more sophisticated interception technologies [74].

Notable Examples of Hypersonic Cruise Missiles

1. **Russia's 3M22 Zircon (Tsirkon)**:
    - One of the most advanced hypersonic cruise missiles, Zircon is capable of reaching speeds of Mach 8 and has a range of around 1,000 kilometres (620 miles). Zircon is designed primarily as an anti-ship missile, capable of targeting large naval vessels such as aircraft carriers.
    - Zircon is powered by a scramjet engine and is considered a major component of Russia's Anti-Access/Area Denial (A2/AD) strategy, which seeks to keep foreign military forces, particularly naval fleets, out of strategically important regions like the Black Sea or Arctic.
2. **India-Russia BrahMos-II**:
    - The **BrahMos-II** is a joint project between India and Russia, intended to be a follow-up to the supersonic BrahMos missile. The BrahMos-II will be a hypersonic cruise missile capable of reaching speeds of Mach 7.
    - The missile will likely be powered by a scramjet engine and is designed for anti-ship and land-attack missions, with a range of up to 600 kilometres (373 miles).
3. **United States Hypersonic Air-breathing Weapon Concept (HAWC)**:
    - The HAWC program is part of the Defense Advanced Research Projects Agency (DARPA) initiative to develop hypersonic cruise missiles. The missile is being developed by Lockheed Martin and Raytheon, with the goal of producing an air-breathing hypersonic missile capable of striking distant targets with precision.
    - HAWC aims to achieve Mach 5+ speeds using a scramjet engine, with the capability to perform evasive manoeuvres during flight.

> This missile is part of the U.S.'s broader effort to maintain technological parity with Russia and China in the hypersonic arms race.

The United States Hypersonic Air-breathing Weapon Concept (HAWC) represents a significant advancement in military technology, focusing on the development of air-breathing hypersonic vehicles capable of sustained flight at speeds exceeding Mach 5. This capability is crucial for modern warfare, as hypersonic weapons can penetrate advanced missile defence systems and strike high-value targets with minimal warning [5]. The HAWC program aims to leverage air-breathing propulsion systems, specifically scramjet engines, which are designed to operate efficiently at hypersonic speeds by utilizing atmospheric oxygen for combustion, thus reducing the need for onboard oxidizers and enhancing range and payload capacity [69, 75].

The HAWC project is part of a broader trend in hypersonic weapon development, which includes various international efforts, notably from Russia and China, to create similar capabilities [5]. The strategic implications of hypersonic weapons are profound, as they can alter the balance of power by enabling rapid global strike capabilities. The U.S. military envisions that such systems will not only enhance deterrence but also provide a tactical advantage in conflict scenarios where speed and precision are paramount [76, 77].

Research indicates that the design and optimization of hypersonic vehicles, such as those under the HAWC program, involve complex aerodynamic considerations and advanced control systems to ensure stability and manoeuvrability during flight [13, 78, 79]. The integration of sophisticated guidance algorithms is critical for intercepting fast-moving targets and executing precise trajectories, which are essential for the effectiveness of hypersonic weapons [52]. Furthermore, the development of these systems necessitates overcoming significant engineering challenges, including thermal management and structural integrity at extreme speeds [80, 81].

Hypersonic Glide Vehicles (HGVs)

Hypersonic glide vehicles (HGVs) differ in that they are boosted to high altitudes by rockets and then glide towards their targets, manoeuvring during flight to evade detection and interception. Notable examples include China's Dongfeng-17 and Russia's Avangard, which are designed to exploit their manoeuvrability and speed to penetrate advanced missile defence systems [5, 52]. U.S. officials have expressed concerns that China's

HGV technology may be more advanced than that of the United States, indicating a competitive landscape in hypersonic capabilities [28].

Hypersonic Glide Vehicles (HGVs) are an advanced class of military weaponry that travel at hypersonic speeds—typically faster than Mach 5 (approximately 3,800 mph or 6,174 km/h)—and possess the ability to manoeuvre unpredictably during flight. Unlike traditional ballistic missiles, which follow a predictable and parabolic trajectory after launch, HGVs glide along the atmosphere and can change their trajectory during flight, making them much harder to detect and intercept. These characteristics make them a critical asset in modern military strategy, capable of evading existing missile defence systems and delivering precision strikes on high-value targets.

HGVs are a type of boost-glide weapon that, once launched by a rocket booster, achieve high altitudes and then glide back through the atmosphere at hypersonic speeds toward their target. Unlike ballistic missiles that ascend and fall back toward Earth in a simple ballistic arc, HGVs use their manoeuvrability to evade detection and interception, giving them a strategic advantage over conventional missiles. They are capable of traveling at extremely high speeds, often exceeding Mach 20, and are designed for long-range precision strikes.

Hypersonic glide vehicles (HGVs) represent a significant advancement in military technology, characterized by their ability to achieve high speeds, manoeuvrability, and resilience in extreme conditions. The development of HGVs is underpinned by several core technologies, each contributing to their operational capabilities.

Boost-Glide System: The boost-glide system is fundamental to the operation of HGVs. Initially, a conventional rocket booster propels the vehicle to hypersonic speeds and high altitudes, typically in the upper atmosphere, before releasing the glide vehicle. This separation marks the transition into the glide phase, where the HGV utilizes its aerodynamic shape to travel vast distances without powered flight. The initial boost phase is critical, as it enables the vehicle to reach speeds exceeding Mach 5, which is essential for its subsequent glide phase [13, 14, 18]. The glide phase distinguishes HGVs from traditional ballistic missiles, as they rely on aerodynamic forces rather than continuous propulsion to maintain high velocities [16, 40].

Typically, the vehicle is launched by a conventional rocket that accelerates the system to hypersonic speeds and delivers it to high altitudes, typically in the upper atmosphere. Once the rocket reaches the desired altitude and speed, the HGV is released. From there, the glide vehicle follows a long, flattened trajectory, using its aerodynamic shape to glide over vast distances, rather than free-falling like a traditional ballistic missile.

- **Rocket Booster**: The initial boost phase relies on a **rocket booster** (either liquid or solid fuel) to launch the glide vehicle into the upper atmosphere. The rocket phase is short but essential, as it propels the HGV to hypersonic speeds and high altitudes.

- **Glide Phase**: After the separation from the booster, the HGV enters the glide phase, where it "glides" through the atmosphere at hypersonic speeds toward its target. The glide phase is what makes HGVs unique, as they use aerodynamic forces rather than a powered engine to continue traveling at high velocities.

Manoeuvrability: A defining feature of HGVs is their enhanced manoeuvrability during flight. Unlike conventional ballistic missiles that follow predictable trajectories, HGVs can adjust their flight path mid-course through the use of aerodynamic control surfaces and changes in their angle of attack. This capability allows them to perform evasive manoeuvres, complicating interception efforts by missile defence systems [3, 20]. The ability to alter direction and altitude unpredictably increases the survivability of HGVs against modern defence mechanisms, making them a formidable challenge for adversaries [38].

HGVs use aerodynamic control surfaces and adjust their angle of attack during the glide phase, allowing them to change direction, altitude, and speed. This makes them extremely difficult for missile defence systems to track and intercept, as they can evade defensive measures and unpredictably alter their flight path.

- **In-flight Manoeuvring**: HGVs can perform evasive manoeuvres during their descent, allowing them to bypass missile defence systems. This capability to adjust their flight path mid-flight helps avoid detection by early-warning radar systems and evade interceptor missiles.

- **Unpredictable Trajectories**: Because of their ability to change direction and altitude during flight, HGVs present a significant challenge to missile defence systems that rely on predicting the target's flight path. This unpredictability increases their survivability against modern missile defences.

Hypersonic Speeds: HGVs operate at speeds greater than Mach 5, with some designs capable of reaching Mach 20. This extreme velocity enables them to cover vast distances in mere minutes, significantly reducing the reaction time available to defence systems [82, 83]. The kinetic energy generated at such speeds is immense, allowing HGVs to inflict substantial damage upon impact, even without explosive warheads. This capability is particularly advantageous for precision strikes on high-value targets, enhancing their strategic utility in military operations [25, 84].

Kinetic Energy of Hypersonic Glide Vehicles (HGVs)

One of the most impactful features of Hypersonic Glide Vehicles (HGVs) is their immense kinetic energy, generated by their extreme speeds. As HGVs travel at velocities greater than Mach 5 (more than 3,800 mph or 6,174 km/h), they accumulate vast amounts of kinetic energy, which is a function of both their mass and velocity. The equation for kinetic energy, $KE = 0.5 * mass * velocity^2$, highlights the exponential impact that speed has on energy. This means that even relatively small HGVs can carry a significant amount of kinetic energy due to their high speeds.

When an HGV strikes a target, this kinetic energy is transferred upon impact, causing devastating physical damage. This energy alone can be sufficient to destroy or severely damage critical infrastructure, vehicles, or military assets. The sheer force of impact from an HGV at hypersonic speeds can penetrate hardened targets, such as underground bunkers, fortified military bases, or missile silos, without the need for a large explosive warhead. In fact, in some designs, the HGV may rely almost entirely on its kinetic energy, using the force of the impact to achieve its mission objectives, a strategy known as a "kinetic kill."

This capability gives HGVs a significant advantage in scenarios where conventional explosive warheads might be less effective, especially against heavily armoured or fortified targets. For example, military installations that are designed to withstand traditional ballistic or cruise missile strikes might still be vulnerable to the high-speed impact of an HGV, as the kinetic force can cause deep structural damage. This characteristic also makes HGVs particularly useful for disabling enemy defences such as air defence systems, missile launch platforms, or command centres, which are often heavily protected.

Long-Range Capabilities of Hypersonic Glide Vehicles

The long-range capabilities of HGVs significantly enhance their strategic value. HGVs can be launched from intercontinental distances, making them suitable for global strike missions. Unlike traditional missiles, which may have limited range and require mid-air refuelling or multiple stages to reach distant targets, HGVs can cover thousands of kilometres in a single glide phase. This allows them to reach targets across continents, making them a versatile tool for military operations that require rapid, long-distance strikes.

HGVs are typically launched from a rocket booster that propels them to high altitudes and hypersonic speeds. Once they are released from the booster, they enter the glide phase in the upper atmosphere, where they maintain their speed while gliding toward their target. This ability to travel long distances without the need for constant propulsion or refuelling sets them apart from other missile systems, offering greater operational flexibility and fewer logistical challenges.

The ability to strike targets located thousands of kilometres away means that HGVs can be used for pre-emptive global strike missions, neutralizing high-value targets such as military installations, radar systems, or command and control centres. Moreover, their intercontinental range allows them to be launched from remote or secure locations, far from enemy territory, reducing the risk to the launch platform and personnel. This also means that adversaries must consider a wider area when planning their defences, as the potential launch points for HGVs are much more diverse than traditional missile systems.

Thermal Protection Systems: The extreme heat generated during hypersonic flight poses a significant challenge for HGV design. As the vehicle travels at high speeds, aerodynamic friction can create temperatures exceeding 2,000°C (3,632°F), which can damage critical components. To mitigate this, HGVs are equipped with advanced thermal protection systems, including ablative materials that dissipate heat by vaporizing and ceramic matrix composites that withstand high temperatures without melting [85, 86]. These materials are essential for ensuring the structural integrity of the vehicle throughout its flight [43].

Ablative Materials in Hypersonic Glide Vehicles (HGVs)

One of the primary challenges faced by Hypersonic Glide Vehicles (HGVs) is the extreme heat generated by aerodynamic friction as they travel at hypersonic speeds through the atmosphere. As the vehicle moves at speeds greater than Mach 5, the friction between the air molecules and the surface of the vehicle creates intense heat, which can exceed temperatures of 2,000°C (3,632°F). Without adequate protection, this heat could compromise the structural integrity of the vehicle, leading to catastrophic failure.

To mitigate this, HGVs are often coated with ablative materials, which are specifically designed to manage the immense heat generated during flight.

Ablative materials function by slowly vaporizing or charring as they are exposed to high temperatures. This process of gradual erosion allows the heat to be carried away as the material burns off, thereby preventing the underlying structure of the vehicle from absorbing excessive thermal energy. The vaporized material creates a protective layer of gas between the surface of the vehicle and the high-temperature air, reducing further heat transfer. This process is known as ablation.

The advantage of using ablative materials is that they are self-sacrificing, meaning that they are intentionally designed to erode during flight in order to protect the vehicle. This method of heat management is highly effective for short-duration, high-temperature events like re-entry into the atmosphere or high-speed gliding. By dissipating heat in this controlled manner, ablative coatings ensure that the HGV can withstand the extreme thermal stresses encountered during flight without overheating critical components, such as guidance systems, sensors, and the vehicle's airframe.

Ablative materials have been used in various aerospace applications, including the heat shields of spacecraft during re-entry into the Earth's atmosphere. However, in the case of HGVs, the materials must be carefully engineered to withstand both extreme heat and the mechanical stresses associated with manoeuvring at hypersonic speeds. This makes the design and selection of ablative materials crucial to the success of hypersonic glide vehicles.

Ceramic Composites in Hypersonic Glide Vehicles

In addition to ablative materials, ceramic matrix composites (CMCs) play a critical role in the construction of HGVs, particularly in providing long-term thermal protection and maintaining structural integrity at high speeds. CMCs are advanced materials made from ceramic fibres embedded in a ceramic matrix. These materials are ideal for applications where extreme heat resistance and mechanical strength are required, such as in the body and control surfaces of hypersonic vehicles.

One of the key properties of ceramic composites is their ability to withstand extreme temperatures—often exceeding 1,500°C (2,732°F)—without melting or losing their strength. Unlike metals, which can deform, melt, or lose structural integrity at high temperatures, CMCs remain stable, making them suitable for high-speed atmospheric flight where the heat generated by friction is significant. This allows the HGV to retain its aerodynamic shape and control surfaces, which are critical for manoeuvrability and flight stability.

CMCs also possess a high level of thermal conductivity, meaning they can effectively dissipate heat over their surface area, preventing hot spots from forming on the vehicle. This even distribution of heat reduces the likelihood of thermal damage to specific parts of the vehicle, ensuring that it can survive long-duration flights through the atmosphere at hypersonic speeds.

In addition to their thermal properties, ceramic composites are extremely lightweight compared to traditional heat-resistant materials like metals. This weight reduction is crucial in the design of HGVs, as it allows the vehicle to achieve higher speeds and greater manoeuvrability without being weighed down by heavy protective materials. The lightweight nature of ceramic composites, combined with their heat resistance, makes them a key material in the construction of modern hypersonic glide vehicles.

Furthermore, ceramic composites are highly resistant to mechanical wear and corrosion, which enhances their durability during repeated exposure to extreme environmental conditions. As HGVs are designed to travel through varying layers of the atmosphere, where air pressure and temperature can change dramatically, CMCs provide a material solution that can endure these harsh conditions while maintaining optimal performance.

Both ablative materials and ceramic matrix composites (CMCs) are essential technologies in the construction of Hypersonic Glide Vehicles (HGVs), each playing a crucial role in managing the intense heat generated during hypersonic flight. Ablative materials provide a self-sacrificing layer that dissipates heat through vaporization, protecting the vehicle's structure during short, high-temperature flight phases. Meanwhile, ceramic composites offer long-term heat resistance, mechanical strength, and weight efficiency, ensuring that the HGV remains functional and manoeuvrable even in the most extreme conditions. Together, these materials enable HGVs to achieve high speeds and precision targeting while surviving the harsh environment of hypersonic flight.

Guidance and Navigation Systems: Guiding an HGV to its target with precision is a complex task, particularly at hypersonic speeds where traditional GPS signals can be disrupted by the plasma sheath surrounding the vehicle. As a result, HGVs predominantly rely on inertial navigation systems (INS) that utilize internal sensors to track position and orientation without external signals [21, 87]. Additionally, advanced algorithms and artificial intelligence are increasingly employed to optimize flight paths in real-time, enhancing the vehicle's ability to evade defences and reach its target accurately [88, 89].

Inertial Navigation Systems (INS) in Hypersonic Glide Vehicles (HGVs)

Inertial Navigation Systems (INS) are a critical component of Hypersonic Glide Vehicles (HGVs), providing essential navigation capabilities that allow the vehicle to maintain precise control over its flight path, even at hypersonic speeds. INS technology calculates the vehicle's position, velocity, and orientation using internal sensors such as gyroscopes and accelerometers. These sensors measure the HGV's movement and changes in direction relative to its initial launch conditions, enabling the system to track the vehicle's trajectory without the need for external signals like GPS.

The main advantage of INS lies in its self-contained nature. Unlike satellite-based navigation systems, which can be disrupted or jammed by enemy countermeasures, INS operates independently, relying solely on the internal sensors to provide continuous data on the vehicle's motion. This makes INS particularly valuable in contested environments where an adversary might attempt to interfere with GPS or other external navigation systems through electronic warfare tactics. As HGVs travel at extreme speeds through the atmosphere, they often generate plasma clouds around the vehicle, which can block or distort radio signals, including those used by GPS. INS bypasses this issue, offering a reliable and robust solution for maintaining navigation accuracy under challenging conditions.

However, one limitation of traditional INS is that it can experience drift over time. Since INS calculates movement based on the integration of sensor data, small errors in the measurements can accumulate over long distances, leading to a gradual degradation in accuracy. To mitigate this, some HGVs use hybrid systems that combine INS with satellite navigation (when available) to periodically correct any drift and recalibrate the system. This hybrid approach ensures that the HGV can maintain high accuracy even during extended flights across intercontinental distances.

In the absence of external correction, INS remains an essential tool for high-speed navigation, particularly in situations where precision targeting is required in GPS-denied environments. The ability to continuously track the vehicle's position based solely on internal measurements is crucial for ensuring that the HGV stays on course, enabling it to evade enemy defences and strike its intended target with precision.

Artificial Intelligence (AI) and Predictive Algorithms in Hypersonic Glide Vehicles

In modern hypersonic weapon systems, Artificial Intelligence (AI) and predictive algorithms are increasingly being integrated to enhance the manoeuvrability and targeting efficiency of Hypersonic Glide Vehicles (HGVs). Given the extreme speeds at which HGVs travel, often exceeding Mach 5, traditional navigation and control systems can struggle to keep up with the rapid decision-making required for real-time adjustments during flight. This is where AI and advanced algorithms come into play, offering real-time optimization and course correction capabilities.

AI-based systems in HGVs are designed to analyse sensor data from the vehicle's internal systems and predict potential threats or flight path deviations. Using advanced algorithms, the AI can optimize the HGV's flight path to ensure maximum efficiency, adjusting the vehicle's trajectory based on real-time conditions such as atmospheric changes, detected threats, or unexpected obstacles. For instance, if the HGV detects incoming missile defence systems or hostile radar tracking, the AI can calculate the most efficient evasive manoeuvres to avoid detection or interception.

Predictive algorithms allow the HGV to anticipate future flight conditions and make adjustments proactively rather than reactively. For example, the system can predict the aerodynamic forces the vehicle will encounter based on its speed and altitude, ensuring that any required course corrections are made smoothly and efficiently. This level of predictive control is especially important for maintaining manoeuvrability at hypersonic speeds, where even small errors in trajectory can lead to significant deviations from the target.

Moreover, the integration of AI also helps the HGV to adapt to changing mission parameters. If the target location is altered mid-flight or if the threat environment changes (such as the activation of enemy air defences), the AI can dynamically adjust the vehicle's flight plan to ensure the mission's success. These real-time updates make the HGV more responsive and capable of handling complex, evolving scenarios, enhancing its overall operational effectiveness.

AI-based navigation and control systems also reduce the cognitive load on human operators. Rather than relying on manual intervention or pre-programmed flight paths, the AI system can autonomously manage most aspects of the vehicle's flight, leaving human controllers free to focus on broader mission objectives. This level of autonomy is particularly valuable in situations where rapid decisions are required, such as first-strike scenarios or missions in contested environments.

In summary, the combination of Inertial Navigation Systems (INS) and Artificial Intelligence (AI) provides HGVs with a powerful set of tools for maintaining precision, manoeuvrability, and adaptability during high-speed flight. INS ensures that the vehicle can track its position accurately even in GPS-denied environments, while AI and predictive algorithms allow the HGV to adjust its course in real-time, optimizing its flight path for maximum efficiency and evasion of defences. Together, these technologies make HGVs formidable assets in modern warfare, capable of delivering precision strikes while overcoming the challenges posed by hypersonic flight and hostile environments.

Warhead Options: HGVs can be equipped with various warhead types, including conventional and nuclear payloads. The combination of high speed and kinetic energy allows even conventional warheads to deliver devastating impacts, making HGVs suitable for precision strikes against critical infrastructure [23, 90]. The potential for nuclear payloads further enhances their strategic value, particularly in deterrence scenarios, as their ability to evade missile defences complicates adversaries' response strategies [89, 91].

Conventional Warheads in Hypersonic Glide Vehicles (HGVs)

Conventional warheads designed for Hypersonic Glide Vehicles (HGVs) are typically used for precision strikes on high-value targets, such as military infrastructure, radar installations, missile defence systems, and command centres. These warheads are intended to destroy or incapacitate key enemy assets with high accuracy, making them a crucial tool in modern warfare. However, what distinguishes HGVs from other missile systems is their ability to travel at hypersonic speeds—speeds greater than Mach 5—which adds a significant advantage when deploying conventional warheads.

At hypersonic velocities, HGVs carry a tremendous amount of kinetic energy, which is the energy of motion. As an HGV approaches its target, its kinetic energy alone can amplify the destructive power of the warhead. This is particularly important because, at such extreme speeds, the vehicle's impact generates a massive force, which can cause significant damage even before the warhead explodes. In some scenarios, the sheer kinetic energy of the HGV can penetrate hardened targets, such as underground bunkers, missile silos, or reinforced military structures. This enables the HGV to strike targets that are typically resistant to conventional missile systems.

The combination of speed and precision targeting allows HGVs with conventional warheads to conduct prompt precision strikes against time-sensitive or mobile assets. For example, targets such as aircraft carriers, mobile missile launchers, or enemy command vehicles can be neutralized quickly, often before enemy forces have the opportunity to react or move. Additionally, because HGVs are difficult to detect and intercept, the risk of the warhead being neutralized before reaching its target is significantly reduced. This capability makes HGVs an important asset in achieving air superiority, disabling air defence systems, and weakening the enemy's military infrastructure in the early stages of a conflict.

In summary, conventional warheads in HGVs are designed for precision destruction of high-priority targets. The kinetic energy generated by hypersonic speeds increases the lethality of the strike, making these warheads highly effective in penetrating defences and causing widespread damage, even to hardened targets. The combination of speed, precision, and kinetic energy makes conventional warheads in HGVs a powerful tool for modern military operations, particularly in first-strike or rapid response scenarios.

Nuclear Warheads in Hypersonic Glide Vehicles (HGVs)

Some Hypersonic Glide Vehicles (HGVs) are specifically designed to carry nuclear warheads, which positions them as a critical component of nuclear deterrence strategies. Nuclear warheads on HGVs offer several strategic advantages, particularly in the context of missile defence evasion and global strike capabilities. The ability of HGVs to travel at hypersonic speeds and manoeuvre unpredictably during flight makes them exceptionally difficult to detect and intercept. This, in turn, enhances their effectiveness as nuclear delivery platforms.

In the realm of nuclear deterrence, HGVs equipped with nuclear warheads provide a reliable second-strike capability. In the event of a nuclear conflict, HGVs could be launched to retaliate against an adversary, ensuring that any pre-emptive nuclear attack does not go unanswered. The ability of HGVs to evade missile defence systems adds to their value in such scenarios. Traditional ballistic missiles, while powerful, follow predictable trajectories, allowing missile defence systems to track and potentially intercept them. In contrast, HGVs can alter their flight path, making their trajectory much harder to predict and reducing the time available for defensive systems to respond. This characteristic makes HGVs particularly valuable in bypassing even the most advanced missile defence systems, thereby ensuring the successful delivery of their nuclear payload.

Moreover, nuclear-armed HGVs extend the range of nuclear deterrence by offering a global strike capability. Launched from long distances, these vehicles can reach any target across the globe in a matter of minutes, providing nations with the ability to deliver swift and devastating nuclear strikes. This capability enhances the credibility of a nation's nuclear deterrent, as potential adversaries must account for the possibility that their defences could be bypassed by an HGV carrying a nuclear warhead.

The presence of nuclear warheads on HGVs also raises the stakes in geopolitical conflicts. Because of their ability to evade missile defences, nuclear-armed HGVs complicate nuclear arms control and strategic stability. Adversaries may feel increased pressure to upgrade their missile defence systems or develop their own hypersonic weapons to maintain a strategic balance. This could potentially lead to arms races in hypersonic technologies, as nations compete to ensure that their nuclear deterrent remains effective.

Hypersonic Glide Vehicle Advantages

Hypersonic Glide Vehicles (HGVs) represent a significant advancement in military technology, offering several strategic advantages over traditional missile systems. This synthesis discusses the primary benefits of HGVs, including their ability to penetrate missile defence systems, their first-strike capability, long-range precision, and enhanced survivability.

One of the most notable advantages of HGVs is their capability to penetrate advanced missile defence systems. Traditional missile defence mechanisms are primarily designed to intercept ballistic missiles, which follow predictable trajectories. In contrast, HGVs can manoeuvre during flight, making their paths unpredictable and significantly complicating interception efforts. Studies have shown that the high manoeuvrability of HGVs allows them to evade detection and interception by existing defence systems, thereby enhancing their effectiveness as a strike weapon [92]. The aerodynamic characteristics of HGVs, combined with their operational speed exceeding Mach 5, contribute to their ability to bypass these defences, posing a substantial challenge to current military capabilities [93, 94].

Furthermore, HGVs provide a strategic advantage through their first-strike capability. Their high speed and precision enable them to conduct pre-emptive strikes on critical targets before adversaries can respond effectively. This rapid response capability is crucial in modern warfare, where the ability to neutralize threats quickly can determine the outcome of conflicts. HGVs can reach any location on the globe within a couple of hours, making them ideal for executing surprise attacks on high-value military assets or infrastructure [20, 93]. This ability to strike swiftly reduces the enemy's reaction time to nearly zero, thereby enhancing the strategic value of HGVs in military operations [14].

In addition to their first-strike capabilities, HGVs are designed for long-range precision strikes. They can travel intercontinental distances while maintaining high accuracy, which is essential for prompt global strike operations. The combination of speed and precision allows HGVs to engage targets that are otherwise difficult to reach with conventional systems [21, 41]. Their operational range and accuracy make them invaluable for military planners seeking to project power globally and respond to emerging threats in real-time [27].

Lastly, the survivability of HGVs in contested environments is significantly enhanced due to their operational characteristics. The high speed and manoeuvrability of HGVs make them difficult to detect and intercept, increasing their chances of successfully reaching their targets. Unlike traditional ballistic missiles, which follow predictable paths, HGVs can alter their trajectories mid-flight, further complicating interception efforts [42, 92]. This capability is particularly important in modern warfare, where adversaries may deploy advanced detection and interception technologies [43]. The ability of HGVs to operate at varying altitudes and speeds also contributes to their survivability, allowing them to evade radar and other detection systems effectively [83].

Notable Examples of Hypersonic Glide Vehicles

- **Russia's Avangard**: One of the most well-known HGVs is the Avangard, developed by Russia. This system can reportedly reach speeds of Mach 20 and is capable of carrying both conventional and nuclear warheads. It is designed to be launched from an intercontinental ballistic missile **(ICBM)** and glide to its target at hypersonic speeds.

- **China's DF-ZF**: The **DF-ZF** is a Chinese HGV that has undergone extensive testing. It is believed to be part of China's broader strategy to develop hypersonic weapons capable of evading U.S. missile defence systems. The DF-ZF is launched from a medium-range or intercontinental missile and glides at hypersonic speeds toward its target.

- **United States' Common Hypersonic Glide Body (C-HGB)**: The U.S. is developing the **C-HGB** as part of its hypersonic weapons programs. The **Long-Range Hypersonic Weapon (LRHW)**, a U.S. Army project, uses the C-HGB and is expected to be operational in the coming years. The U.S. Navy's Conventional Prompt Strike (CPS) program will also utilize the C-HGB for its hypersonic systems.

Speed, Range, and Flight Dynamics

Speed

The most defining and critical feature of hypersonic weapons is their speed, which allows them to travel at velocities between Mach 5 and Mach 25, corresponding to a range of 3,800 to 19,000 miles per hour (6,174 to 30,600 kilometres per hour). This incredible speed is what distinguishes hypersonic weapons from conventional missiles or aircraft, providing them with a unique advantage in terms of reaction time, target engagement, and evasion of defence systems. The ability to travel at such high speeds makes hypersonic weapons highly lethal, as they reduce the window for detection, tracking, and interception to mere seconds, offering a near-instantaneous strike capability.

At the lower end of the hypersonic spectrum, Mach 5 is five times the speed of sound, which is roughly equivalent to 1.7 kilometres per second. However, the exact speed of sound varies depending on altitude, temperature, and atmospheric conditions. For example, at sea level, the speed of sound is approximately 767 miles per hour (1,235 kilometres per hour), but this value decreases at higher altitudes where the air is thinner and temperatures are lower. In the context of hypersonic flight, vehicles must sustain speeds above Mach 5 across much of their flight path to maintain their classification as hypersonic. These speeds are not just extremely fast; they exist in a realm that places unique demands on the materials, guidance systems, and propulsion technologies used to operate these weapons.

Achieving and maintaining Mach 5 and above throughout the flight trajectory places hypersonic weapons in a highly challenging operational environment. The friction generated by moving through the atmosphere at such velocities creates intense heat—temperatures can exceed 2,000°C (3,632°F). This requires the use of advanced heat-resistant materials, such as ceramics or special alloys, to prevent the vehicle from burning up. The ability to withstand and operate under such extreme thermal stress while maintaining structural integrity is one of the key technical challenges that hypersonic weapons systems must overcome. Additionally, as these vehicles travel at hypersonic speeds, they experience rapid changes in altitude and atmospheric pressure, further complicating their operation and guidance.

Hypersonic speeds allow these weapons to significantly reduce the reaction time available to adversaries' defence systems. Traditional missile defence systems, which are designed to detect and engage slower, predictable ballistic missile trajectories, are often unable to respond quickly enough to intercept hypersonic weapons. Even if the weapon is detected, the sheer speed at which it travels can overwhelm current tracking and interception technologies. For example, a hypersonic missile traveling at Mach 10 (about 7,600 mph / 12,250 km/h) can cross a distance of 1,000 kilometres (620 miles) in approximately five minutes. This gives defence systems very little time to assess the threat, coordinate a response, and attempt to intercept the weapon.

The near-instantaneous strike capability of hypersonic weapons also offers a strategic advantage in terms of pre-emptive or precision strikes. Given their speed, these weapons can be used to target critical infrastructure, military assets, or command centres with minimal warning. By the time a hypersonic weapon is detected, it may already be too late to defend against it. This makes hypersonic weapons particularly effective for striking time-sensitive or high-value targets, such as naval ships, missile defence installations, or underground bunkers. Furthermore, because of their high speeds, hypersonic weapons are capable of overcoming many existing missile defence shields, which are optimized for intercepting slower-moving ballistic missiles.

Altitude

Hypersonic weapons are unique not only because of their speed and manoeuvrability but also because of their ability to operate at various altitudes, depending on their type. This capability allows them to exploit different parts of the atmosphere to maximize range, effectiveness, and evasiveness. Two primary types of hypersonic weapons—Hypersonic Glide Vehicles (HGVs) and Hypersonic Cruise Missiles (HCMs)—differ significantly in terms of the altitudes at which they travel, and these altitude differences offer distinct tactical advantages in warfare.

Hypersonic Glide Vehicles (HGVs) typically operate at high altitudes in the upper atmosphere, usually between 20 to 100 kilometres (about 12 to 62 miles). These altitudes lie above the typical operational range of most traditional aircraft and below the low-Earth orbit range of satellites. After being launched by a rocket into this high-altitude zone, the HGV separates from its booster and enters the glide phase. At these altitudes, the thin atmosphere presents less air resistance, allowing the vehicle to maintain its hypersonic speeds while covering vast distances. This high-altitude glide phase gives HGVs a strategic advantage in terms of both distance and evasion.

Operating at such altitudes makes HGVs challenging to track for traditional ground-based radar systems, which are designed to monitor missile trajectories in lower sections of the atmosphere. The curvature of the Earth also contributes to this difficulty in detection. Radar systems, limited by their line-of-sight range, often struggle to track objects at these high altitudes, especially when they are moving at hypersonic speeds and constantly changing course. This makes it difficult for missile defence systems to engage or intercept the weapon before it descends for the final strike. Furthermore, the high altitude gives HGVs the ability to bypass missile defence systems that are typically optimized for lower-flying or ballistic threats, increasing the weapon's survivability during its flight.

In contrast, Hypersonic Cruise Missiles (HCMs) fly at lower altitudes, typically within the Earth's atmosphere, and can operate as low as 30,000 feet (9 kilometres) or even lower, depending on the mission profile. Unlike HGVs, which use the glide mechanism for propulsion, HCMs are powered throughout their flight by engines such as scramjets. These air-breathing engines require the vehicle to stay within the denser layers of the atmosphere, as they rely on the intake of atmospheric oxygen to maintain propulsion at hypersonic speeds.

This lower altitude profile gives HCMs a different set of tactical advantages. By flying closer to the Earth's surface, these missiles can exploit the terrain and curvature of the Earth to hide from radar systems. Ground-based radar, as well as space-based tracking systems, have a harder time detecting objects flying at low altitudes, especially those moving at high speeds. The missile's ability to "hug the ground" reduces the radar cross-section and makes it difficult for detection systems to get a lock on it. This low-altitude flight path enhances the stealth of HCMs, making them harder to detect and engage.

Additionally, the lower altitude allows HCMs to conduct precision strikes against specific ground or sea-based targets, such as naval ships or fortified military installations. By flying lower, HCMs can manoeuvre more easily through complex airspaces, using terrain features like mountains and valleys to avoid detection by enemy radar. The ability to operate at lower altitudes and maintain sustained flight throughout their mission gives HCMs an advantage in reaching tactical and operational targets with extreme accuracy.

Range and Endurance

The effectiveness of hypersonic weapons in military operations is significantly influenced by their range and endurance, which are primarily determined by their design and propulsion systems. Hypersonic weapons are categorized into two main types: Hypersonic Glide Vehicles (HGVs) and Hypersonic Cruise Missiles (HCMs). Each type possesses unique capabilities regarding range and endurance, aligning with different strategic military objectives. Both HGVs and HCMs are designed to travel at speeds exceeding Mach 5, yet their operational ranges differ markedly due to their propulsion methods and flight dynamics.

HGVs generally exhibit longer ranges than HCMs. They are launched atop a rocket or ballistic missile, allowing them to ascend into the upper atmosphere before entering a glide phase. This separation from the booster enables HGVs to cover extensive distances, often exceeding thousands of kilometres, without the need for continuous propulsion, which is a limitation for HCMs. The glide capability of HGVs, combined with their high-altitude flight that minimizes air resistance, allows them to maintain speed over long distances. This characteristic makes HGVs particularly effective for intercontinental missions, as they can strike targets across vast distances while remaining challenging to detect and intercept [2, 5].

For instance, the Russian Avangard system exemplifies the extended range of HGVs, reportedly capable of reaching distances over 6,000 kilometres (3,728 miles). This intercontinental reach allows the Avangard to target locations far from its launch point, enhancing its utility for long-range precision strikes and strategic deterrence. The ability of HGVs to manoeuvre unpredictably at hypersonic speeds further complicates interception efforts by missile defence systems, thereby providing a significant strategic advantage [2, 5].

Conversely, HCMs typically have shorter operational ranges, often limited to around 1,500 kilometres (930 miles) or less. These missiles utilize air-breathing engines, such as scramjets, to sustain hypersonic speeds throughout their flight within the atmosphere. However, their reliance on continuous propulsion constrains their range, as it is directly tied to fuel capacity and engine endurance. Despite this limitation, HCMs are highly effective for regional or tactical missions, allowing military forces to engage nearby threats or high-value targets within a specific operational theatre [2, 5].

The shorter range of HCMs makes them particularly suited for targeting regional threats or conducting rapid strike missions. For example, an HCM launched from a naval vessel or aircraft can effectively engage enemy installations or military infrastructure within a confined area, such as coastal defences or airbases. The high-speed capabilities of HCMs, combined with their operational altitudes that reduce radar detection, enhance their stealth and make them difficult to defend against in regional combat scenarios [2, 5].

In terms of endurance, HGVs typically demonstrate greater endurance due to their ability to glide long distances without requiring continuous propulsion. Once in the glide phase, HGVs can travel significant distances with minimal energy expenditure, allowing them to remain operational for extended periods. This endurance, coupled with their range, makes HGVs particularly suitable for strategic deterrence missions, as they can remain undetected for much of their flight and deliver surprise attacks from great distances [2, 5].

In contrast, while HCMs can sustain high-speed flight, their endurance is limited by fuel supply and the technical constraints of their propulsion systems. Consequently, HCMs are best utilized for missions requiring rapid response over shorter distances, where their speed and precision can be leveraged effectively [2, 5].

Guidance and Navigation

Guidance and navigation are critical components of hypersonic weapons, as these systems must maintain extreme precision while the weapon travels at speeds of Mach 5 or above. At such high velocities, even small deviations from the intended flight path can lead to significant errors over long distances, potentially causing the weapon to miss its target. This challenge is further compounded by the need for hypersonic weapons to evade interception by manoeuvring unpredictably. To address these issues, hypersonic weapons rely on a combination of advanced inertial navigation systems (INS), satellite guidance, and increasingly, artificial intelligence (AI) and predictive algorithms to ensure accuracy and effectiveness.

One of the most reliable methods of navigation for hypersonic weapons is the Inertial Navigation System (INS). This system is particularly valuable because it does not depend on external signals like GPS, which can become unreliable or lost at hypersonic speeds. Instead, INS calculates the vehicle's position, velocity, and orientation using internal sensors such as accelerometers and gyroscopes. These sensors continuously track changes in the vehicle's motion, allowing the system to determine its current position relative to its starting point. Because INS is self-contained, it is immune to jamming or signal loss, which is especially important when a hypersonic weapon must navigate through contested or hostile environments where GPS signals might be blocked or degraded by enemy forces.

However, while INS provides a significant level of autonomy and reliability, it is not without limitations. Over time and distance, small errors in the sensors' readings can accumulate, causing drift—a gradual deviation from the intended course. This drift can reduce the precision of the weapon if not corrected periodically. To address this issue, INS is often paired with other guidance systems to recalibrate the navigation and correct any accumulated drift. This is where satellite-based positioning systems, such as GPS (Global Positioning System) or GLONASS (Russia's Global Navigation Satellite System), come into play.

Satellite guidance is widely used in modern weapons systems, including hypersonic weapons, to provide additional precision. These systems rely on signals from a network of satellites to determine the weapon's exact position on Earth. By receiving signals from multiple satellites, the navigation system can triangulate the weapon's location with high accuracy, ensuring that it stays on course toward its target. However, at hypersonic speeds, signal delays and loss of connection become significant challenges. The speed at which a hypersonic vehicle travels means that even slight delays in receiving satellite signals can affect the system's ability to calculate the precise location in real time. Moreover, at the high altitudes or rapid manoeuvring typical of hypersonic weapons, satellite signals can be disrupted, further complicating navigation.

To overcome these issues, hypersonic weapons often integrate multiple guidance systems to ensure continuous accuracy. For instance, while INS provides a stable baseline, satellite signals can be used intermittently to correct any drift in the system. This hybrid approach ensures that the weapon remains on course even when one system is compromised, enhancing the overall reliability of the guidance and navigation system.

As hypersonic weapons evolve, the use of Artificial Intelligence (AI) and predictive algorithms is becoming increasingly important in their guidance systems. Hypersonic vehicles must not only maintain precision but also manoeuvre evasively to avoid interception by missile defence systems. This requires real-time path optimization, which can be achieved through the integration of AI. By analysing the surrounding environment, potential threats, and the current position of the weapon, AI systems can calculate the fastest and most accurate route to the target while dynamically adjusting to avoid detection or interception.

AI enhances the weapon's ability to make split-second decisions, factoring in changes in the flight path, atmospheric conditions, or enemy responses. For example, if an incoming interceptor is detected, the AI can predict the trajectory of the threat and adjust the weapon's course to avoid it, all while continuing toward the target. This ability to calculate multiple variables in real-time makes AI an essential tool for maintaining the effectiveness of hypersonic weapons, especially in contested environments where the weapon must evade defences while staying on target.

Predictive algorithms further enhance the capability of hypersonic weapons by anticipating future changes in the flight path or environment. These algorithms can be programmed to forecast how external factors, such as wind patterns, temperature fluctuations, or the curvature of the Earth, will affect the weapon's trajectory over time. By incorporating these predictions into the navigation system, the weapon can pre-emptively adjust its course to ensure it remains on the most efficient path to the target. This level of foresight is crucial for hypersonic weapons, as the speed at which they travel requires them to react to changes much more quickly than traditional weapons systems.

The Physics of Hypersonic Flight

Aerodynamics of Hypersonic Speed

Hypersonic Speed

The term hypersonic speed refers to any velocity greater than Mach 5, which is five times the speed of sound. This threshold marks a significant point in aerospace engineering because at such speeds, several physical effects begin to emerge, posing complex challenges. Hypersonic objects traveling through the dense layers of Earth's atmosphere experience massive heat flux, which subjects the vehicle's surface to extreme temperatures. The resulting aerodynamic heating, combined with the need to maintain stability and control at such speeds, makes the development of hypersonic vehicles both difficult and expensive. Achieving and sustaining hypersonic speeds requires advanced materials, propulsion systems, and thermal management, as the physical stresses increase exponentially beyond Mach 5 [64].

However, defining hypersonic speed as anything above Mach 5 introduces complications, particularly because the speed of sound—which forms the basis for the Mach scale—varies depending on altitude. The speed of sound, or Mach 1, is not constant; it depends on factors like the chemical composition of the air and, crucially, temperature. As the US Standard Atmosphere model shows, temperature fluctuates with altitude, which means that the actual speed corresponding to Mach 5 in kilometres per hour will change depending on the altitude at which a vehicle is traveling. For example, at lower altitudes where the air is denser,

Mach 5 will represent a different speed than at higher altitudes where the atmosphere is thinner and colder. This variation introduces ambiguity in defining hypersonic speeds and makes it harder to set clear benchmarks for what qualifies as hypersonic flight [64].

There's another issue with using Mach numbers as a reference at high altitudes where the air thins out. The Mach scale becomes less meaningful as the surrounding gas density decreases with altitude. While scientists and engineers can still use Mach numbers to define speeds at altitudes around 30 kilometres (where weather balloons and specialized planes can reach), the atmosphere becomes too thin at much higher altitudes—around 300 kilometres—where satellites orbit the Earth. At such altitudes, there is insufficient air for the concept of Mach numbers to apply. This thinning atmosphere makes it difficult to determine a clear altitude limit for when Mach numbers should still be used, creating challenges for defining the boundaries of hypersonic flight and, by extension, hypersonic missiles [64].

Given that hypersonic speed refers to velocities greater than Mach 5, which is five times the speed of sound, or approximately 3,800 mph (6,174 km/h). At these extreme speeds, the aerodynamic behaviour of vehicles differs significantly from that of subsonic or supersonic flight. The forces and physical phenomena experienced by objects moving at hypersonic speeds present unique challenges to vehicle design, materials, and stability, making the study of hypersonic aerodynamics a critical area in aerospace engineering, especially for Hypersonic Glide Vehicles (HGVs), missiles, and spacecraft re-entry vehicles.

Characteristics of Hypersonic Flow

At hypersonic speeds (Mach 5 and above), the behaviour of airflow around a vehicle changes significantly compared to subsonic or supersonic flight. While there is no strict boundary between supersonic and hypersonic regimes, hypersonic flow is marked by distinct physical phenomena that cannot be discounted analytically, such as shock waves, aerodynamic heating, and real gas effects. These phenomena introduce significant challenges for designing vehicles that can survive and perform at such extreme speeds.

The terms "subsonic" and "supersonic" are traditionally used to describe speeds below and above the local speed of sound, respectively. However, in aerodynamics, these terms refer to specific ranges of Mach values. For instance, when an aircraft approaches transonic speeds (around Mach 1), it enters a regime where the usual Navier–Stokes equations begin to break down. This is because, even though the freestream flow may not be supersonic, parts of the airflow around the aircraft can exceed Mach 1, requiring more sophisticated methods to manage the complexities of transonic flow.

The supersonic regime typically includes a range of Mach numbers where linearized theories can still apply. In these conditions, the airflow is not chemically reacting, and heat transfer between the vehicle and the surrounding air can be reasonably neglected. NASA defines high hypersonic speeds as between Mach 10 and 25, and re-entry speeds as anything above Mach 25. Spacecraft such as the Soyuz, Dragon capsules, the Space Shuttle, and newer reusable spacecraft like the SpaceX Starship and Rocket Lab Electron operate in these regimes.

Subsonic speeds (below Mach 0.8, or 614 mph) are typically achieved by commercial aircraft and other propeller-driven planes. These planes have high-aspect-ratio wings and rounded features, such as their noses and leading edges, to ensure efficient airflow at subsonic velocities.

In the transonic regime (Mach 0.8–1.2), aircraft experience airflow that transitions between subsonic and supersonic conditions across different parts of the body. Swept wings and supercritical designs are often employed to delay the onset of wave drag, as seen in aircraft like the Northrop X-4 Bantam.

At supersonic speeds (Mach 1.2–5), all of the airflow over an aircraft is supersonic. These aircraft are designed with sharp edges, thin aerofoil sections, and all-moving tailplanes or canards. Examples include the XB-70 Valkyrie, SR-71 Blackbird, and Concorde, which operate in this range.

Hypersonic speeds (between Mach 5 and 10) introduce additional design complexities. Vehicles operating in this range, such as the X-15, NASA X-43, and the BrahMos-II, must account for extreme aerodynamic heating and interference effects between components. These craft are highly integrated, with cooled nickel or titanium skins, and their designs often require a complete understanding of how air flows around each component, making small design changes impactful across the entire vehicle.

In the high-hypersonic regime (Mach 10–25), such as with the HTV-2, Agni-V, and Avangard, thermal control becomes a dominant design factor. Vehicles in this regime must contend with chemically reacting flows, as free atomic oxygen and high temperatures cause corrosion and radiative heat transfer. These vehicles are often blunt-shaped to reduce aerodynamic heating.

Finally, re-entry speeds (above Mach 25) involve vehicles like the Space Shuttle and Boeing X-37, which must use ablative heat shields to survive the extreme temperatures experienced during atmospheric re-entry.

As Mach numbers increase, different flow regimes require specialized design adaptations:

1. Subsonic: Streamlined shapes and rounded edges optimize airflow.

2. Transonic: Swept and supercritical wings manage drag and transition zones.

3. Supersonic: Sharp edges and delta wings balance high-speed aerodynamics with low-speed handling.

4. Hypersonic: Integrated designs with materials that withstand extreme temperatures.

5. High-Hypersonic and Re-entry: Blunt shapes and ablative heat shields counteract the severe thermal environment.

These Mach regimes illustrate the wide variety of challenges faced at different speeds, particularly as vehicles transition into hypersonic and re-entry conditions, where aerodynamic and thermal stresses reach their extremes.

The following provides an explanation of the primary characteristics of hypersonic flow:

At hypersonic speeds, vehicles generate intense shock waves as they move through the atmosphere. The shock layer is the region between the vehicle's surface and the bow shock, which forms in front of the vehicle. In this region, the air is highly compressed and heated as it interacts with the vehicle's surface. The shock layer is much thinner in hypersonic flow than at lower speeds due to the sharp compression of air. This results in strong pressure, density, and temperature gradients across the shock layer, which significantly affect the vehicle's aerodynamic performance and thermal management.

One of the most challenging aspects of hypersonic flight is aerodynamic heating, caused by the friction between the vehicle and the surrounding air. At hypersonic speeds, air molecules cannot move out of the way fast enough, leading to extreme compression and heating of the air as it is forced to flow around the vehicle. This process generates temperatures of 2,000°C to 3,000°C (3,632°F to 5,432°F) or even higher, which can severely damage or destroy a vehicle unless it is protected by advanced thermal management systems, such as ablative materials or ceramic composites.

Aerodynamic heating also affects structural materials and internal systems, which must be able to withstand these extreme temperatures. Thermal protection becomes a major design consideration, particularly for reusable vehicles or those designed for long-duration hypersonic flight.

The entropy layer forms when the air passing through the shock wave experiences a sharp increase in entropy due to the intense compression and heating. As the Mach number increases, the change in entropy across the shock becomes more significant. This creates a highly vortical flow with a strong entropy gradient that interacts with the boundary layer around the vehicle. The mixing of the entropy layer with the boundary layer adds to the complexity of the flow and can influence the overall stability and control of the vehicle at hypersonic speeds.

At hypersonic velocities, the air can no longer be treated as an ideal gas. The extreme temperatures resulting from shock compression and aerodynamic heating cause air molecules to undergo dissociation and ionization. These real gas effects introduce additional challenges to hypersonic flow because the properties of the air change dramatically.

For example, diatomic gases such as nitrogen and oxygen begin to dissociate into individual atoms when exposed to high temperatures, which alters the flow properties. As speeds increase further, these atoms can become ionized, creating a plasma around the vehicle. This plasma affects heat transfer, radiation, and communication with the vehicle, as the ionized gases can block radio signals.

At high altitudes, where hypersonic vehicles often operate, the air density is much lower. This creates a unique flow environment where the air's interaction with the vehicle becomes rarified, meaning that the assumptions of continuum flow—where the air behaves as a continuous medium—no longer apply. In low-density environments, the number of collisions between air molecules decreases, and the flow becomes dominated by molecular interactions rather than fluid dynamics. This requires specialized models to account for rarefied gas behaviour and thermal management in hypersonic vehicles.

At hypersonic speeds, aerodynamic coefficients such as drag coefficient (Cd) and lift coefficient (Cl) become less dependent on the vehicle's Mach number. In lower-speed regimes, these coefficients typically change as the speed increases. However, in the hypersonic regime, the flow around the vehicle stabilizes to a point where these coefficients remain relatively constant over a wide range of Mach numbers, especially beyond Mach 10. This behaviour allows for more predictable modelling of vehicle performance at extreme speeds.

As the Mach number increases, the density behind the bow shock increases, leading to a decrease in volume behind the shock due to the conservation of mass. This results in a smaller distance between the bow shock and the surface of the vehicle. At higher Mach numbers, the shock stand-off distance becomes very small, creating sharp gradients in pressure and temperature that can directly impact the vehicle's structural integrity. This small stand-off distance presents challenges for thermal management, as the heat generated by the shock layer must be quickly dissipated.

As the vehicle travels at high Mach numbers, the increase in entropy across the shock leads to a strong entropy gradient. This gradient interacts with the boundary layer, the thin region of slower-moving air in contact with the vehicle's surface. At hypersonic speeds, a significant portion of the vehicle's kinetic energy is converted into internal energy due to viscous effects, which manifest as a sharp increase in temperature within the boundary layer. The density decreases in the boundary layer due to the temperature rise, causing the boundary layer to expand. In some cases, the boundary layer can grow so thick that it merges with the shock wave near the leading edge of the vehicle, further complicating the aerodynamics and heat management.

At hypersonic speeds, the flow around the vehicle reaches extremely high temperatures, resulting in non-equilibrium chemical reactions such as vibrational excitation, dissociation, and ionization of the air molecules. These high temperatures cause changes in the flow's chemical composition, leading to convective and radiative heat transfer that affects both the structural materials and the aerodynamic properties of the vehicle. These reactions also lead to heat flux from both convective and radiative sources, which further complicates thermal protection strategies.

The characteristics of hypersonic flow introduce significant challenges for vehicle design, including shock layers, aerodynamic heating, entropy gradients, and real gas effects. As Mach numbers increase, the shock stand-off distance decreases, leading to intense pressure and temperature gradients that require specialized thermal management systems and advanced materials to protect the vehicle. Understanding and managing these phenomena are critical for developing hypersonic vehicles that can survive the extreme conditions of high-speed atmospheric flight.

Mach and Reynolds Numbers

In the study of aerodynamics, Mach number and Reynolds number are two fundamental dimensionless parameters that help categorize fluid flows, particularly in the context of compressibility and the relative importance of inertial and viscous forces. These parameters are often introduced through dimensional analysis and are central to understanding the behaviour of different aerodynamic flows [95].

Mach Number

The Mach number is defined as the ratio of the local flow velocity to the local speed of sound. It is used to quantify the compressibility effects in a flow and is expressed mathematically as [95]:

$$M = \frac{V}{a}$$

Where:

- V is the flow velocity.
- a is the local speed of sound.

The Mach number plays a crucial role in determining whether a flow is subsonic, transonic, supersonic, or hypersonic. At subsonic speeds (Mach < 1), the compressibility effects of the gas are minimal. As the Mach number increases and approaches transonic speeds (Mach ≈ 1), these effects become more pronounced, and shock waves can form, leading to complex changes in flow behaviour. Supersonic flows (Mach > 1) introduce shock waves and wave drag, which greatly affect the aerodynamics of the vehicle.

The speed of sound itself depends on the local temperature of the gas and can be determined using the equation:

$$a = \sqrt{\gamma RT}$$

Where:

- γ is the ratio of specific heats.
- R is the gas constant.
- T is the absolute temperature.

Thus, the Mach number varies throughout a flow, depending on both the flow velocity and temperature changes caused by compressibility effects. At high Mach numbers (Mach > 5), or hypersonic speeds, additional phenomena such as kinetic heating and chemical dissociation of gases become critical, making the flow difficult to model and predict.

Reynolds Number

The Reynolds number (Re) is another critical dimensionless parameter in aerodynamics that measures the ratio of inertial forces to viscous forces in a fluid flow. It is defined as [95]:

$$Re = \frac{\rho V L}{\mu}$$

Where:

- ρ is the fluid density.
- V is the velocity of the flow.
- L is the characteristic length (e.g., chord length of a wing).
- μ is the dynamic viscosity of the fluid.

The Reynolds number helps determine whether a flow is laminar or turbulent. Laminar flows, characterized by smooth streamlines, occur at lower Reynolds numbers, where viscous forces dominate. At higher Reynolds numbers, turbulent flow develops, leading to chaotic mixing and increased skin friction drag.

In aerodynamic design, increasing the Reynolds number can enhance aerodynamic efficiency by reducing viscous drag. For example, a higher Reynolds number results in thinner boundary layers, which means lower drag over an aircraft's surfaces. However, at very high Reynolds numbers, turbulent boundary layers form, contributing to more drag but also better stability in controlling flow separation.

Physical Interpretation and Interaction

The interaction between Mach and Reynolds numbers is vital in understanding compressible and incompressible flows. The Mach number is primarily concerned with compressibility, while the Reynolds number focuses on viscous effects. In many practical aerodynamic applications, both parameters need to be considered to understand the flow characteristics [95].

For instance, at low Mach numbers (subsonic flows), the Reynolds number is more important for predicting the drag and lift characteristics of a wing. As the Mach number increases to supersonic and hypersonic speeds, the importance of the Mach number grows because shock waves, compressibility effects, and wave drag become dominant.

Mach Number Regimes

The Mach number is used to classify different regimes of flight [95]:

- **Subsonic (Mach < 1)**: Air flows smoothly over the surface, and compressibility effects are negligible.
- **Transonic (Mach 0.8 – 1.2)**: Compressibility effects start to appear, with shock waves forming on the surface of the aircraft.
- **Supersonic (Mach 1.2 – 5)**: Shock waves become a dominant feature, causing wave drag and a need for streamlined designs.
- **Hypersonic (Mach > 5)**: Extreme kinetic heating and molecular dissociation occur, complicating vehicle design.

Compressible Flows and Shock Waves

As an aircraft accelerates to supersonic speeds, disturbances in the flow cannot propagate upstream, leading to the formation of shock waves. At Mach 1, a shock wave forms perpendicular to the aircraft's direction, and as the speed increases, the Mach angle decreases, confining the shock waves to a narrow cone behind the aircraft. The steep angle of shock waves at hypersonic speeds also leads to intense wave drag and heating effects [95].

Reynolds Number and Flow Stability

The Reynolds number is critical for understanding boundary layer behaviour. A high Reynolds number generally results in thinner boundary layers, while a low Reynolds number can cause early flow separation and increased drag. The transition from laminar to turbulent flow is particularly important in determining the aerodynamic performance of airfoils and wings. For example, in subsonic flows, laminar boundary layers minimize drag, but at supersonic speeds, the turbulent boundary layers can prevent flow separation and reduce drag under certain conditions [95].

Both the Mach and Reynolds numbers play crucial roles in understanding aerodynamic flows. The Mach number helps determine the compressibility and shock wave formation, while the Reynolds number helps assess the balance between inertia and viscous forces. Together, these parameters enable engineers to predict flow behaviour and optimize the performance of aircraft, missiles, and spacecraft across a wide range of speeds and conditions. Understanding their implications helps solve aerodynamic problems, from improving the efficiency of airfoils to designing vehicles that operate at supersonic and hypersonic speeds [95].

Shock Waves and Flow Regimes

When an object travels at hypersonic speeds, the aerodynamic behaviour of the air around it changes drastically. The vehicle experiences intense shock waves, which form because the air cannot move out of the way fast enough to allow the object to pass smoothly. These shock waves create sharp variations in pressure, temperature, and density around the vehicle, posing significant challenges to its structural integrity, aerodynamic stability, and control.

Shock waves are a hallmark of hypersonic flight and play a critical role in determining the overall performance and survivability of the vehicle. Two primary types of shock waves are formed at these speeds: bow shocks and oblique shocks.

Shock waves are indeed a hallmark of hypersonic flight, significantly influencing the performance and survivability of hypersonic vehicles. At hypersonic speeds, defined as speeds greater than Mach 5, two primary types of shock waves are generated: bow shocks and oblique shocks. The formation and characteristics of these shock waves are critical to understanding the aerodynamics of hypersonic vehicles.

Bow shocks are formed when a hypersonic vehicle travels through the atmosphere, creating a region of compressed air in front of it. This phenomenon is primarily due to the friction between the vehicle and the atmosphere, resulting in substantial heat generation. Zhao et al. [96] emphasize that the bow shock is a direct consequence of this interaction, leading to extreme thermal environments that can challenge vehicle integrity. Furthermore, the strength of the bow shock can be influenced by the vehicle's geometry, as noted by Kumar and Thornber [97], who observed that variations in cowl lip radii can affect the intensity of the bow shock compared to oblique shocks.

Oblique shocks, on the other hand, occur when the flow is deflected at an angle, typically at compression surfaces or ramps on the vehicle. These shocks are crucial for the operation of scramjet engines, which rely on the compression of incoming air for efficient combustion. Jiang et al. [98] highlight the significance of oblique shock wave interactions in the context of shock wave/turbulent boundary layer interactions, which are essential for the aerodynamic performance of hypersonic vehicles. The interaction between these oblique shocks and the boundary layer can lead to complex flow phenomena that must be carefully managed to ensure vehicle stability and performance.

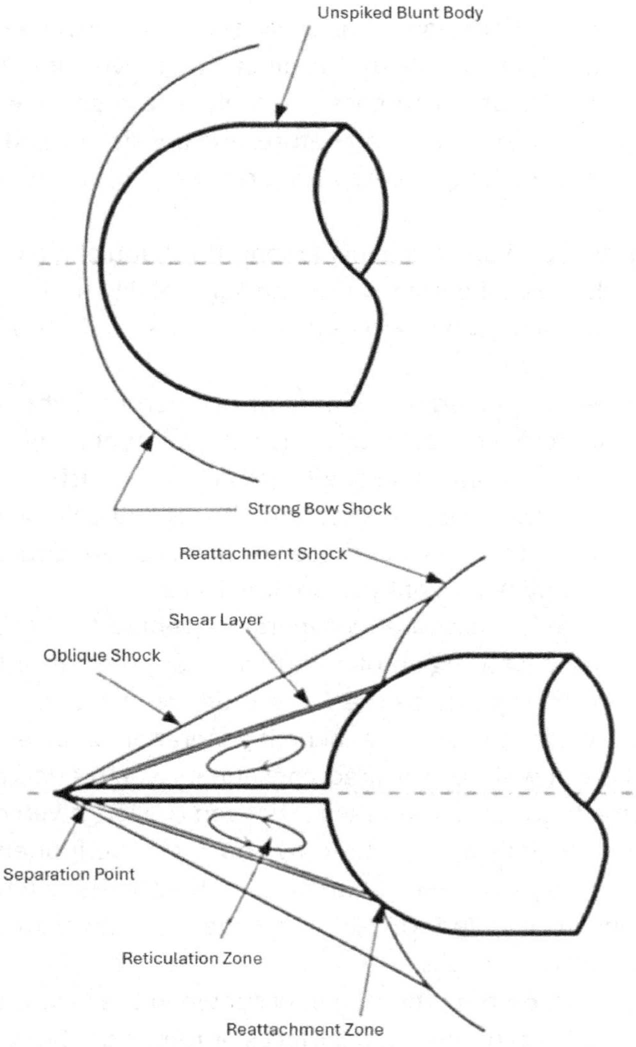

Figure 8: Features of flowfield around a blunt body with and without a spike during hypersonic flights.

The implications of shock waves extend beyond mere aerodynamic performance; they also play a vital role in thermal management. The intense heating associated with shock waves can lead to severe aerodynamic heating, which poses risks of surface ablation and structural failure. Wang et al. [99] discuss the challenges posed by aerodynamic heating in hypersonic flight, emphasizing the need for advanced thermal protection systems to mitigate these effects. Additionally, the interaction of shock waves with boundary layers can amplify temperature fluctuations and pressure variations, further complicating the thermal environment around the vehicle [100].

Bow Shocks at Hypersonic Speeds

A bow shock forms in front of the vehicle as it travels through the atmosphere at hypersonic speeds. This occurs because the air in front of the vehicle is compressed so quickly that it cannot flow smoothly around the body. Instead, the air is forced into a highly compressed and heated region, creating a shock wave that wraps around the front of the vehicle, much like a bow wave in front of a ship traveling through water [97, 101].

- **Compression and Heating**: As the air is compressed by the vehicle's movement, its temperature rises dramatically, reaching thousands of degrees Celsius. This creates a layer of superheated air in front of the vehicle, generating intense heat that is transferred to the vehicle's surface. Managing this heat is one of the most significant challenges in hypersonic vehicle design. The thermal protection systems must be capable of withstanding these extreme conditions to prevent damage or failure.

- **Increased Drag**: The bow shock also increases aerodynamic drag, which is the resistance the vehicle faces as it moves through the air. The drag force generated by the bow shock is much higher than what is encountered in subsonic or supersonic flight. At hypersonic speeds, this drag can significantly reduce the vehicle's speed if not properly managed. To minimize drag, hypersonic vehicles often feature aerodynamically efficient shapes, such as sharp noses and wedge-like surfaces, designed to reduce the intensity of the bow shock.

The bow shock is a key factor in the aerodynamic environment of hypersonic vehicles. It directly impacts lift, drag, and heat transfer, which are critical to maintaining the vehicle's structural integrity and overall performance during flight.

The bow shock is a fundamental aspect of the aerodynamic environment surrounding hypersonic vehicles. It plays a pivotal role in determining the vehicle's aerodynamic performance, influencing both lift and drag while also posing significant challenges related to thermal management. The design of hypersonic vehicles must therefore carefully consider the implications of bow shock formation, ensuring that thermal protection systems are robust enough to handle the extreme conditions encountered during flight. The interplay between aerodynamic forces and thermal stresses necessitates a multidisciplinary approach to vehicle design, integrating insights from fluid dynamics, materials science, and structural engineering [102, 103].

Oblique Shocks at Hypersonic Speeds

In addition to the bow shock that forms in front of the vehicle, oblique shocks occur along the sides and body of the vehicle. Unlike bow shocks, which are perpendicular to the vehicle's motion, oblique shocks are angled relative to the surface of the vehicle. These shocks occur as the compressed air flows around the vehicle's body and over its aerodynamic surfaces.

- **Pressure and Temperature Gradients**: One of the primary effects of oblique shocks is the generation of pressure and temperature gradients along the vehicle's surface. These gradients result in sharp variations in pressure and temperature, which can create high-pressure zones in certain areas and low-pressure zones in others. Such uneven pressure distributions are crucial for the stability and control of the vehicle during flight. If these gradients are not effectively managed, they can lead to instability, complicating the vehicle's ability to maintain a stable flight path at hypersonic speeds [97, 104]. The interaction of oblique shocks with the boundary layer can also result in significant total pressure losses, which are known to be high in stream tubes crossing strong shocks, with losses proportional to the square of the Mach number [104].

- **Flow Separation**: Moreover, oblique shocks can induce flow separation, a condition where the airflow no longer adheres to the vehicle's surface. This phenomenon can drastically affect the vehicle's aerodynamic performance, leading to increased drag and loss of control [105]. The design of hypersonic vehicles must therefore incorporate strategies to mitigate the adverse effects of oblique shocks. This includes optimizing the placement and strength of these shocks through careful design of aerodynamic surfaces, such as employing sharp leading edges and angled surfaces to guide airflow effectively [106]. For instance, the use of waverider configurations has been shown to leverage the properties of oblique shocks to enhance performance across a range of speeds [106].

Interaction of Shock Waves with Vehicle Aerodynamics

The presence of bow shocks and oblique shocks at hypersonic speeds leads to strong pressure gradients around the vehicle, which drastically affect its aerodynamic performance. Air at hypersonic speeds behaves like a compressible fluid, meaning that its density changes significantly as it is compressed by the vehicle. This compressibility causes the formation of high-pressure zones behind the shock waves, particularly in regions where the air is most compressed, such as along the front and underside of the vehicle.

- **Lift and Stability**: The high-pressure zones behind shock waves contribute to the generation of lift, but they also create significant aerodynamic instability. Hypersonic vehicles must carefully balance the lift generated by these pressure zones with the stabilizing forces provided by control surfaces, such as fins or wings. Because of the intense pressures and rapid airflow changes, maintaining precise control over the vehicle becomes more challenging as speed increases. This dynamic environment can lead to complex interactions between aerodynamic forces and structural responses, potentially resulting in phenomena such as flutter or limit cycle oscillations [107, 108].

- **Drag and Heat**: Both bow and oblique shocks contribute to the overall drag force acting on the vehicle. At hypersonic speeds, the drag increases exponentially, requiring efficient aerodynamic designs to minimize resistance. Additionally, the heat transfer from the compressed and superheated air to the vehicle's surface can cause material degradation or failure if not properly managed. To mitigate these risks, advanced thermal protection systems, including ablative materials and ceramic composites, are employed to shield the vehicle from the extreme thermal loads encountered during hypersonic flight [109-111].

The aerodynamic behaviour of vehicles traveling at hypersonic speeds is dominated by the formation of intense shock waves, including bow shocks in front of the vehicle and oblique shocks along its surface. These shock waves create sharp changes in pressure, temperature, and density around the vehicle, posing significant challenges for structural integrity, heat management, and aerodynamic control.

The interaction of these shocks with the vehicle's surface results in strong pressure gradients that affect the lift, drag, and stability of the vehicle during flight. To overcome these challenges, hypersonic vehicles are designed with streamlined shapes and advanced materials to withstand the extreme conditions generated by shock waves, allowing them to maintain high-speed flight while minimizing drag and heat transfer.

In summary, the aerodynamic performance of hypersonic vehicles is profoundly affected by the presence of shock waves, which create strong pressure gradients and high-pressure zones. These factors contribute to both lift generation and drag, while also introducing significant challenges in terms of stability and thermal management. The design of hypersonic vehicles must therefore integrate advanced aerodynamic principles with robust thermal protection strategies to ensure safe and effective operation at these extreme speeds [112, 113].

Spiked Blunt Bodies for Hypersonic Flights

Since the introduction of hypersonic vehicles in the late 1940s, continuous research has been dedicated to overcoming the formidable challenges associated with high-speed atmospheric flight. Two primary objectives drive this research: reducing aerodynamic heating during re-entry and minimizing drag during atmospheric operations. Achieving these goals necessitates altering the flowfield pattern around the vehicle's blunt body, specifically by eliminating the strong detached bow shock that typically forms. A myriad of techniques have been explored and implemented to modify the flowfield, including the use of spikes, forward-facing cavities, aero discs, plasma torches, focused gas jets, laser or microwave beams directed upstream of the nose stagnation point, arc discharges, DC corona discharges, and non-ablative thermal protection systems. Among these diverse approaches, the spiked blunt body has emerged as the most promising due to its simplicity, efficiency, and effectiveness [114].

The spiked blunt body concept involves attaching a slender rod, or spike, at the stagnation point of a blunt body. This seemingly simple addition brings about two significant modifications to the vehicle's flowfield. Firstly, the spike transforms the strong, detached bow shock typically generated in front of the blunt nose into a series of weaker oblique shock waves. Secondly, the spike acts as a flow separator, enforcing boundary layer separation and creating a shear layer. These changes result in the formation of a recirculation zone behind the spike, which plays a crucial role in reducing both drag and aerodynamic heating [114].

In the absence of a spike, a blunt body traveling at hypersonic speeds generates a strong, detached bow shock wave ahead of it. As the vehicle moves, fluid particles encounter this shock wave, experiencing a sudden deceleration. This leads to a significant increase in pressure and temperature behind the shock, contributing to high levels of pressure drag and intense aerodynamic heat transfer to the vehicle's surface. To mitigate these effects, designers often opt for a larger nose radius, as the wall heat flux is inversely proportional to the square root of the nose radius. A larger radius effectively reduces heat transfer to the vehicle by scaling back the heat and dissipating it into the surrounding air, thereby protecting the vehicle from extreme thermal stresses [114].

Introducing a spike at the stagnation point fundamentally alters the hypersonic flow around the vehicle. The slender rod replaces the strong bow shock with a foreshock, a system of weaker oblique shock waves. This transition significantly lowers the rise in pressure and temperature at the front of the vehicle. Additionally, the spike induces flow separation in the boundary layer, leading to the creation of a shear layer. This shear layer forms a recirculation zone of low-pressure, low-temperature air behind the spike, effectively shielding the blunt forebody from the direct impact of the high-pressure, high-temperature air flow. Consequently, this recirculation zone results in a substantial reduction in both drag and aerodynamic heating [114].

Despite the advantages, the introduction of a spike creates new aerodynamic challenges, particularly at the point where the shear layer reattaches to the blunt body. Here, the pressure and heat flux on the vehicle's surface experience a sudden increase, leading to a reattachment shock. The strength of this shock is influenced by the flow turning angle, which depends on the location of the reattachment point on the nose surface. The position of the reattachment point can be adjusted by altering the nose radius and spike length, thereby influencing the overall aerodynamic performance. The reattachment shock must be carefully managed to prevent excessive pressure spikes and thermal loads on the vehicle's structure [114].

At certain flow conditions, especially with specific nose shapes and spike dimensions, the flow around a spiked blunt body can become unstable. These instabilities manifest as either pulsatory or oscillatory modes. In the pulsatory mode, the shape of the foreshock and the recirculation zone undergo drastic changes, characterized by high-frequency oscillations and significant fluctuations in surface pressure and heat flux. This mode can lead to increased acoustic and structural loads, spike bending, and flight perturbations, complicating flight control. In contrast, the oscillatory mode involves smaller, more regular changes in the flow pattern, with lower oscillation frequencies and reduced pressure amplitudes. Managing these flow instabilities is crucial for maintaining structural integrity and flight stability. Research indicates that flow stability is highly dependent on the nose contour and spike geometry, necessitating the optimization of these parameters for each specific flight condition to ensure optimal aerothermodynamic performance and structural stability [114].

The aerodynamic benefits of a spiked blunt body are significantly influenced by various geometric parameters, including nose shape, spike length, and spike tip geometry.

Spike Length: Increasing the spike length generally enhances the performance of the vehicle by expanding the recirculation zone and reducing the reattachment shock strength. As the spike length increases, the reattachment point moves toward the shoulder of the blunt nose, and the recirculation region covers a larger area of the nose. This shift reduces the angle required to deflect the flow outside the shear layer, thereby decreasing the pressure increase on the nose surface. However, beyond a certain critical spike length, the recirculation zone can split into two bubbles, with a shock wave forming between them, leading to an increase in overall drag. Therefore, an optimal spike length exists where drag reduction and aerodynamic heating are minimized without causing flow separation issues.

Spike Tip Geometry: The geometry of the spike tip plays a pivotal role in the aerodynamic performance of the spiked blunt body. Various spike tip shapes, such as sharp points, flat-head aero discs, hemispherical tips, and biconical shapes, have been experimented with. Among these, aerodisks have consistently shown superior performance in terms of thermal protection and drag reduction compared to pointed spikes. Double-headed profiles also offer better drag reduction than single-profile spikes. Furthermore, flat aerodisks generate the least drag compared to biconical, hemispherical, or pointed spikes when the spike length is fixed. This improved performance is attributed to the aerodisks' ability to more effectively disrupt the strong bow shock and enhance the recirculation zone.

Disk Diameter: While spike length is crucial, the disk diameter at the spike tip also influences aerodynamic performance. Increasing the disk diameter shifts the reattachment point towards the nose shoulder, thereby expanding the recirculation zone and reducing surface pressure. However, beyond a critical disk diameter, the bowed foreshock becomes stronger, which can negate the benefits gained from the expanded recirculation zone, leading to increased drag. Therefore, an optimal disk diameter is essential and is typically inversely proportional to the spike length. This means that the spike length should be greater than the disk diameter to ensure stable flow, reduced drag, and lower aerodynamic heating. Additionally, regardless of the spike tip shape, drag increases with the angle of attack over a broad range of Mach numbers. To address this, some researchers have proposed the concept of a pivoting spike, where the spike remains aligned with the freestream direction while the vehicle body aligns with the flight path. This alignment helps maintain consistent drag levels even when the vehicle experiences changes in angle of attack.

The integration of a spike on a blunt body in hypersonic vehicles offers a promising solution to the dual challenges of reducing aerodynamic heating and drag. By transforming strong bow shocks into weaker oblique shocks and creating a recirculation zone that shields the forebody, spiked blunt bodies achieve significant reductions in both pressure drag and heat flux. However, the effectiveness of this approach is highly dependent on precise geometric design and flow stability. Optimal spike length, tip geometry, and disk diameter are critical parameters that must be carefully balanced to maximize performance while mitigating potential flow instabilities. Continued research and experimentation in these areas hold the key to advancing hypersonic vehicle technology, enabling safer and more efficient high-speed atmospheric and re-entry operations [114].

Drag Forces at Hypersonic Speeds

At hypersonic speeds, defined as Mach 5 and above, aerodynamic drag becomes a critical challenge for vehicle design. The nature of drag at these velocities is complex, as it increases significantly with speed, necessitating innovative engineering solutions to manage it effectively. The primary contributors to drag in hypersonic flight are wave drag and skin friction drag, both of which pose significant threats to vehicle performance, energy efficiency, and stability.

Wave drag, which arises from the formation of shock waves as the vehicle travels through the atmosphere, is particularly pronounced at hypersonic speeds. Research indicates that shock wave drag can account for approximately two-thirds of the total drag experienced by hypersonic vehicles [115]. This phenomenon necessitates careful design considerations to minimize its impact. For instance, the integration of aerodynamic features such as aerospikes and counterflow jets has been shown to effectively reduce wave drag by altering the flow characteristics around the vehicle [116, 117]. These methods not only mitigate drag but also help in managing the thermal loads experienced during hypersonic flight, as the reduction in drag correlates with a decrease in heat transfer to the vehicle's surface [115, 116].

Skin friction drag, while less dominant than wave drag at hypersonic speeds, still plays a significant role. Studies have demonstrated that turbulent boundary layers can contribute to about 30% of the total drag even in hypersonic conditions [118]. Techniques such as heated wall blowing have been explored to control the boundary layer and reduce skin friction drag, highlighting the multifaceted approach required to tackle drag in hypersonic flight [118]. Furthermore, the overall aerodynamic performance of hypersonic vehicles is heavily influenced by their geometric configurations, which must be optimized to achieve favourable lift-to-drag ratios [119, 120].

The management of aerodynamic heating is another critical aspect intertwined with drag reduction. As vehicles encounter extreme speeds, the aerodynamic heating can lead to structural failures if not adequately addressed. The design of thermal protection systems (TPS) is thus essential, and advancements in drag reduction techniques can also contribute to thermal management by lowering the heat flux experienced by the vehicle [121]. The coupling of fluid dynamics and thermal analysis is crucial in the design process, as it allows for a holistic approach to vehicle performance optimization [121].

The challenges posed by aerodynamic drag at hypersonic speeds require a comprehensive understanding of both wave and skin friction drag, alongside effective thermal management strategies. The integration of innovative design features and advanced computational methods is essential for the development of efficient and stable hypersonic vehicles.

Wave Drag

Wave drag is a type of aerodynamic drag that arises from the formation of shock waves around the vehicle as it moves through the atmosphere at hypersonic speeds. When an object travels at these velocities, the air in front of the vehicle becomes compressed so rapidly that it cannot move out of the way smoothly. This compression creates a bow shock wave in front of the vehicle, which leads to a sharp increase in pressure.

The intense pressure differential between the front and the rear of the shock wave results in significant drag forces. The region where the bow shock meets the vehicle's surface experiences the most pressure, and this pressure difference directly contributes to wave drag. Wave drag is a dominant force at hypersonic speeds, and it can severely diminish the vehicle's momentum and energy if not managed properly. As such, reducing wave drag is essential for maintaining the vehicle's speed and stability, and much of the vehicle's design is focused on controlling this effect.

Skin Friction Drag

In addition to wave drag, skin friction drag plays a crucial role at hypersonic speeds. Skin friction drag occurs due to the boundary layer, a thin layer of air that comes into direct contact with the surface of the vehicle. Within this layer, shear forces develop as the air molecules rub against the surface, creating friction.

At hypersonic speeds, the air molecules in the boundary layer move at extremely high velocities, often leading to turbulent flow. Turbulence within the boundary layer increases friction, which in turn raises the overall drag experienced by the vehicle. The high-energy turbulent flow, combined with the heat generated by aerodynamic friction, exacerbates skin friction drag, making it another major contributor to the total drag forces acting on the vehicle.

While wave drag is more significant in hypersonic conditions, skin friction drag cannot be ignored. Its impact, especially when combined with shock waves generated by the vehicle, creates additional resistance that must be accounted for in the vehicle's design.

Mitigating Drag at Hypersonic Speeds

To address the severe drag forces at hypersonic speeds, vehicle designs must incorporate aerodynamic efficiency while balancing stability and thermal protection. One of the primary strategies for m tigating drag is the use of streamlined, conical, or wedge-shaped bodies. These shapes are favoured because they minimize the size of the bow shock and reduce the pressure build-up at the leading edges, which helps in controlling wave drag.

A conical shape directs airflow more smoothly over the surface of the vehicle, minimizing the intensity of shock waves. The streamlined design also allows the vehicle to cut through the air more effectively, reducing the overall drag forces. Wedge-shaped bodies are similarly designed to deflect airflow in a way that reduces drag and improves flight stability.

Additionally, advanced materials and thermal coatings are often employed to manage the heating effects caused by skin friction drag and to prevent structural damage due to the high temperatures generated during flight. The material choices for hypersonic vehicles are critical to withstanding the immense thermal stress while maintaining the vehicle's integrity.

Extreme Heating and Aerodynamic Heating

Hypersonic flight presents unique challenges primarily due to extreme aerodynamic heating, which is a consequence of both friction and shock wave formation as vehicles travel at speeds exceeding Mach 5. The interaction between the vehicle and the atmosphere leads to significant thermal stresses that can compromise structural integrity if not effectively managed.

One of the primary sources of heat during hypersonic flight is friction between the vehicle's surface and the high-speed air. As the vehicle moves through the atmosphere, the air is compressed, generating enormous amounts of heat due to viscous dissipation. This phenomenon can elevate surface temperatures to several thousand degrees Celsius, necessitating robust thermal protection systems (TPS) to prevent damage to the vehicle's structure and internal systems [122]. The structural materials must not only withstand high temperatures but also maintain their mechanical properties under extreme thermal conditions. For instance, the use of ablative coatings and ceramic matrix composites (CMCs) has been shown to be effective in dissipating heat and protecting the underlying structures from thermal degradation [123].

Additionally, shock waves play a critical role in aerodynamic heating. The formation of bow shocks leads to adiabatic compression of air, which significantly raises its temperature. The intense heat generated in the region behind the shock wave can lead to thermal failure if not adequately managed [109]. The coupling of thermal and mechanical stresses due to aerodynamic heating complicates the design of hypersonic vehicles, as local failures such as panel buckling can occur from the combined effects of thermal and aerodynamic loads [124, 125].

To address these challenges, advanced materials and designs are being explored. For example, multilayer thermal insulations and bio-inspired structures have been proposed to enhance the thermal protection capabilities of hypersonic vehicles [126, 127]. These materials are designed to absorb and dissipate heat effectively while minimizing weight, which is crucial for maintaining flight performance [128]. Furthermore, the integration of innovative cooling techniques, such as transpiration cooling, is being researched to manage the extreme thermal environments encountered during flight [129].

Lift and Control at Hypersonic Speeds

At hypersonic speeds, the challenges of achieving and maintaining lift and control are exacerbated by the unique aerodynamic environment that these vehicles encounter. The design of aerodynamic surfaces, including wings and control fins, must be fundamentally reimagined to cope with extreme conditions such as high-pressure gradients, shock interactions, and intense aerodynamic forces. Traditional aerodynamic principles, which govern lift generation at subsonic and supersonic speeds, become less effective at hypersonic velocities due to the dominance of shock waves and the resulting changes in airflow dynamics [29, 119].

In hypersonic flight, the formation of shock waves around the vehicle plays a critical role in lift generation. As the vehicle moves through the atmosphere at speeds exceeding Mach 5, the air compresses and decelerates, creating a high-pressure zone beneath the vehicle. This phenomenon allows for sufficient lift to be generated even when conventional wing designs are less effective [116, 130]. To optimize lift in this regime, vehicle designs often incorporate shallow-angled surfaces that enhance the maintenance of high-pressure zones while minimizing aerodynamic heating and drag, which are significant concerns at these speeds [131, 132].

Control and stability at hypersonic speeds present additional challenges due to the complex flow dynamics and the intense aerodynamic forces acting on the vehicle. Control surfaces such as fins and canards must be designed to operate effectively in environments dominated by shock waves and high-pressure regions, which can produce unpredictable aerodynamic forces [133, 134]. The intense drag forces encountered can destabilize flight if not properly managed, necessitating control surfaces that can withstand extreme thermal loads while providing sufficient manoeuvrability for trajectory adjustments [135, 136]. Moreover, thrust vectoring is often employed to enhance control, allowing for adjustments in trajectory through the manipulation of exhaust flow, which is particularly beneficial in the thinner upper atmosphere where traditional control surfaces may be less effective [137].

Advanced guidance systems are also crucial for the successful operation of hypersonic vehicles. Inertial Navigation Systems (INS) are vital as they do not rely on external signals, making them reliable in environments where GPS may be compromised by the plasma cloud generated around the vehicle [138]. Additionally, the integration of AI-based predictive algorithms enables real-time adjustments to the vehicle's flight path, optimizing trajectory based on current conditions and enhancing the vehicle's ability to respond to unexpected challenges during flight [139, 140]. This level of automation is essential for maintaining stability and control at the extreme speeds characteristic of hypersonic flight.

Plasma Effects

At hypersonic speeds, vehicles experience plasma effects due to the extreme heating of the air surrounding them. As the vehicle moves through the atmosphere at such high velocities, the intense heat causes the air molecules to ionize, turning into a plasma sheath that envelops the vehicle [141, 142]. This phenomenon is a direct result of the vehicle's interaction with the atmosphere, where temperatures can soar to thousands of degrees Celsius. The formation of this plasma layer presents significant challenges in terms of communication, navigation, and aerodynamic performance.

Communication and Navigation Disruption

One of the major challenges caused by the plasma sheath is its impact on radio communications and navigation signals. As the vehicle travels at hypersonic speeds, the ionized air forms a barrier that can block or distort radio waves, making it difficult to send or receive signals. This phenomenon is often referred to as a communication blackout, a well-known issue during the re-entry of space capsules and other high-speed vehicles.

For hypersonic vehicles, this communication blackout can disrupt GPS signals and make it difficult to track the vehicle's position or remotely control its flight. The plasma sheath effectively acts as a shield that interferes with electromagnetic waves, which include the signals used for navigation and telemetry. To mitigate these effects, advanced systems, such as plasma-resistant antennas or communication relays that operate in different frequency bands, are required. Additionally, onboard systems like inertial navigation systems (INS) are crucial, as they do not rely on external signals and can help guide the vehicle even when external communication is compromised.

Aerodynamic Impact of Plasma

In addition to its effects on communication, the plasma layer also significantly influences the vehicle's aerodynamic properties. As the air around the vehicle ionizes, it changes the nature of the airflow, which can lead to unpredictable drag forces and potentially destabilize the vehicle. The interaction between the plasma and the vehicle's surface can alter the distribution of pressure around the body, affecting lift and drag. These changes in aerodynamic forces make it more challenging to maintain stability and control at hypersonic speeds.

The plasma's interaction with the vehicle's surface creates a dynamic and complex flow environment, which requires specialized designs to ensure that the vehicle can maintain predictable flight behaviour. Materials that can withstand the extreme heat of ionized air, while also minimizing the impact of plasma-induced drag, are essential. Some advanced vehicles use ceramic composites and ablative materials to manage the intense heat while maintaining structural integrity. In some cases, electromagnetic field generators have been proposed as a way to control or reduce plasma formation, thus minimizing its impact on both aerodynamics and communication systems.

Managing Plasma Effects

Given the challenges posed by plasma at hypersonic speeds, designing a vehicle that can effectively manage these effects is critical. Thermal protection systems are essential to handle the extreme temperatures caused by air friction and ionization, ensuring that the vehicle's structure and sensitive electronics are not damaged. The development of heat-resistant materials that can also reduce the impact of plasma-induced drag plays a key role in maintaining vehicle stability.

Recent studies have shown that the plasma sheath surrounding hypersonic vehicles exhibits turbulent characteristics, which complicates the propagation of electromagnetic signals [143]. This turbulence can lead to communication disruptions, commonly referred to as "communication blackout," where the vehicle loses contact with ground control or other systems [144]. To mitigate these issues, researchers are exploring alternative communication technologies that can penetrate the plasma sheath. For instance, using different frequency bands that are less affected by ionization or deploying relay satellites during critical flight phases has been proposed as potential solutions [145, 146]. These innovations are essential for maintaining continuous communication, which is vital for the safe operation of both manned and unmanned hypersonic vehicles [144].

Moreover, the aerodynamic design of hypersonic vehicles must account for the complex interactions between the vehicle's structure and the surrounding flow field. The integration of advanced flight control methods, such as the use of plasma actuators, has been identified as a promising approach to enhance control during hypersonic flight [147, 148]. These methods can help manage the aerodynamic forces acting on the vehicle, thereby improving stability and manoeuvrability under extreme conditions. Additionally, the implementation of drag reduction techniques, such as the use of opposing jets or aerodynamic spikes, can significantly lower the thermal loads experienced by the vehicle, further enhancing the effectiveness of the thermal protection systems [149].

Thermodynamics and Heat Challenges

Heat and Thermal Management

At hypersonic speeds, vehicles encounter extreme heat due to the intense atmospheric friction they experience as they travel through the air at velocities of Mach 5 or higher. This friction generates temperatures ranging from 2,000°C to 3,000°C (about 3,632°F to 5,432°F), which creates significant challenges for both the materials used to construct the vehicle and the systems designed to manage these high temperatures. Effective heat and thermal management is essential to ensuring the survivability and functionality of hypersonic vehicles, as exposure to such extreme temperatures can cause structural damage, system failures, and even complete destruction if not properly mitigated. Several specialized techniques and materials are employed to address the unique thermal challenges posed by hypersonic flight.

One of the primary methods of managing heat in hypersonic vehicles is through the use of ablative coatings. Ablative coatings are designed to protect the surface of the vehicle by absorbing and dissipating heat as they undergo a controlled vaporization process. As the vehicle heats up during flight, the outer layer of the ablative material gradually erodes or vaporizes, carrying away heat and preventing it from reaching the underlying structure. This process acts as a thermal barrier, ensuring that the most critical components of the vehicle, such as guidance systems, payloads, or propulsion units, remain protected from the intense temperatures generated during flight. Ablative coatings are particularly useful for Hypersonic Glide Vehicles (HGVs), which experience sustained high temperatures throughout the glide phase. By sacrificing the outer layer of material, the integrity of the vehicle is maintained for the duration of its mission, even at the cost of surface damage.

Ablative coatings are often made from specialized polymers, composites, or resin-based materials that are engineered to handle both the high heat flux and the aerodynamic stresses of hypersonic flight. These materials are selected for their ability to withstand rapid temperature changes and continue functioning even as they erode. This self-sacrificing property is crucial because it prevents heat from building up to levels that could cause catastrophic failure, such as the melting or warping of key components. While the use of ablative coatings is effective, it does impose certain design limitations, as the coatings are consumed during flight and cannot be reused after a mission. Nevertheless, they remain one of the most effective solutions for protecting hypersonic vehicles from extreme thermal conditions.

In addition to ablative coatings, ceramic matrix composites (CMCs) play a vital role in managing the extreme heat encountered during hypersonic flight. CMCs are advanced materials composed of ceramic fibres embedded in a ceramic matrix, which gives them exceptional heat resistance and structural integrity at high temperatures. Unlike metals, which tend to soften, melt, or lose strength when exposed to intense heat, ceramics retain their properties even at the extreme temperatures generated by hypersonic speeds. This makes them ideal for use in structural components that need to maintain their shape and functionality throughout the vehicle's high-speed journey.

Ceramic composites are particularly important for leading edges and other parts of the vehicle that bear the brunt of the heat load. These areas, which include the nose cone, wing edges, and engine components, are exposed to the most intense heat as they directly confront the airflow during hypersonic flight. By using ceramic composites in these areas, the vehicle can maintain its aerodynamic performance and structural stability without suffering from heat-induced deformation or failure. In addition to their thermal properties, CMCs offer advantages in terms of weight and durability, as they are often lighter than traditional metal alloys, further enhancing the performance of hypersonic systems by reducing overall vehicle mass.

Another critical component of heat management in hypersonic systems is the use of thermal shields and insulation systems. These systems are designed to protect sensitive electronic systems, guidance systems, and payloads from the extreme heat that permeates the vehicle during flight. Thermal shields work by reflecting or absorbing heat, preventing it from penetrating into the interior compartments of the vehicle. This is particularly important for the survival of electronics, which are highly sensitive to temperature fluctuations and can malfunction or fail entirely if exposed to excessive heat.

Thermal shields are often made from materials with high thermal resistance, such as carbon-carbon composites, which can withstand both high heat and the mechanical stresses associated with hypersonic flight. These shields are strategically placed throughout the vehicle to ensure that critical systems remain insulated from the extreme temperatures outside. In some cases, active cooling systems may be integrated into the thermal shielding to provide additional protection, although these systems are generally reserved for particularly high-heat applications due to their complexity and weight. By combining both passive (insulation and shielding) and active (cooling) methods, thermal management systems ensure that hypersonic vehicles can maintain operational performance without overheating.

The importance of thermal management cannot be overstated, as the failure to properly address the extreme heat generated during hypersonic flight can lead to catastrophic consequences. The combination of ablative coatings, ceramic composites, and thermal shields ensures that hypersonic vehicles can withstand the punishing conditions of their missions. By leveraging the unique properties of these materials and technologies, engineers are able to create systems that not only survive but thrive in the extreme environments where hypersonic weapons operate.

Heat Generation and Aerodynamic Heating

At hypersonic speeds, the phenomenon of aerodynamic heating becomes a critical concern for vehicle design and operation. As a hypersonic vehicle travels through the atmosphere, it compresses the air in front of it, generating shock waves that lead to a significant increase in temperature due to the principles of thermodynamics, particularly the first law, which states that energy conservation results in increased thermal energy when a gas is compressed [109]. This compression heating can elevate the temperature of the air to several thousand degrees Celsius, necessitating consideration of real gas effects, such as vibrational excitation, dissociation, and ionization, which become prominent at these extreme conditions [110].

The intense aerodynamic heating experienced by hypersonic vehicles is primarily due to the interaction between shock waves and boundary layers, creating a high-temperature environment around the vehicle [109]. This heating is not uniform; it is particularly pronounced near critical components, such as optical domes, where the heat transfer can jeopardize sensitive electronic equipment [110]. The heat generated by aerodynamic forces is transferred to the vehicle through conduction, convection, and radiation, leading to severe thermal stresses on the vehicle's surface. The management of this heat is a significant challenge, necessitating advanced thermal protection systems (TPS) that can withstand prolonged exposure to high heat fluxes while maintaining structural integrity [122].

To mitigate the effects of aerodynamic heating, various strategies are employed in the design of hypersonic vehicles. For instance, the use of high-temperature-resistant materials and composite structures is essential to endure the harsh thermal environments encountered during flight [150]. Additionally, innovative cooling techniques, such as active cooling channels and heat pipes, are being explored to manage heat distribution effectively [151]. The design of TPS must also consider the aerodynamic shape of the vehicle, as this influences the heat transfer rates and overall thermal performance [152]. Furthermore, the integration of lightweight materials is crucial to minimize the impact on flight velocity and fuel consumption, which are critical factors in hypersonic flight [150].

The design and operation of hypersonic vehicles face significant challenges due to the extreme thermal conditions encountered during flight. Aerodynamic heating at hypersonic speeds, often exceeding Mach 5, generates intense heat that can severely impact the structural integrity and performance of the vehicle. To mitigate these effects, several advanced techniques and strategies are employed. Here's a detailed explanation of the key approaches used to manage aerodynamic heating in hypersonic vehicles:

1. High-Temperature-Resistant Materials and Composite Structures

One of the most fundamental strategies to combat aerodynamic heating is the use of high-temperature-resistant materials. At hypersonic speeds, the heat generated by air friction and shock waves can reach thousands of degrees Celsius, requiring materials that can maintain their structural integrity under such conditions. Ceramic matrix composites (CMCs) and carbon-carbon composites are often employed due to their ability to withstand extreme temperatures without degrading.

- **Ceramic Matrix Composites (CMCs)**: These materials are highly resistant to heat and provide excellent thermal insulation, making them suitable for critical areas such as the leading edges, nose cone, and other surfaces exposed to the highest heat loads.

- **Metallic Alloys and Superalloys**: In some parts of the vehicle, high-strength metallic alloys such as Inconel or titanium alloys are used. These materials have high melting points and good mechanical properties at elevated temperatures.

- **Ablative Materials**: In specific cases, ablative coatings are used, where the outer layer of the material gradually vaporizes, taking heat away from the vehicle and protecting the underlying structure.

2. Cooling Techniques: Active Cooling Channels and Heat Pipes

To further manage heat, cooling systems are integrated into the design of hypersonic vehicles. Two of the most promising methods are active cooling channels and heat pipes.

- **Active Cooling Channels**: This technique involves the circulation of a cooling fluid through channels embedded in the structure of the vehicle. The fluid absorbs heat from the vehicle's skin and carries it away, preventing excessive temperature buildup. In hypersonic vehicles, the use of liquid coolants in closed-loop systems allows for continuous heat extraction during flight. This is particularly useful in areas like the engine, where the heat generated by the scramjet or rocket engines is most intense.

- **Heat Pipes**: Heat pipes are another cooling method used to manage localized heat loads. These devices use a phase-change process, where a fluid inside the pipe evaporates at the hot end, absorbs heat, and then condenses at the cooler end, releasing heat. This method allows for effective heat distribution across the surface of the vehicle without the need for complex mechanical systems. Heat pipes are commonly used in aerospace applications to manage thermal loads efficiently.

3. Thermal Protection System (TPS) Design and Aerodynamic Shape

The design of the Thermal Protection System (TPS) is another critical element in managing aerodynamic heating. The shape of the vehicle influences how heat is distributed across its surface. For example, blunt bodies generate stronger shock waves that cause higher heat loads but are more effective at spreading this heat over a larger area, reducing the peak temperature in any one location.

- **Aerodynamic Shaping**: The design of the vehicle's surface plays a crucial role in minimizing heat buildup. Shallow-angled surfaces or wedge shapes are often used to generate oblique shock waves rather than strong bow shocks. This design reduces the intensity of aerodynamic heating by spreading the energy over a larger area.

- **Leading-Edge Blunting**: While sharp edges tend to concentrate heat and result in higher temperatures, blunt edges help distribute heat more evenly. By blunting the leading edges, the TPS can reduce the peak heat load on the vehicle, thus decreasing the risk of thermal failure.

4. Lightweight Materials for Minimizing Flight Velocity Impact

In hypersonic flight, the weight of the vehicle is a critical factor. Heavier vehicles require more fuel and experience higher aerodynamic drag, which can reduce their speed and range. Therefore, the use of lightweight materials is essential for maintaining the high velocities needed for hypersonic flight.

- **Carbon Fiber Reinforced Polymers (CFRPs)**: These materials offer a high strength-to-weight ratio and are often used in structural components that do not experience the highest temperatures. CFRPs reduce the overall weight of the vehicle without sacrificing structural integrity.

- **Lightweight Metal Alloys**: Alloys such as aluminium-lithium or magnesium alloys are used in areas where extreme heat is not the primary concern but where weight reduction is critical for fuel efficiency and maintaining high flight speeds.

5. Heat Distribution Management

Effective heat distribution is essential to prevent localized thermal hotspots that could compromise the vehicle's integrity. Advanced thermal protection designs focus on managing the heat flow to spread it more evenly across the surface of the vehicle.

- **Multi-Layer Insulation**: Hypersonic vehicles may employ multi-layer thermal insulation, where multiple layers of heat-resistant materials and insulating layers are used to slow the transfer of heat to the inner structure. This design also helps to dissipate heat more evenly across the surface.

- **Phase Change Materials (PCMs)**: Some hypersonic vehicles integrate materials that can absorb and store heat during flight by changing their phase from solid to liquid. These phase change materials prevent sudden temperature spikes and help regulate the heat flow over time.

The strategies employed to mitigate aerodynamic heating in hypersonic vehicles combine advanced materials, innovative cooling technologies, and precise aerodynamic designs. High-temperature-resistant materials like CMCs, along with active cooling channels, help manage extreme heat loads, while lightweight materials ensure the vehicle can maintain high speeds and minimize fuel consumption. Additionally, the aerodynamic shape of the vehicle plays a crucial role in heat distribution and drag reduction, while thermal protection systems ensure the vehicle remains structurally sound even in the most extreme conditions. These combined technologies are essential to overcoming the thermodynamic challenges posed by hypersonic flight, enabling sustained performance at speeds greater than Mach 5.

Real Gas Effects

At hypersonic speeds, the behaviour of air deviates significantly from the ideal gas law due to the extreme temperatures and pressures encountered. This results in complex thermodynamic phenomena, including vibrational excitation, dissociation, and ionization, which are critical to understanding the dynamics of hypersonic flight.

Vibrational Excitation

Vibrational Excitation occurs when air molecules, primarily nitrogen (N_2) and oxygen (O_2), absorb energy from shock waves, leading to an increase in their vibrational modes. At temperatures exceeding 1,000°C, the vibrational energy contributes significantly to the internal energy of the gas, thereby raising its overall temperature. Studies have shown that vibrational degrees of freedom play a crucial role in the heat transfer characteristics of polyatomic gases, indicating that the energy distribution among vibrational modes is essential for understanding thermal behaviours in high-temperature environments [153]. Additionally, the interactions between vibrationally excited states and translational energy can lead to enhanced thermal conductivity in gases, further complicating the thermodynamic modelling of hypersonic flows [154].

In the realm of hypersonic flight, vibrational excitation plays a significant role in understanding and managing the thermal environment encountered by vehicles traveling at extremely high speeds. This phenomenon primarily involves air molecules, particularly nitrogen (N_2) and oxygen (O_2), absorbing energy from shock waves, which leads to an increase in their vibrational energy states. At temperatures above 1,000°C, the vibrational energy contributes significantly to the internal energy of the gas, which in turn raises its overall temperature. This has critical implications for the design of hypersonic weapons and vehicles, as it influences heat transfer, thermal management, and overall vehicle performance in high-speed atmospheric flight.

<u>Vibrational Excitation and its Effects in Hypersonic Flight</u>

At hypersonic speeds, the extreme heat generated by air friction and shock waves causes air molecules to undergo **vibrational excitation**. In this state, the energy absorbed by the gas molecules is not only distributed among their translational and rotational modes but also excites the vibrational modes of polyatomic molecules like nitrogen and oxygen.

- **Energy Distribution in Vibrational Modes**: As the molecules absorb energy from the intense shock waves formed around the hypersonic vehicle, they shift into higher vibrational energy states. This shift becomes more pronounced as temperatures exceed 1,000°C, which is common in hypersonic environments. The energy contained within these vibrational modes contributes significantly to the overall internal energy of the gas, leading to higher local temperatures.

- **Thermal Conductivity and Heat Transfer**: Vibrational excitation also alters the heat transfer characteristics of the gas. The interactions between vibrationally excited molecules and their translational energy can lead to increased thermal conductivity, further intensifying the heat transfer to the vehicle's surface. This added complexity makes it difficult to predict and model the thermodynamic behaviour of the gas flow around the hypersonic vehicle.

Implications for Hypersonic Vehicle Design

Incorporating the effects of vibrational excitation is essential in the design of hypersonic weapons and vehicles. This phenomenon significantly impacts how heat is managed and how thermal protection systems (TPS) are developed. The design must account for the additional energy stored in the vibrational modes and its influence on heat transfer and overall thermal load on the vehicle.

- **Thermal Protection Systems (TPS)**: The increased heat transfer due to vibrational excitation demands that thermal protection systems be highly efficient. Materials used in the TPS must be capable of withstanding not only the intense convective heating from the airflow but also the enhanced thermal conductivity caused by the vibrationally excited molecules. This is one of the reasons why advanced ablative materials and ceramic matrix composites (CMCs) are used in hypersonic vehicle designs. These materials are designed to endure the elevated temperatures by either dissipating heat through ablation or maintaining structural integrity at high temperatures.

- **Heat Transfer Modelling**: The presence of vibrational excitation complicates the **thermodynamic modelling** of hypersonic flows. Designers and engineers must account for how vibrational energy interacts with translational and rotational energy in the gas flow, which requires sophisticated computational fluid dynamics (CFD) models that incorporate vibrational degrees of freedom. These models help predict the heat flux and temperature distribution across the vehicle's surface, allowing for optimized thermal management solutions.

- **Impact on Vehicle Surface Materials**: The elevated temperatures and enhanced thermal conductivity associated with vibrational excitation influence the choice of surface materials. Designers must ensure that materials can withstand not only the intense heat generated by the shock waves but also the sustained heat load due to increased thermal conductivity. This is critical in protecting sensitive onboard electronics and maintaining the vehicle's structural integrity throughout its flight path.

Interaction Between Vibrational and Translational Energy

The interaction between vibrationally excited states and translational energy further complicates the thermodynamics of hypersonic flight. When air molecules are vibrationally excited, they can transfer energy back into translational modes (and vice versa), which impacts the overall energy balance in the gas. This interplay affects the shock layer surrounding the vehicle, altering the temperature distribution and potentially leading to non-equilibrium thermodynamics, where different parts of the gas flow exhibit varying temperature and energy distributions.

In conclusion, vibrational excitation is a critical factor in understanding the thermal environment of hypersonic vehicles. At temperatures exceeding 1,000°C, the vibrational energy of air molecules significantly influences heat transfer characteristics and requires advanced thermal protection systems to manage the additional heat load. Designers of hypersonic weapons and vehicles must account for these effects by employing high-temperature-resistant materials, advanced cooling systems, and sophisticated thermodynamic models to ensure the vehicle can withstand the extreme conditions of hypersonic flight.

Dissociation

Dissociation becomes prominent at temperatures above 2,000°C, where the kinetic energy of air molecules is sufficient to break molecular bonds. This process transforms diatomic gases like N_2 and O_2 into atomic gases, which not only increases the internal energy but also reduces the density of the air. The energy required for dissociation is derived from the kinetic energy of the flow, leading to a complex interplay between energy conservation and molecular dynamics. Research indicates that as gas temperatures rise, the likelihood of molecular dissociation increases, which is critical for accurately modelling hypersonic flight conditions [155]. Moreover, the dissociation process contributes to the thermal non-equilibrium state of the gas, complicating the thermodynamic analysis further [155].

At temperatures above 2,000°C, molecular dissociation becomes a significant factor in the thermodynamics of hypersonic flight. As air molecules gain sufficient kinetic energy, the bonds in diatomic gases like nitrogen (N_2) and oxygen (O_2) begin to break apart, transforming them into atomic gases (N and O). This process not only increases the internal energy of the gas but also reduces its density, introducing a host of complex thermodynamic challenges for designers of hypersonic vehicles and weapons. Understanding and accounting for dissociation is critical in modelling the behaviour of air at hypersonic speeds, where the extreme heat and energy dynamics affect the vehicle's performance, thermal protection systems, and overall stability.

<u>Molecular Dissociation at High Temperatures</u>

In hypersonic flight, the high-speed flow of air around a vehicle generates extreme heat through aerodynamic friction, causing the temperature to exceed 2,000°C in many cases. At these temperatures, the energy from the flow is enough to overcome the molecular bonds in diatomic gases, particularly nitrogen and oxygen, leading to dissociation.

- **Energy Redistribution**: The dissociation process involves taking kinetic energy from the flow to break molecular bonds, effectively reducing the energy available for other processes like raising temperature or generating shock waves. As a result, while dissociation increases the internal energy of the gas, it can also create a more complex thermal environment, as the gas transitions into a lower-density atomic state.

- **Reduction in Air Density**: Dissociation lowers the density of the gas surrounding the vehicle, which directly impacts the aerodynamic forces acting on the hypersonic object. Reduced density can lower drag forces, but it also influences how heat is transferred to the vehicle's surface. Designers must carefully balance the effects of dissociation on density with the need to control drag and manage heat distribution.

<u>Impact on Thermodynamic Modelling</u>

Dissociation is a key factor that complicates the thermodynamic modelling of hypersonic vehicles. Traditional models that assume a chemically inert gas or an ideal gas fail to account for the energy exchange between kinetic motion and molecular dissociation, making them inadequate for capturing the full dynamics of high-temperature airflows.

- **Non-Equilibrium Thermodynamics**: The dissociation process introduces thermal non-equilibrium conditions, where different parts of the gas exhibit varying temperatures and energy states. In a non-equilibrium state, energy is distributed unevenly between translational, rotational, and vibrational modes, and a portion of the energy goes into breaking molecular bonds. This necessitates the use of non-equilibrium thermodynamic models that can account for the variations in energy distribution and the presence of dissociated species in the flow.

- **Complex Interplay Between Energy and Molecular Dynamics**: The energy needed for dissociation comes from the flow's kinetic energy, creating a feedback loop where the flow loses energy to molecular dissociation, altering its overall thermodynamic behaviour. Accurate modelling must capture this complex interplay, especially in regions with strong shock waves, where the gas experiences rapid compression and heating.

<u>Dissociation's Role in Thermal Protection Systems</u>

For hypersonic vehicles, managing the extreme heat generated by high-speed flight is one of the most critical challenges. The dissociation of air molecules plays a dual role in this regard: while it increases the gas's internal energy, it can also reduce the temperature rise that would otherwise occur if all the energy remained in the translational or vibrational modes. This dynamic has important implications for the design of thermal protection systems (TPS).

- **Heat Absorption and Ablation**: The TPS must account for the heat generated by both the kinetic energy of the flow and the additional internal energy from dissociated gases. **Ablative materials**, which absorb heat by slowly vaporizing, are commonly used to dissipate heat away from the vehicle. As the material ablates, it also sheds some of the heat generated by dissociated gases, providing a buffer against the intense thermal environment.

- **Material Selection**: The materials used in TPS must be able to withstand not only high temperatures but also the chemical effects of dissociated atomic gases. Atomic oxygen, in particular, can react with vehicle surfaces, causing chemical erosion or weakening of the structure. Designers often use ceramic composites and other advanced materials that can resist both thermal and chemical degradation.

Design Implications for Hypersonic Weapons

The dissociation of air molecules has several implications for the design and operation of hypersonic weapons, particularly those that operate at extremely high altitudes or through atmospheric re-entry. The changes in air density and energy distribution caused by dissociation affect everything from vehicle stability to drag reduction and heat management.

- **Drag and Stability**: As dissociation reduces the air density around the vehicle, it can lower drag forces, helping to maintain high speeds. However, this also affects the vehicle's stability, as lower density means fewer aerodynamic forces are available to counterbalance the vehicle's motion. Control surfaces and thrust vectoring must be precisely designed to operate effectively in this altered environment.

- **Hypersonic Glide Vehicles (HGVs)**: For vehicles like Hypersonic Glide Vehicles, which travel at extreme speeds through the upper atmosphere, dissociation can become a dominant factor. Designers must carefully model how the gas will behave as the vehicle moves through different temperature regimes, ensuring that the vehicle can maintain lift and control while managing heat and shock wave interactions.

Dissociation and Advanced Computational Fluid Dynamics (CFD)

To accurately predict the behaviour of hypersonic vehicles, advanced computational fluid dynamics (CFD) tools must incorporate the effects of molecular dissociation. These models need to account for the complex interactions between the flow's kinetic energy, molecular dynamics, and thermal effects.

- **CFD Simulations**: **Non-equilibrium CFD models** are essential for predicting how dissociated gases will impact heat transfer, drag, and overall vehicle performance. These models simulate the flow of air around the vehicle, including the breakdown of diatomic gases and the redistribution of energy within the shock layer.

Dissociation is a crucial factor in the thermodynamics of hypersonic flight. As temperatures exceed 2,000°C, air molecules begin to break apart, leading to changes in density, energy distribution, and heat transfer characteristics. This process must be carefully accounted for in the design of hypersonic vehicles, with particular attention to thermal protection, material selection, and aerodynamic stability. Advanced computational tools that incorporate non-equilibrium thermodynamics and molecular dissociation are essential for ensuring that hypersonic weapons can operate effectively in the extreme conditions they encounter.

Ionization

Ionization occurs at even higher temperatures, typically above 5,000°C, where the energy is sufficient to strip electrons from atoms, resulting in the formation of a plasma state. This ionization introduces charged particles into the flow, which can significantly affect heat transfer, flow dynamics, and electromagnetic interactions. The presence of plasma in hypersonic flows creates a unique environment that alters the thermodynamic properties of the gas, leading to phenomena such as the formation of a plasma sheath around hypersonic vehicles [156]. The interactions within this plasma state can also influence the effectiveness of various flow control techniques, such as magnetohydrodynamic (MHD) methods, which leverage the properties of ionized gases to manipulate flow characteristics [98].

Ionization becomes a critical factor at temperatures exceeding 5,000°C, where air molecules experience sufficient energy to strip electrons from atoms, resulting in the creation of a plasma state. This ionized gas, composed of charged particles such as electrons and ions, profoundly impacts the behaviour of the flow around hypersonic vehicles, altering key thermodynamic properties, heat transfer rates, and electromagnetic interactions. The formation of a plasma sheath is one of the most significant effects at these extreme temperatures, and it plays a crucial role in both the design and operational performance of hypersonic weapons.

<u>Plasma Formation and Thermodynamic Impacts</u>

As air molecules are exposed to the intense kinetic energy generated during hypersonic flight, they undergo ionization, transforming into a plasma. This plasma environment significantly changes the thermodynamic properties of the gas.

- **Energy Redistribution**: During ionization, a portion of the flow's kinetic energy is absorbed in the process of stripping electrons from atoms, which adds to the internal energy of the gas. This redistribution of energy impacts heat transfer, as ionized particles increase the complexity of thermal dynamics in the boundary layer surrounding the vehicle.

- **Impact on Heat Transfer**: The introduction of charged particles affects the way heat is transferred to the surface of the vehicle. As the plasma sheath forms, it can act as an insulator, reducing the conductive heat transfer to the vehicle. However, the plasma can also emit radiation, which increases radiative heat transfer, requiring careful consideration in the thermal protection system (TPS) design.

Plasma Sheath and Its Effects on Communication and Control

One of the most well-known effects of ionization in hypersonic flight is the formation of a plasma sheath around the vehicle. This sheath can block radio frequency (RF) signals, creating a "communication blackout" during critical phases of flight, such as atmospheric re-entry or high-speed manoeuvres.

- **Communication Disruptions**: The plasma sheath can interfere with GPS, telemetry, and communication signals, making it difficult for ground control to maintain contact with the vehicle. Hypersonic weapon designs must account for this challenge by integrating solutions such as plasma-resistant communication systems, advanced signal processing techniques, or alternative navigation methods like inertial navigation systems (INS).

- **Electromagnetic Interactions**: The presence of ionized gas also introduces new challenges in terms of electromagnetic interactions. The charged particles in the plasma can reflect or absorb electromagnetic waves, affecting radar cross-section (RCS) and electronic warfare systems. This necessitates the use of plasma-tolerant electronics and carefully designed sensor systems that can function effectively within this ionized environment.

Magnetohydrodynamic (MHD) Flow Control

Magnetohydrodynamic (MHD) methods leverage the unique properties of ionized gases to manipulate the flow around hypersonic vehicles. MHD flow control is particularly useful in hypersonic weapons because the charged particles in plasma can be influenced by magnetic and electric fields, allowing for the adjustment of flow characteristics in real-time.

- **Flow Manipulation**: In MHD flow control, magnetic fields are applied to the plasma flow around the vehicle to change its behaviour. By controlling the flow of ionized gas, MHD techniques can reduce drag, manage heat distribution, and even enhance lift. This is particularly valuable in hypersonic flight, where managing the intense aerodynamic forces is essential for maintaining control and stability.

- **Thermal Management**: MHD flow control can also be used to manipulate how heat is distributed across the surface of the vehicle. By directing the flow of plasma away from certain areas, MHD systems can protect critical components from overheating and improve the overall performance of the vehicle's TPS.

Design Implications for Hypersonic Weapons

The ionization process and the subsequent formation of plasma around hypersonic weapons present several unique challenges that must be accounted for in the design phase.

- **Thermal Protection Systems**: The plasma sheath and its effects on heat transfer require robust thermal protection systems. Materials must be chosen to withstand not only the conductive and convective heat from the flow but also the radiative heat emitted by the ionized gas. Ablative materials and ceramic composites are commonly used to dissipate heat and protect the vehicle from the extreme thermal environment.

- **Communication Systems**: Since plasma can interfere with RF signals, communication systems in hypersonic weapons must be adapted to ensure reliability during flight. One approach is to use plasma-resistant antennas or frequency-agile communication systems that can operate across a broader spectrum to avoid signal loss. In some cases, hypersonic weapons may rely more heavily on inertial navigation systems (INS) during high-temperature phases, where GPS signals are blocked by plasma.

- **Structural Integrity and Stability**: The high temperatures and complex flow dynamics associated with plasma formation also place immense stress on the vehicle's structural components. Hypersonic weapon designs must use high-strength, heat-resistant materials to ensure that the vehicle remains intact and functional under these extreme conditions. Additionally, plasma flow control techniques, such as MHD systems, may be employed to manage drag and maintain stability at ultra-high speeds.

Challenges in Thermodynamic Modelling

Ionization introduces further complexity to the thermodynamic modelling of hypersonic weapons. The presence of charged particles and the interactions between the plasma, shock waves, and the vehicle's surface create a non-equilibrium thermodynamic state that must be carefully modelled to predict the vehicle's behaviour accurately.

- **Non-Equilibrium Models**: Standard equilibrium models of thermodynamics are insufficient for hypersonic vehicles, as they do not account for the varying energy states of ionized particles. Advanced non-equilibrium computational fluid dynamics (CFD) models must be used to simulate the behaviour of the plasma and its effects on heat transfer, drag, and control dynamics. These models incorporate ionization, recombination, and the interactions between charged particles and electromagnetic fields.

- **Plasma Diagnostics**: To further refine hypersonic designs, plasma diagnostic tools are often used during testing to gather data on ionization levels, plasma density, and electromagnetic interactions. This data is then fed back into the design process to improve the performance and survivability of the vehicle.

Ionization at hypersonic speeds creates a unique set of challenges for the design and operation of hypersonic weapons. The formation of plasma significantly affects heat transfer, flow dynamics, and communication systems, necessitating the use of advanced materials, thermal protection systems, and innovative flow control techniques such as MHD. As researchers continue to explore the behaviour of ionized gases at extreme temperatures, the development of non-equilibrium models and plasma-tolerant technologies will be critical for ensuring the success and reliability of hypersonic vehicles in military applications.

Shock Waves and Thermodynamic Changes

Shock waves are a critical aspect of the thermodynamics associated with hypersonic flight, particularly for vehicles traveling at speeds exceeding Mach 5. These shock waves arise due to the inability of air to move out of the way quickly enough, resulting in abrupt changes in thermodynamic properties such as pressure, temperature, and density. The formation of bow shocks ahead of the vehicle and oblique shocks along its body significantly influences the aerodynamic behaviour and thermal environment surrounding the vehicle. As noted by Bertin and Cummings, the high-temperature gas effects at hypersonic speeds strongly affect the forces acting on the vehicle's surface, including pressure and skin friction, as well as energy fluxes related to convective and radiative heating [157]. This interaction is crucial for understanding the heat transfer and aerodynamic drag experienced by hypersonic vehicles.

The Rankine-Hugoniot relations provide a framework for understanding the pressure and temperature increases across shock waves. As air molecules are compressed when passing through these waves, there is a significant rise in pressure, which is particularly pronounced in hypersonic conditions due to the high velocities involved [158]. This compression leads to a rapid deceleration of the air, resulting in increased thermal energy as kinetic energy is converted into heat. Wang et al. highlight that the extreme temperatures in the shock layer can exceed 10,000 K, leading to complex phenomena such as dissociation and ionization of gas molecules, which further complicates the thermodynamic environment around the vehicle [158]. The thermal protection systems (TPS) must be designed to withstand these extreme conditions, as they are critical for vehicle integrity during flight [159].

Moreover, the increase in entropy associated with shock waves is a key factor in the wave drag experienced by hypersonic vehicles. Shock waves are irreversible processes that lead to energy dissipation as heat, which contributes to increased entropy and represents a loss of mechanical energy that could otherwise be utilized for propulsion [157]. This entropy increase directly correlates with wave drag, a specific form of aerodynamic drag that significantly impacts the efficiency and energy consumption of hypersonic flight [160]. The design of hypersonic vehicles must therefore account for these thermodynamic losses to optimize performance.

The interaction between shock waves and the boundary layer also introduces thermodynamic instabilities, particularly at hypersonic speeds where heat and pressure gradients are extreme. These instabilities can lead to flow separation and turbulent mixing, which complicate heat transfer and aerodynamic performance [161]. The fluctuating nature of the boundary layer can create localized hot spots on the vehicle's surface, exacerbating thermal challenges [162]. Effective management of these instabilities is essential for maintaining aerodynamic control and ensuring the structural integrity of the vehicle during flight.

Rankine-Hugoniot Equations

The Rankine-Hugoniot relations are a set of equations that describe the behaviour of a fluid (such as air) when it passes through a shock wave. These relations are derived from the conservation laws of mass, momentum, and energy and are critical in understanding how pressure, temperature, density, and velocity change across a shock front, particularly in high-speed, compressible flows like those found in supersonic and hypersonic flight.

When a vehicle travels at supersonic or hypersonic speeds, it generates shock waves due to the sudden compression of air. Shock waves are thin regions where the fluid properties—velocity, pressure, temperature, and density—change almost instantaneously. These rapid changes in the fluid's state are governed by the Rankine-Hugoniot relations, which quantify how the properties of the flow upstream of the shock differ from those downstream.

The Rankine-Hugoniot relations are built upon three fundamental conservation principles:

1. **Conservation of Mass**: The mass flow rate must be conserved across the shock wave. This means that the amount of mass flowing into the shock must equal the amount flowing out.

2. **Conservation of Momentum**: The momentum (mass times velocity) of the fluid must also be conserved. This accounts for the forces exerted by pressure changes across the shock wave.

3. **Conservation of Energy**: The total energy of the fluid (including kinetic, thermal, and internal energy) must remain constant across the shock wave, assuming no heat is added or removed from the system.

These conservation laws form the basis of the Rankine-Hugoniot relations, which are expressed mathematically to describe the jump in physical properties from the upstream side of the shock (denoted as region 1) to the downstream side (denoted as region 2).

For a shock wave moving through a compressible fluid, the Rankine-Hugoniot relations can be written as a set of equations that relate the conditions in front of the shock (region 1) to the conditions behind the shock (region 2):

Conservation of Mass:

$$\rho_1 u_1 = \rho_2 u_2$$

Where ρ_1 and ρ_2 are the densities, and u_1 and u_2 are the flow velocities upstream and downstream of the shock, respectively. This equation states that the mass flux (density times velocity) is the same on both sides of the shock.

Conservation of Momentum:

$$P_1 + \rho_1 u_1^2 = P_2 + \rho_2 u_2^2$$

Where ρ_1 and ρ_2 are the pressures upstream and downstream of the shock. This equation expresses the balance of forces, considering both the pressure and momentum flux.

Conservation of Energy:

$$h_1 + \frac{u_1^2}{2} = h_2 + \frac{u_2^2}{2}$$

Where h_1 and h_2 are the specific enthalpies upstream and downstream of the shock. This equation ensures that the total energy (including both kinetic and thermal energy) remains constant across the shock.

Key Parameters and Variables

To understand the Rankine-Hugoniot relations, we must define several key variables:

- **Density** (ρ): The mass per unit volume of the fluid.
- **Velocity** (u): The speed of the fluid in the direction of the shock wave.
- **Pressure** (P): The force exerted by the fluid per unit area.
- **Specific Enthalpy** (h): The sum of the fluid's internal energy and the work required to move it, expressed as $h = \frac{\gamma}{\gamma - 1}\frac{P}{\rho}$ for an ideal gas.
- **Mach Number** (M): The ratio of the fluid velocity to the local speed of sound, which determines whether the flow is subsonic, supersonic, or hypersonic.

Types of Shocks: Normal and Oblique

The Rankine-Hugoniot relations are typically applied to two types of shocks:

1. **Normal Shock**: The shock front is perpendicular to the flow direction. Normal shocks are common in supersonic inlets and diffusers. In a normal shock, the velocity of the fluid decreases significantly, and the flow transitions from supersonic to subsonic downstream of the shock.

2. **Oblique Shock**: The shock front is inclined at an angle to the flow direction. Oblique shocks occur when the flow is deflected by a surface, such as the nose of a vehicle or an aircraft wing. Unlike normal shocks, the velocity remains supersonic behind an oblique shock, though it is reduced in magnitude.

<u>Effects of Shock Waves on Flow Properties</u>

The Rankine-Hugoniot relations describe several critical changes in flow properties across a shock:

1. **Pressure Increase**: The pressure downstream of the shock is always higher than upstream. This increase is particularly pronounced in strong shock waves (e.g., in hypersonic flight).

2. **Temperature Increase**: The temperature downstream of the shock also rises, as part of the fluid's kinetic energy is converted into thermal energy.

3. **Velocity Decrease**: The velocity of the fluid decreases significantly after passing through a shock, particularly in normal shocks where the flow often transitions from supersonic to subsonic.

4. **Density Increase**: The fluid becomes denser after the shock due to the compression of air as it passes through the shock front.

5. **Entropy Increase**: Shock waves are inherently irreversible processes, meaning that entropy increases across the shock, reflecting a loss of useful energy and an increase in disorder in the system. This increase in entropy leads to energy losses and contributes to wave drag, which is a major challenge in hypersonic vehicle design.

<u>Applications in Hypersonic Vehicles</u>

In the context of hypersonic flight, the Rankine-Hugoniot relations are crucial for predicting the behaviour of shock waves and designing systems to manage the resulting pressure and temperature changes. Hypersonic vehicles generate intense bow shocks that lead to significant increases in temperature and pressure. Understanding these changes helps engineers design thermal protection systems (TPS), manage aerodynamic heating, and minimize wave drag.

Moreover, the relations help in the design of air-breathing propulsion systems, like scramjets, by predicting the behaviour of air as it compresses and heats up during the intake process. The efficient management of shock waves is critical to ensuring the proper function of engines and maintaining vehicle stability at extreme speeds.

The Rankine-Hugoniot relations provide a mathematical framework for understanding how fluid properties change across a shock wave. These relations are essential in high-speed, compressible flow regimes, such as supersonic and hypersonic flight. They describe how the fluid's velocity, pressure, temperature, and density change, and they form the foundation for designing effective hypersonic vehicles and managing the intense heat and pressure generated by shock waves.

Boundary Layers and Heat Transfer

The study of boundary layers and heat transfer is critical in understanding the thermal dynamics experienced by vehicles, especially at hypersonic speeds. The boundary layer, which is the thin layer of fluid in immediate contact with the surface of a vehicle, undergoes significant changes due to the extreme conditions present at such velocities.

Thermal Boundary Layer Dynamics

At hypersonic speeds, the thermal boundary layer becomes significantly heated due to viscous heating, which occurs as a result of friction between the vehicle's surface and the surrounding air. This phenomenon is exacerbated by the increase in temperature as the vehicle accelerates, leading to a thickening of the boundary layer. The transition from laminar to turbulent flow within this layer can dramatically enhance heat transfer rates to the vehicle's surface. Schlichting and Gersten [163] provide a comprehensive overview of boundary layer theory, detailing how the characteristics of the thermal boundary layer evolve under varying conditions, including the onset of turbulence. Additionally, the effects of internal heat generation and thermal radiation on boundary layers have been explored, showing that these factors can significantly influence thermal diffusion and the overall heat transfer process [164].

Mechanisms of Heat Transfer

The primary mechanisms of heat transfer in the context of hypersonic vehicles are convection and radiation. Convective heat transfer occurs when the hot air in the boundary layer transfers heat to the vehicle's surface, a process that is influenced by the flow characteristics and temperature gradients present at the wall [165]. Dong et al. emphasize the importance of understanding convective heat transfer in the boundary layer, particularly for applications such as anti-icing systems in airframes, where precise control of temperature is critical [165].

Radiative heat transfer also plays a significant role, especially at elevated temperatures where molecular dissociation and ionization can occur. This aspect of heat transfer becomes increasingly relevant as the vehicle approaches hypersonic speeds, where the thermal radiation emitted by the hot gases can contribute substantially to the thermal load experienced by the vehicle [166]. The interplay between convective and radiative heat transfer mechanisms is crucial for accurately predicting the thermal environment around hypersonic vehicles, as highlighted in various studies focusing on boundary layer flows under extreme conditions [167, 168].

Thermal Protection Systems (TPS)

Thermal Protection Systems (TPS) are essential for hypersonic vehicles due to the extreme heat generated during flight at hypersonic speeds. The thermodynamic conditions in such high-speed environments can lead to temperatures exceeding thousands of degrees Celsius, which pose significant risks to the structural integrity of the vehicle. Without adequate thermal protection, the intense heat can cause materials to degrade or fail, leading to catastrophic consequences. To combat these challenges, advanced TPS designs are incorporated, using a range of passive and active strategies.

One of the most widely used methods of thermal protection in hypersonic vehicles is ablative coatings. Ablative materials function by slowly vaporizing or charring when exposed to high temperatures. As the material ablates, it carries away the heat, preventing it from reaching the underlying structural components. The vaporized material forms a protective boundary layer that insulates the vehicle from further heating.

This sacrificial process is particularly effective for short-duration, high-heat environments, such as atmospheric re-entry or brief hypersonic flight. Ablative coatings are commonly applied to areas of the vehicle that experience the highest heat loads, such as the nose cone and leading edges, which encounter strong shock waves and aerodynamic heating. The self-destruction of the ablative material limits the temperature buildup, making it a vital aspect of TPS in many hypersonic designs.

Another critical component of TPS in hypersonic vehicles is the use of ceramic matrix composites (CMCs). These materials are composed of ceramic fibres embedded within a ceramic matrix, resulting in materials that can withstand extreme temperatures without melting or structurally degrading.

Ceramic composites are ideal for regions of the vehicle exposed to the most intense heat, such as the nose, wing leading edges, and other areas subject to direct shockwave interaction. CMCs exhibit high resistance to thermal and mechanical stresses, maintaining their integrity even at temperatures exceeding 2,000°C. Additionally, ceramics have low thermal conductivity, which helps prevent the transfer of heat to the internal components of the vehicle, enhancing the overall thermal management system.

While passive systems like ablative coatings and ceramic composites provide effective thermal protection, active cooling systems are also employed in some hypersonic vehicle designs. Active cooling involves the circulation of a coolant (often liquid or gas) through channels or cavities within the vehicle's structure to absorb and redistribute heat away from critical components.

This technique is particularly useful in sustaining long-duration hypersonic flights, where the vehicle may experience sustained heating over time. By actively removing heat from the vehicle's surface, active cooling systems help to maintain structural integrity and prevent thermal degradation. These systems are often integrated into propulsion units, such as scramjets, where managing heat is crucial for maintaining engine performance.

Active cooling systems can involve complex engineering solutions, including regenerative cooling, where fuel is used as a coolant before it is combusted, absorbing heat from the surface and lowering temperatures. This type of system is vital for protecting internal systems, avionics, and other sensitive equipment from extreme thermal environments.

Thermal protection is one of the most critical challenges in hypersonic flight due to the severe heat loads caused by atmospheric friction and shockwaves. The TPS of hypersonic vehicles is designed using a combination of ablative coatings, ceramic composites, and active cooling systems to protect the vehicle's structure and ensure its survivability at extreme speeds. Each of these strategies plays a vital role in managing the heat generated during hypersonic flight, ensuring the vehicle remains operational while maintaining structural integrity under intense thermal conditions.

Entropy Layer and Viscous Interaction

At hypersonic speeds, the interaction between the boundary layer and shock waves significantly influences the aerodynamic and thermal environment surrounding a vehicle. The boundary layer, a thin region of slower-moving air adjacent to the vehicle's surface, interacts with shock waves generated during flight, leading to the formation of an entropy layer characterized by steep gradients in pressure, temperature, and density. This interaction is crucial as it introduces complex aerodynamic behaviours that affect heat transfer, drag, and overall vehicle stability.

In the context of hypersonic flight, entropy refers to a thermodynamic property that measures the level of disorder or randomness in a system. It is also a crucial factor in determining how energy is distributed and lost during the rapid, high-energy interactions that occur at hypersonic speeds. Entropy is especially important in hypersonic flight because the shock waves generated around the vehicle and the extreme thermal and aerodynamic conditions significantly affect the energy states and flow characteristics of the surrounding air.

At hypersonic speeds (Mach 5 and above), vehicles generate strong shock waves that cause sudden and dramatic changes in the properties of the air, including pressure, temperature, and density. These shock waves are highly irreversible processes, meaning that some of the mechanical energy of the flow is lost as heat, and this results in an increase in entropy. As the air passes through the shock waves, the energy that was previously used to propel the vehicle is partially dissipated into the atmosphere in the form of heat, and this increases the randomness or disorder of the air molecules—an increase in entropy.

Key aspects of entropy in hypersonic flight:

- **Shock Waves and Entropy Increase**: When air flows around a hypersonic vehicle, it passes through strong shock waves. These shock waves are very sharp transitions where the air is rapidly compressed and heated. According to the Rankine-Hugoniot relations, the pressure, temperature, and density all increase dramatically across a shock wave. The process is irreversible, meaning that some of the energy in the air is converted into heat, and the flow becomes more disordered, increasing entropy. This increase in entropy indicates a loss of useful mechanical energy (which could otherwise contribute to lift or thrust) and a corresponding increase in heat. This is one of the reasons why wave drag (drag caused by shock waves) is a dominant force at hypersonic speeds—because the increase in entropy represents the energy lost in overcoming drag.

- **Heat Transfer and Entropy**: The high temperatures experienced in hypersonic flight—often exceeding 2,000°C (3,632°F)—mean that the air molecules absorb a tremendous amount of energy. As the kinetic energy of the air molecules increases, the vibrational, rotational, and translational motion of the molecules increases, contributing to a higher entropy state. The vehicle's surface, subjected to this extreme heat, must manage the heat load effectively to avoid structural damage. The management of entropy increase through thermal protection systems (TPS) is a critical aspect of vehicle design.

- **Viscous Interactions and Entropy**: In hypersonic flight, viscous effects (the resistance of air due to its internal friction) become much more pronounced. As the vehicle moves through the atmosphere, the boundary layer of air—the thin layer of air in direct contact with the vehicle's surface—experiences significant shear forces. These forces cause the kinetic energy in the flow to be converted into heat, increasing the entropy within the boundary layer. This rise in entropy indicates energy losses, complicating the vehicle's ability to maintain aerodynamic efficiency.

- **Entropy and Thermodynamic Efficiency**: In general, an increase in entropy represents a loss of thermodynamic efficiency. For hypersonic vehicles, managing this entropy increase is vital because it directly affects performance. High entropy means more energy is lost to heat and less is available to perform useful work, such as generating lift or reducing drag. Thermal protection systems, aerodynamic designs that minimize shock wave formation, and materials that can handle heat loads are all part of mitigating the effects of entropy in hypersonic flight.

- **Entropy and Non-Equilibrium Flow**: In hypersonic flight, the flow is often in a non-equilibrium state because the chemical reactions in the air (such as dissociation and ionization) may not have time to complete before conditions change. This introduces additional complexities in entropy calculation since the gas does not follow traditional equilibrium thermodynamics. Non-equilibrium conditions can lead to uneven entropy distributions in the flow, which complicates the modelling and prediction of heat transfer and drag forces on the vehicle.

Entropy plays a crucial role in the thermodynamics of hypersonic flight. It represents the irreversible losses of mechanical energy due to the extreme aerodynamic conditions and shock waves generated at these speeds. The rise in entropy through shock waves, viscous effects, and heat transfer contributes to increased drag and heat loads, requiring careful management through vehicle design. Understanding and accounting for entropy in hypersonic flight is essential for optimizing performance, minimizing drag, and protecting the vehicle from the harsh thermal environments it encounters.

The formation of the entropy layer at hypersonic speeds is primarily due to the abrupt changes in thermodynamic properties as air molecules traverse shock waves. When a shock wave compresses the air in front of a hypersonic vehicle, the entropy of the flow increases sharply, resulting in a high-entropy gradient near the vehicle's surface. This phenomenon is particularly pronounced in regions where strong shock waves, such as bow shocks, interact with the boundary layer, leading to vortical flows that disrupt the smooth airflow over the vehicle's surface. Such disruptions can cause unpredictable aerodynamic behaviours and increase heat loads on critical surfaces, complicating vehicle control and stability [169-171].

Viscous interaction plays a pivotal role in the dynamics of hypersonic flight. At these extreme velocities, the kinetic energy of the airflow is converted into internal energy through viscous effects, resulting in significant heating of the boundary layer. This viscous heating, combined with the already intense aerodynamic heating from atmospheric friction, poses challenges for thermal management systems. The friction between the air and the vehicle surface leads to a substantial increase in the temperature of the boundary layer, complicating the thermodynamic behaviour of the flow and necessitating advanced thermal protection systems (TPS) to prevent structural failure [125, 172, 173].

Boundary layer thickening is another critical consequence of viscous interaction at hypersonic speeds. As the vehicle moves through the atmosphere, the boundary layer thickens due to the increased temperatures, which reduce air density and cause the layer to expand. This thickening can result in the boundary layer merging with the shock wave, creating complex interactions known as shock-boundary layer interactions (SBLI). Such interactions can lead to increased drag and heat transfer rates, further complicating vehicle dynamics and control. In extreme cases, the boundary layer may transition to turbulence, exacerbating drag and heat transfer issues, and necessitating sophisticated computational fluid dynamics (CFD) simulations and wind tunnel testing to accurately model these effects [174-177].

The entropy layer and viscous interactions are critical challenges in the design and operation of hypersonic vehicles. The high entropy gradients near shock waves, coupled with viscous heating, contribute to increased drag and heat transfer, posing significant risks to vehicle performance and structural integrity. The thickening of the boundary layer at hypersonic speeds further complicates these issues, underscoring the need for robust thermal protection systems and careful aerodynamic shaping to mitigate these complex interactions [178-180].

Non-Equilibrium Flow

In hypersonic flight, the flow of air around the vehicle often exists in a non-equilibrium state, primarily due to the extreme speeds and temperatures encountered at these velocities. This non-equilibrium condition arises because the chemical reactions, such as dissociation (the breaking apart of molecular bonds) and ionization (the stripping of electrons from atoms), require time to occur. However, the timescales for these reactions are often longer than the rapid changes in the flow conditions experienced by a hypersonic vehicle. As the vehicle moves at speeds exceeding Mach 5, the air molecules experience abrupt and extreme changes in pressure, temperature, and density, causing the gas to behave in ways that deviate from the assumptions of thermodynamic equilibrium.

Non-Equilibrium in Chemical Reactions

At hypersonic speeds, the kinetic energy of air molecules is so high that they begin to break apart through dissociation, particularly for diatomic gases like nitrogen (N_2) and oxygen (O_2). This dissociation process absorbs a significant amount of energy from the flow, which would otherwise contribute to heat and pressure. However, because the vehicle is moving so quickly, the flow conditions change before the dissociation process can fully reach equilibrium. This means that the gas may still contain a mixture of dissociated and undissociated molecules, leading to a non-equilibrium chemical state.

Similarly, ionization occurs at even higher temperatures, when the energy levels are sufficient to strip electrons from atoms, creating a plasma of charged particles. Ionization introduces additional complexities in the flow, such as changes in electrical conductivity and electromagnetic interactions. However, as with dissociation, the timescale for ionization may be longer than the time available for the gas to fully adjust to the changing flow conditions. This results in a partially ionized gas, further complicating the thermodynamic analysis of the flow around the hypersonic vehicle.

Impact on Heat Transfer

In non-equilibrium flow, heat transfer becomes more challenging to predict and control. The process of dissociation and ionization absorbs a significant portion of the energy that would otherwise contribute to the temperature of the gas, meaning that the traditional methods of calculating heat transfer—based on equilibrium thermodynamics—are no longer accurate. The gas may be at a lower temperature than expected in regions where dissociation and ionization are occurring, but as these processes are incomplete, the overall energy distribution is difficult to model precisely.

This impacts the design of thermal protection systems (TPS), as engineers must account for the fact that the heat load experienced by the vehicle may not follow the standard predictions based on equilibrium conditions. The TPS must be robust enough to handle not only the extreme temperatures associated with hypersonic flight but also the uneven heat distribution caused by the ongoing chemical reactions in the flow.

Effect on Pressure and Drag

The presence of a non-equilibrium flow also affects pressure and drag. As the air molecules undergo dissociation and ionization, they absorb energy, which would otherwise contribute to the pressure on the vehicle's surface. This means that the pressure distribution around the vehicle can vary in unpredictable ways, making it more difficult to maintain stable flight. The gas may be less dense than expected due to dissociation, reducing the pressure, but the overall drag on the vehicle could still be high because of the complex interactions between the non-equilibrium gas and the vehicle's surface.

Furthermore, the shock waves generated by the vehicle at hypersonic speeds complicate the flow dynamics even more. Shock waves create sharp increases in pressure and temperature, accelerating the rate of chemical reactions. However, since these reactions cannot keep pace with the rapid changes in flow conditions, the gas behind the shock wave remains in a non-equilibrium state. This contributes to wave drag, a major source of resistance in hypersonic flight, as the vehicle must push through the highly compressed and heated air in front of the shock wave.

Thermodynamic Modelling Challenges

Traditional thermodynamic models assume that the gas is in equilibrium, meaning that its temperature, pressure, and chemical composition can be described using standard equations of state. However, in hypersonic flows, the non-equilibrium nature of the gas means that these assumptions no longer hold. The gas does not reach a steady state, and its behaviour depends on the specific timescales of the chemical reactions and the rate at which the flow conditions are changing.

To address this, non-equilibrium thermodynamic models must be used. These models account for the fact that the gas is in a transient state, with ongoing chemical reactions that alter its properties over time. This makes the analysis of hypersonic flows much more complex, requiring advanced computational fluid dynamics (CFD) simulations that incorporate the kinetics of dissociation, ionization, and other chemical processes. These models are essential for predicting how the flow will behave around a hypersonic vehicle, allowing engineers to design vehicles that can withstand the extreme conditions without losing stability or efficiency.

The non-equilibrium nature of hypersonic flows presents significant challenges in the design and analysis of hypersonic vehicles. The slow timescales of chemical reactions, such as dissociation and ionization, relative to the rapid changes in flow conditions, create a complex environment where the gas does not behave according to standard thermodynamic principles. This affects heat transfer, pressure distribution, and drag, all of which are critical factors in the performance of hypersonic vehicles. Understanding and modelling these non-equilibrium effects is essential for the development of effective thermal protection systems, aerodynamic designs, and flight control strategies for hypersonic vehicles.

Materials and Structural Requirements

Designing and developing Hypersonic Glide Vehicles (HGVs) presents numerous challenges, as these vehicles are required to independently navigate, manoeuvre, and reach their targets at extreme speeds. Despite not needing a main propulsion system like a traditional missile, HGVs must integrate a range of advanced subsystems to ensure they can carry out their missions. These subsystems include guidance and control systems, a lightweight but heat-resistant airframe, and the payload, which can vary based on mission objectives. Each of these components must work seamlessly together, which is no easy task given the constraints of space, weight, and the extreme conditions of hypersonic flight [64].

The guidance and control system of an HGV is a particularly complex subsystem. It requires multiple components, such as inertial navigation sensors, computers, power sources, and actuators that control aerodynamic surfaces or gas thrusters. These components enable the HGV to make real-time adjustments to its flight path, allowing it to manoeuvre at hypersonic speeds. However, adding these components increases the vehicle's overall weight and volume, making it difficult to maintain the desired balance between size, weight, and functionality. As more systems are added—such as wiring, fasteners, and connectors—the available space for the payload becomes increasingly constrained, limiting the vehicle's overall capabilities. This requires careful trade-offs between the performance of the guidance system and the vehicle's ability to carry effective payloads [64].

Another major challenge is the thermal management of HGVs, which must endure extreme heat as they travel through the denser layers of the atmosphere. At hypersonic speeds, the airframe is subjected to massive heat loads due to friction, creating temperatures that can reach thousands of degrees Celsius. The airframe must be made of lightweight materials that can withstand these high temperatures without melting or degrading. In addition to heat, the mechanical stresses generated by rapid manoeuvres at such speeds further strain the airframe. Materials like thermal shielding are essential for maintaining the structural integrity of the vehicle while also keeping internal components, such as guidance systems and payloads, protected from the extreme heat [64].

A particularly difficult challenge arises from the plasma cloud that forms around the HGV at hypersonic speeds, caused by the ionization of air molecules due to extreme heat. This plasma cloud can disrupt the vehicle's sensors and communication systems, making it hard for the vehicle to detect its surroundings or receive external guidance. This means that HGVs must rely heavily on inertial navigation systems (INS), which do not depend on external inputs, to maintain accurate positioning. However, achieving the required level of precision in such extreme conditions is difficult, and the technology required to create accurate inertial sensors that can withstand hypersonic environments is both sophisticated and expensive. The technological challenges faced by HGVs are comparable to those of designing spacecraft for atmospheric re-entry, though HGVs have stricter size and weight limits due to their military applications [64].

Despite these significant technical challenges, several countries are pursuing HGV development because of the strategic advantages these vehicles offer. The ability to manoeuvre at hypersonic speeds while evading missile defence systems makes HGVs a highly desirable tool in modern warfare. However, the complexity of designing reliable HGVs also means that these systems are expensive to develop, with high testing requirements and access to advanced materials and technologies often being limiting factors [64].

Hypersonic Cruise Missiles (HCMs) share many of the same technical requirements as HGVs but face additional challenges due to the need for propulsion systems. Unlike HGVs, which are launched by rocket boosters and glide unpowered toward their targets, HCMs must carry and sustain propulsion throughout their flight to maintain hypersonic speeds. This propulsion is typically provided by scramjet engines (supersonic combustion ramjets), which are capable of propelling the missile at hypersonic speeds by compressing and igniting atmospheric air. Scramjets, however, are highly sophisticated engines that present extreme engineering challenges [64].

For instance, scramjet engines need to maintain combustion at hypersonic speeds, where air enters the engine at such high velocity that stable ignition is difficult to achieve. The technology needed to maintain continuous air-breathing propulsion at these speeds has yet to be fully demonstrated in operational systems. Current prototypes, such as the US X-51A Waverider, have been able to achieve hypersonic flight for a few minutes, but the vehicle was not equipped to carry a payload, and the available space was entirely filled with engine components. This highlights one of the main challenges for HCMs—fitting a highly efficient propulsion system, along with guidance, thermal shielding, and a payload, into a small, lightweight airframe. The weight and volume constraints severely limit the potential size and power of the payload that an HCM can carry [64].

Additionally, HCMs face many of the same thermal and mechanical stresses as HGVs. The high speeds generate intense heat, which requires thermal shielding to protect internal components. Like HGVs, HCMs must be made of lightweight, heat-resistant materials that can maintain their structural integrity during the entire flight. They must also have highly accurate guidance systems that can work in conjunction with the propulsion system to ensure precise targeting while manoeuvring at hypersonic speeds. This requires a delicate balance between speed, manoeuvrability, and precision, all while ensuring the missile can survive the harsh thermal environment [64].

Both HGVs and HCMs are designed to reach and maintain speeds greater than Mach 5, but their development is limited by the sophistication of the technology, the design choices, and the engineering trade-offs made by the countries developing them. The technical challenges involved in creating these systems go far beyond just reaching high speeds; they include overcoming issues related to heat management, guidance, control, and propulsion. Furthermore, the ability of hypersonic missiles to evade detection and interception makes them valuable in military applications, but it also raises concerns about the stability of global missile defence systems [64].

Materials Used in Hypersonic Weapons

Hypersonic weapons, defined as those capable of traveling at speeds exceeding Mach 5, face significant challenges due to the extreme conditions they encounter during flight. These conditions include high temperatures, intense aerodynamic forces, and substantial structural stresses, which necessitate the use of advanced materials that can endure such harsh environments without degrading. The selection of materials for hypersonic weapon systems is critical, as they must exhibit high temperature resistance, mechanical strength, and durability under prolonged exposure to extreme conditions.

The aerodynamic heating experienced by hypersonic vehicles is a major concern, as it can lead to significant thermal loads that affect the structural integrity of the materials used. Hu et al. discuss the importance of understanding aerodynamic heating in the context of hypersonic rocket sled tests, which are essential for evaluating the performance of weapon system materials under high aerodynamic pressures and thermal environments [181]. This understanding is crucial for developing materials that can withstand the extreme heating associated with hypersonic speeds.

Advanced materials such as ultrahigh-temperature ceramics and refractory metals are being explored for their potential in hypersonic applications. Lü et al. [182] highlight the significance of these materials, noting that they can operate effectively at temperatures exceeding 2000 °C, making them suitable for the demanding conditions faced by hypersonic missiles and reusable delivery systems. Additionally, Xie et al. [183] emphasize the role of zirconium diboride ceramics, which possess exceptional properties for ultrahigh temperature applications, including thermal protection systems in hypersonic vehicles. These materials are engineered to maintain structural integrity and performance even under extreme thermal and mechanical loads.

Moreover, the integration of thermal protection systems (TPS) is vital for ensuring the operational capability of hypersonic weapons. The TPS must be lightweight yet robust enough to handle the intense heat generated during flight. Chang et al. [184] discuss the development of superelastic carbon aerogels, which are emerging as promising materials for advanced thermal protection due to their lightweight nature and ability to withstand extreme thermal environments. Similarly, the work of Kim et al. [185] indicates that the skin panel structures of hypersonic vehicles must be designed to endure combined aerodynamic, thermal, and acoustic loads, which further underscores the need for innovative material solutions.

The design considerations for hypersonic materials are critical due to the extreme conditions these vehicles encounter during flight. Key aspects include heat resistance, weight reduction, structural integrity, and thermal expansion. Each of these factors plays a significant role in the overall performance and safety of hypersonic vehicles.

Heat Resistance is paramount in hypersonic flight, where vehicles are subjected to extreme temperatures from aerodynamic heating. Materials such as ceramic composites and ablatives are essential for thermal protection systems (TPS) that shield critical surfaces from heat damage. Research indicates that the TPS must effectively manage heat transfer through various mechanisms, including convective, conductive, and radiative cooling [186, 187]. The use of carbon-carbon composites is particularly noted for their high latent heat capacity and strength at elevated temperatures, making them suitable for thermal protection applications [186]. Additionally, the design of leading edges must consider the heat flux, which is inversely proportional to the radius of curvature, emphasizing the need for careful geometric considerations in material selection [188].

Weight Reduction is another critical factor in hypersonic vehicle design. Lightweight materials such as titanium and aluminium-lithium alloys are often employed to minimize drag and enhance fuel efficiency. The integration of advanced materials not only contributes to weight savings but also ensures that the structural components can withstand the intense aerodynamic loads experienced during flight [150, 189]. The development of high-temperature resistant resin matrix composites further supports the need for lightweight structures while maintaining the necessary thermal resistance [150].

Structural Integrity under extreme thermal and aerodynamic loads is essential for the safety and functionality of hypersonic vehicles. The use of refractory metals and nickel-based superalloys is common, as these materials maintain their mechanical properties even at high temperatures [110, 189]. The multidisciplinary approach to hypersonic design necessitates the consideration of structural mechanics alongside aerodynamics and propulsion systems, ensuring that all components can withstand the harsh conditions of hypersonic flight [130, 190].

Thermal Expansion must also be addressed, as hypersonic vehicles experience rapid temperature fluctuations that can lead to material deformation. Materials like carbon-carbon composites are favoured for their low thermal expansion coefficients, which help prevent structural deformation during flight [186]. The coupling of thermal and mechanical effects is a significant consideration in the design of hypersonic structures, as it influences the overall stability and performance of the vehicle [109].

Key Materials Used in Hypersonic Weapons

- **Ceramic Matrix Composites (CMCs)** Ceramic Matrix Composites (CMCs) are widely used in the construction of hypersonic vehicles due to their high-temperature resistance and durability. CMCs can withstand extreme temperatures without melt ng or losing strength, which is critical for components exposed to the highest thermal loads, such as the nose cone, leading edges, and control surfaces.

  - **Advantages**: High thermal resistance (over 2,000°C), excellent structural stability, and resistance to thermal shock.

  - **Examples**: Silicon carbide (SiC) composites, alumina (Al2O3) composites.

- **Ablative Materials** Ablative materials are used in hypersonic weapons to provide protection against extreme heat by gradually vaporizing or "ablating" during flight. As the materia vaporizes, it carries away heat from the surface of the weapon, thereby protecting the underlying structure from damage.

  - **Advantages**: Effective heat management through material sacrificial vaporization, widely used in heat shields and thermal protection systems.

  - **Examples**: Phenolic resin-based ablatives, carbon-carbon composites.

- **Ultra-High-Temperature Ceramics (UHTCs)** Ultra-High-Temperature Ceramics (UHTCs) are a class of materials that can withstand temperatures above 3,000°C. These ceramics are used in hypersonic applications where traditional materials cannot survive the extreme heat generated by atmospheric friction.
    - **Advantages**: Extremely high melting points and thermal stability at temperatures exceeding 3,000°C.
    - **Examples**: Hafnium carbide (HfC), zirconium carbide (ZrC), tantalum carbide (TaC).
- **Titanium Alloys** Titanium alloys are commonly used in hypersonic weapons for structural components because they offer a good balance between strength, weight, and temperature resistance. They are lighter than steel and maintain their mechanical properties at high temperatures.
    - **Advantages**: High strength-to-weight ratio, good temperature resistance (up to 600°C), corrosion resistance.
    - **Examples**: Ti-6Al-4V (a titanium-aluminium-vanadium alloy).
- **Nickel-Based Superalloys** Nickel-based superalloys are used in hypersonic weapons to provide structural integrity and heat resistance in critical components, such as engine parts and fasteners. These materials are designed to maintain mechanical strength at high temperatures while resisting corrosion and oxidation.
    - **Advantages**: Excellent mechanical properties at high temperatures, resistance to oxidation, and structural integrity at temperatures above 1,000°C.
    - **Examples**: Inconel (a nickel-chromium alloy), Hastelloy.
- **Refractory Metals** Refractory metals are a group of metals known for their exceptional resistance to heat and wear. These materials are used in components exposed to extreme temperatures, such as engine nozzles, heat shields, and leading edges.

- **Advantages**: High melting points (above 2,000°C), resistance to thermal stress, and durability under harsh conditions.
- **Examples**: Molybdenum (Mo), tungsten (W), niobium (Nb), tantalum (Ta).

- **Carbon-Carbon Composites** Carbon-carbon composites are made by reinforcing carbon fibers with a carbon matrix. These materials are extremely lightweight yet capable of withstanding extreme temperatures, making them ideal for thermal protection systems and structural components in hypersonic weapons.
  - **Advantages**: High strength at high temperatures, low weight, and resistance to thermal expansion.
  - **Examples**: Used in nose cones, heat shields, and leading edges.
- **Aluminium-Lithium Alloys** Aluminium-lithium alloys are used in hypersonic weapons for lightweight structural components where lower temperature resistance is acceptable. These alloys reduce the overall weight of the weapon, improving its range and manoeuvrability.
  - **Advantages**: Lightweight, good strength-to-weight ratio, and corrosion resistance.
  - **Examples**: Al-Li alloys used in aircraft and space structures.
- **Thermal Protection Coatings** Hypersonic weapons often use specialized coatings to protect their surfaces from oxidation, corrosion, and wear caused by high-speed flight through the atmosphere. These coatings are applied to surfaces that experience high temperatures and help maintain structural integrity by reducing the rate of heat transfer and material degradation.
  - **Advantages**: Provides thermal insulation, reduces oxidation, and protects against wear.
  - **Examples**: Thermal barrier coatings (TBCs) made of zirconia (ZrO_2) or alumina (Al_2O_3), often stabilized with yttria (Y_2O_3).

Hypersonic weapons require materials that can endure the extreme environments of high-speed flight. The use of ceramic composites, ablative coatings, refractory metals, and advanced alloys ensures that these weapons can withstand the intense heat, pressure, and mechanical stresses they face while traveling at hypersonic speeds. These advanced materials play a critical role in the design, performance, and survivability of hypersonic weapons in modern military applications.

Hypersonic Propulsion Systems

Boost-Glide Systems and Scramjets

The propulsion systems of hypersonic weapons are a key factor in enabling these vehicles to reach and sustain the extreme speeds required to qualify as hypersonic—Mach 5 and above. The technical complexity of achieving such speeds, while maintaining control and effectiveness over long distances, relies heavily on the choice of propulsion system. The two most common types of propulsion systems used in hypersonic weapons are boost-glide systems and scramjet engines. Each of these propulsion methods is suited to different types of hypersonic vehicles, such as Hypersonic Glide Vehicles (HGVs) and Hypersonic Cruise Missiles (HCMs), and provides distinct operational advantages based on their principles of operation.

Boost-glide systems are the most common propulsion method for Hypersonic Glide Vehicles (HGVs). These systems involve the use of conventional rocket boosters to accelerate the weapon to hypersonic speeds during the initial launch phase, known as the "boost phase." During this phase, the rocket carries the glide vehicle into the upper atmosphere, typically at altitudes of 20 to 100 kilometres (12 to 62 miles), and accelerates it to hypersonic speeds. Once the vehicle reaches the desired altitude and velocity, the rocket booster is jettisoned, and the weapon enters the glide phase, where it coasts toward its target at hypersonic speeds without the need for further propulsion.

During the glide phase, the vehicle remains in the upper atmosphere, where the air resistance is minimal, allowing it to maintain its extreme speed over long distances with minimal loss of energy. This glide phase is where the HGV's manoeuvrability becomes critical, as the weapon can make adjustments to its trajectory to avoid interception or adapt to changing target locations. The ability to manoeuvre while gliding at hypersonic speeds makes HGVs particularly challenging to detect and intercept, as their flight paths are unpredictable compared to traditional ballistic missiles. The use of a boost-glide system allows for intercontinental range capabilities, as the vehicle can travel thousands of kilometres once it reaches its gliding phase. For example, the Russian Avangard system is reported to have a range exceeding 6,000 kilometres, demonstrating the vast distances that boost-glide systems can cover.

Boost-glide systems are advantageous because they allow hypersonic vehicles to reach extreme speeds and altitudes quickly, while the glide phase provides efficient long-range travel with reduced propulsion demands. However, these systems rely heavily on the initial rocket launch, which can be vulnerable to detection, although once the vehicle enters the glide phase, it becomes far more challenging to track.

In contrast, scramjet engines (supersonic combustion ramjets) are used to power Hypersonic Cruise Missiles (HCMs), offering a different method of achieving and maintaining hypersonic speeds. Unlike traditional jet engines that rely on subsonic airflow into the engine, scramjets are designed to operate at hypersonic speeds, where the air entering the engine remains supersonic throughout the combustion process. This is a critical distinction that allows scramjets to function efficiently at speeds greater than Mach 5, where traditional engines would become ineffective.

The principle behind a scramjet engine is relatively simple, but its execution is highly complex due to the extreme conditions involved. A scramjet engine intakes air at hypersonic speeds, compresses it, and then ignites the fuel-air mixture to produce thrust. Unlike rocket engines, scramjets do not require onboard oxidizers because they rely on atmospheric oxygen for combustion. This makes scramjets lighter and more efficient than rocket-powered systems, as they can sustain high-speed flight over long distances without the need to carry heavy fuel and oxidizer supplies. The absence of the need for oxidizers also allows for sustained hypersonic flight within the atmosphere, making scramjet-powered hypersonic cruise missiles ideal for regional or theatre-level strikes.

The air-breathing nature of scramjet engines gives hypersonic cruise missiles several tactical advantages. First, they can maintain hypersonic speeds for extended periods, providing the ability to rapidly strike targets over medium-range distances (typically up to 1,500 kilometres or 930 miles). Unlike HGVs, which rely on a glide phase, scramjet-powered missiles are continuously propelled, allowing them to fly at lower altitudes and maintain high speeds throughout their flight. This low-altitude flight profile, combined with their sustained speed, makes them difficult to detect by radar and challenging to intercept by missile defence systems. Additionally, the ability of scramjets to operate within the atmosphere means that Hypersonic Cruise Missiles can leverage terrain features, such as mountains and valleys, to evade detection, further enhancing their stealth.

While scramjet engines provide tremendous advantages in terms of speed and operational efficiency, they also pose significant technical challenges. The extreme temperatures generated by hypersonic flight require advanced materials and cooling systems to prevent the engine and vehicle from overheating. Furthermore, maintaining stable combustion at hypersonic speeds is a complex engineering problem, as the air is moving through the engine so quickly that it must be compressed and ignited within milliseconds. These challenges have made scramjet technology difficult to perfect, but recent advancements have made it a viable propulsion option for hypersonic cruise missiles.

Boost-Glide Systems

A boost-glide system is an advanced propulsion and flight mechanism used primarily in hypersonic weapons, designed to achieve extreme speeds and long-range strike capabilities. It operates by combining two phases: an initial boost phase, where the vehicle is propelled to very high speeds and altitudes by a rocket, followed by a glide phase, where the vehicle descends and glides toward its target at hypersonic speeds. This system takes advantage of the high kinetic energy gained during the boost phase, allowing the vehicle to cover vast distances at speeds typically exceeding Mach 5.

The operation of a Boost-Glide system, particularly in the context of hypersonic glide vehicles (HGVs), can be divided into two critical phases: the boost phase and the glide phase. Each phase plays a significant role in the overall performance and capabilities of the system.

1. The Boost Phase:

In the first phase of a boost-glide system, a rocket booster is used to propel the hypersonic vehicle, often referred to as a hypersonic glide vehicle (HGV), into the upper atmosphere or near space. During this phase, the vehicle is launched almost vertically, similar to a ballistic missile. The rocket provides the necessary thrust to overcome gravity and reach extremely high speeds, generally in the hypersonic range (Mach 5+), and high altitudes, where the air is thinner, reducing drag [191].

This initial phase is crucial because the vehicle must gain sufficient velocity and altitude to enter the glide phase. The altitude reached can be tens of kilometres above the Earth's surface, depending on the specific mission and target. Once the vehicle reaches the required speed and altitude, the booster is jettisoned, and the vehicle transitions to the next phase.

Research has shown that optimizing the ascent trajectory during this phase can lead to significant improvements in fuel efficiency and overall mission success [192]. For instance, the use of advanced optimization algorithms, such as the improved chicken swarm optimization method, has been demonstrated to enhance the ascent trajectory of hypersonic vehicles, thereby reducing fuel consumption [192]. Additionally, the dynamics of the booster separation process are critical; studies indicate that understanding the motion parameters during this phase can optimize the transition to the glide phase [193].

2. The Glide Phase:

After the booster is discarded, the vehicle begins the glide phase, which distinguishes boost-glide systems from traditional ballistic missiles. Unlike ballistic missiles, which follow a predictable parabolic trajectory, hypersonic glide vehicles are capable of controlled, sustained flight within the atmosphere. The glide vehicle re-enters the atmosphere and uses its aerodynamic design to "glide" at hypersonic speeds toward its target.

The glide phase is where the system's real advantages come into play. The vehicle uses the kinetic energy it acquired during the boost phase to maintain high speeds while manoeuvring through the atmosphere. This phase allows for much greater range and flexibility than traditional missiles. Because of the vehicle's ability to manoeuvre, it becomes difficult for missile defence systems to predict its flight path, making it extremely challenging to intercept.

This manoeuvrability is facilitated by the aerodynamic design of the vehicle, which allows it to glide efficiently while maintaining high speeds [13]. The glide phase leverages the kinetic energy gained during the boost phase, enabling the vehicle to cover vast distances with enhanced flexibility compared to conventional missiles.

The ability of HGVs to manoeuvre during the glide phase poses significant challenges for missile defence systems. The unpredictable flight paths make interception exceedingly difficult, as these vehicles can adjust their trajectories in real-time based on external conditions and targets

[13, 38]. Moreover, trajectory optimization techniques, such as the Gauss pseudospectral method, have been applied to enhance the glide phase performance, allowing for more efficient path planning and execution [25]. This optimization is essential for maximizing the operational range and effectiveness of hypersonic glide vehicles.

Boost-glide systems, characterized by their unique operational capabilities, present several key features and advantages that distinguish them from traditional ballistic missile systems. These features include high speeds, manoeuvrability, long-range strike capability, and the ability to evade missile defences.

High Speeds: Boost-glide systems operate at hypersonic speeds, typically exceeding Mach 5. This capability significantly reduces the time required to reach targets, making them ideal for rapid strikes over long distances. The operational advantages of hypersonic vehicles include fast response times and efficient penetration of defences, which are critical in modern warfare scenarios where adversaries have limited time to react [13, 20]. The ability to travel at such high velocities allows these systems to cover vast distances quickly, enhancing their strategic value [40].

Manoeuvrability: Unlike traditional ballistic missiles that follow predictable parabolic trajectories, boost-glide vehicles can manoeuvre during their glide phase. This manoeuvrability complicates interception efforts by missile defence systems, as the vehicles can alter their flight paths mid-flight, making them less predictable and harder to track [13, 40]. The aerodynamic configurations of these vehicles are designed to optimize lift-to-drag ratios, further enhancing their ability to manoeuvre effectively at hypersonic speeds [194].

Long-Range Strike Capability: Boost-glide systems are particularly suited for long-range strike missions due to their ability to glide over extended distances while maintaining hypersonic speeds. This capability allows them to strike targets deep within enemy territory, often across continents, which is a significant advantage over traditional missile systems that may have limited ranges [20]. The aerodynamic efficiency of these vehicles enables them to travel further with less propulsion, making them more effective for strategic operations [120].

Evasion of Missile Defences: One of the most significant advantages of boost-glide systems is their ability to evade existing missile defence systems. Traditional defences are designed to intercept ballistic missiles that follow predictable paths. In contrast, the combination of high speed, unpredictable flight paths, and manoeuvrability of boost-glide systems poses a substantial challenge to these defence systems, making interception increasingly difficult [195]. The unique flight characteristics of boost-glide vehicles allow them to exploit gaps in current missile defence architectures, enhancing their survivability and effectiveness in combat scenarios [20].

Differences from Traditional Ballistic Missiles: Boost-glide systems differ from traditional ballistic missiles in several critical aspects. First, their flight paths are not fixed; they can manoeuvre during flight, making them less predictable compared to the parabolic trajectories of ballistic missiles [13, 40]. Second, the re-entry phase of boost-glide vehicles is more gradual, allowing them to glide through the atmosphere at high speeds, which reduces their vulnerability to interception [120]. Lastly, boost-glide vehicles utilize aerodynamic lift during their glide phase, allowing for greater range and efficiency compared to the propulsion-dependent trajectories of traditional missiles [194].

A missile's ballistic trajectory can be shaped in various ways. The figure below illustrates a simple minimum-energy trajectory (MET), which closely resembles a parabola, primarily influenced by gravity after the missile's rocket booster burns out. To lower the trajectory, a ballistic missile could be launched at a shallower angle compared to the MET missile. However, to achieve the same range as the MET missile, the missile with the depressed trajectory would need a larger rocket booster to attain a higher initial speed.

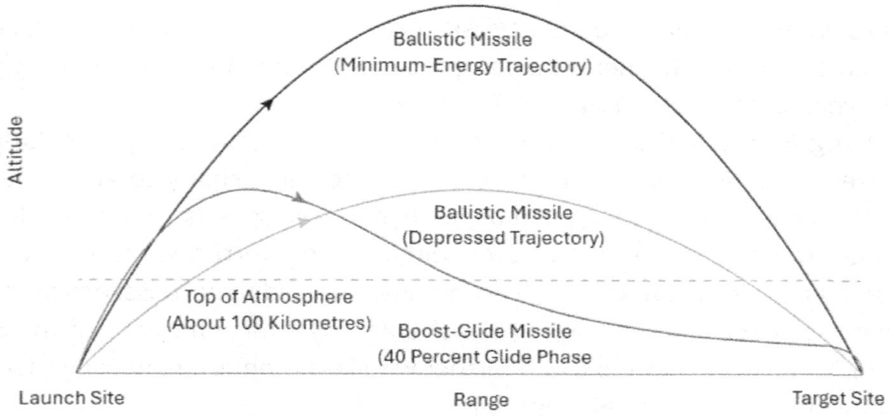

Figure 9: Notional Flight Paths for Different Ballistic and Boost-Glide Missile Trajectories With the Same Range.

In contrast, a boost-glide missile begins on a ballistic trajectory, but during the early phase of its flight, thrusters are used to bring it back toward Earth sooner than it would in a pure ballistic flight. Control surfaces on the glide body are then employed in the second phase to initiate the gliding process. The missile's maximum potential range is determined by its initial speed, the lift-to-drag ratio of the glide body, and the portion of its trajectory spent within the atmosphere.

A boost-glide system used in hypersonic weapons operates through a two-phase propulsion mechanism: the boost phase, powered by a rocket engine, and the glide phase, where the weapon travels at hypersonic speeds through the atmosphere without active propulsion. This unique system leverages the combination of ballistic missile technology and aerodynamic gliding to achieve extremely high speeds and extended ranges, while maintaining manoeuvrability. The system is designed to overcome traditional missile defence systems through speed and unpredictability, making it a significant advancement in missile technology.

In the boost phase, the propulsion system is driven by a powerful rocket engine, typically either a solid or liquid-fuelled rocket. This phase is critical as it provides the necessary velocity and altitude for the weapon to achieve hypersonic flight. The rocket booster generates immense thrust to rapidly accelerate the weapon into the upper layers of the atmosphere. Most boost-glide systems use solid-fuel rockets due to their reliability and compact design, although some may use liquid-fuel rockets for precise control over thrust, particularly in longer-range systems. During the high-speed ascent, the missile climbs at an extremely steep angle, reaching speeds greater than Mach 5 (five times the speed of sound) to ensure it quickly reaches the necessary altitude for the glide phase. Once this altitude and speed are reached, the rocket booster phase ends, and the glide vehicle or warhead separates from the booster, leaving the missile unpowered for the remainder of its flight.

The glide phase begins immediately after booster separation. At this point, the missile or glide vehicle enters an unpowered, high-speed glide through the atmosphere. The propulsion is no longer active, but the glide vehicle relies on the velocity and altitude achieved during the boost phase. The aerodynamic design of the glide body, often sleek and wedge-like, allows it to maintain high velocities while minimizing drag. This aerodynamic glide enables the vehicle to travel long distances while gradually descending through the atmosphere. The performance of the glide phase heavily depends on the vehicle's lift-to-drag ratio (L/D), which dictates how efficiently it can travel. A higher L/D ratio allows the vehicle to travel farther while losing less speed and altitude, enhancing both its range and manoeuvrability.

Unlike ballistic missiles, which follow a predictable, parabolic trajectory, a boost-glide weapon has significant manoeuvrability during its glide phase. Control surfaces on the glide body, such as fins or canards, enable the vehicle to change direction, making it harder to predict and more difficult to intercept with missile defence systems. This ability to manoeuvre is a key advantage of the boost-glide system, allowing the weapon to avoid defences and maintain a less predictable flight path.

The glide phase also presents significant engineering challenges, primarily due to the extreme conditions encountered during hypersonic flight. One of the most pressing challenges is aerodynamic heating. As the vehicle travels at hypersonic speeds, the friction between the vehicle and the atmosphere generates extreme temperatures, reaching several thousand degrees Celsius. To manage this, the vehicle must be equipped with advanced thermal protection systems, such as ablative materials or ceramic composites, to prevent structural failure from the intense heat.

Types of Propulsion Systems in Boost-Glide Hypersonic Weapons

Different hypersonic boost-glide systems may use varying types of rocket propulsion during the boost phase, depending on their mission profile, range, and speed requirements.

- **Solid-Fuel Rockets**: Many boost-glide systems use solid-fuel rockets for the boost phase due to their simplicity and high energy density. Solid rocket motors are easier to store and maintain, and they provide a high thrust-to-weight ratio, making them ideal for weapons that need to quickly reach hypersonic speeds. Research indicates that solid rocket-powered launch vehicles are particularly advantageous for hypersonic glide vehicles (HGVs), offering storage stability and short launch preparation times, making them suitable for ultra-long-range and fast attack missions [21]. Furthermore, studies have demonstrated that solid-fuel rocket scramjets can achieve specific impulses significantly higher than traditional solid rocket motors, enhancing their performance in hypersonic applications [196, 197].

- **Liquid-Fuel Rockets**: In some advanced systems, liquid-fuel rockets may be used for more precise control over the thrust profile. Liquid-fuel engines allow for throttling, which means the missile can vary its thrust during different stages of the boost phase, optimizing performance [198]. However, they are more complex and require careful handling. The ability to adjust thrust dynamically can be critical for advanced hypersonic systems that need to navigate complex flight paths while maintaining high speeds [199]. The flexibility of liquid-fuel propulsion systems makes them suitable for specific applications where precision is paramount, despite their increased complexity compared to solid-fuel systems [5].

- **Hybrid Rocket Systems**: Some boost-glide weapons may incorporate hybrid propulsion systems that combine both solid and liquid propellants. This hybrid approach allows for a balance between the simplicity and reliability of solid-fuel systems with the flexibility and control of liquid-fuel rockets. Research has shown that hybrid rockets can provide significant advantages, including reduced costs and increased reliability, making them an attractive option for future hypersonic vehicles [200]. The versatility of hybrid systems allows for the selection of more sustainable fuels, which can potentially reduce the environmental impact of rocket launches [201]. Moreover, advancements in hybrid rocket technology have led to improved performance metrics, such as thrust and efficiency, which are critical for meeting the demanding requirements of hypersonic flight [202].

Solid Fuel Rockets

A solid-fuel rocket is a type of propulsion system that uses a solid propellant composed of a fuel and an oxidizer mixed together in a stable, solid form. This system has been widely used in various aerospace applications, including space exploration and military weaponry, due to its simplicity, reliability, and ability to generate powerful thrust. In the context of hypersonic flight, solid-fuel rockets are particularly valuable during the boost phase, where they provide the initial velocity and altitude required for the vehicle to reach hypersonic speeds. Understanding how a solid-fuel rocket works, and how it supports hypersonic flight, involves exploring the key processes that occur during ignition, combustion, and thrust generation.

At the core of a solid-fuel rocket's operation is the solid propellant itself, which typically consists of a mixture of a fuel (such as powdered metals like aluminium) and an oxidizer (such as ammonium perchlorate). These two components are intimately mixed in a binder material that holds them together and provides additional energy during combustion. The entire propellant is cast or moulded into a specific shape, often a cylindrical grain with a hollow core, to maximize the surface area exposed to combustion and to control the burn rate of the propellant. Once the solid fuel is ignited, it begins to burn from the inside out, steadily releasing energy in the form of hot gases.

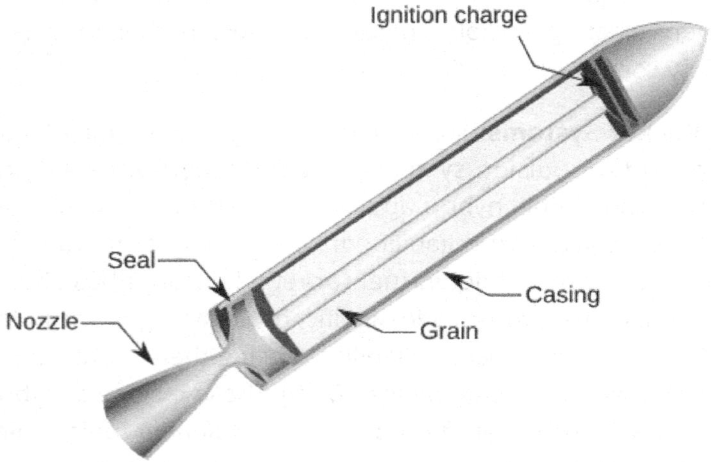

Figure 10: A simplified diagram of a solid-fuel rocket. Pbroks13, CC BY-SA 4.0, via Wikimedia Commons.

The combustion process in a solid-fuel rocket is initiated by an ignition source, which can be a pyrotechnic charge or an electric igniter. Once ignited, the fuel burns continuously along the exposed surfaces of the propellant grain, generating immense heat and high-pressure gases. These gases are forced out of the rocket's nozzle at high speeds, creating the thrust needed to propel the rocket forward. One of the advantages of solid-fuel rockets is that they burn until the entire fuel load is consumed, delivering a constant thrust throughout the burn time. This characteristic makes solid-fuel rockets highly reliable, as there are no moving parts or complex systems required to control the combustion process.

In the context of hypersonic flight, the solid-fuel rocket plays a critical role during the boost phase, where it is responsible for accelerating the hypersonic vehicle to speeds exceeding Mach 5 (five times the speed of sound). Hypersonic vehicles, such as boost-glide systems or scramjet-powered missiles, rely on solid-fuel rockets to provide the initial velocity needed to enter the high-speed flight regime. During this phase, the solid-fuel rocket generates the necessary thrust to lift the vehicle to high altitudes while rapidly accelerating it. The efficiency of the solid-fuel rocket in generating high thrust is essential for overcoming the drag forces encountered at lower altitudes and propelling the vehicle to the point where the glide or scramjet phases can take over.

The design of the solid-fuel rocket's nozzle is also crucial for hypersonic flight. Nozzles are typically designed to expand and accelerate the exhaust gases as they exit the combustion chamber, converting the thermal energy from combustion into kinetic energy, which generates thrust. For hypersonic applications, the nozzle must be optimized to handle extremely high temperatures and pressures while ensuring that the exhaust gases are expelled at hypersonic velocities. This is typically achieved using a convergent-divergent nozzle (also known as a de Laval nozzle), where the gases are compressed in the narrow throat and then expanded in the diverging section, accelerating them to supersonic or hypersonic speeds.

One of the key advantages of solid-fuel rockets in hypersonic applications is their ability to deliver consistent thrust in a compact and lightweight package. Unlike liquid-fueled rockets, which require complex pumps, tanks, and plumbing to deliver the fuel and oxidizer, solid-fuel rockets contain the fuel and oxidizer in a single, solid mass, making them simpler to design and operate. This simplicity is particularly important in military hypersonic weapons, where reliability and rapid deployment are critical. Solid-fuel rockets can be stored for long periods and are ready to launch with minimal preparation, making them ideal for applications where speed and readiness are essential.

However, solid-fuel rockets do have limitations in the context of hypersonic flight. Once ignited, the rocket burns until all the propellant is consumed, and the burn rate cannot be throttled or controlled mid-flight. This lack of flexibility can pose challenges for missions that require precise control over thrust levels or flight duration. Additionally, the high combustion temperatures and pressures generated by solid-fuel rockets place significant thermal stress on the rocket structure, requiring the use of heat-resistant materials in critical areas, such as the nozzle and combustion chamber.

Liquid-Fuel Rockets

A liquid fuel rocket operates by burning liquid propellants—typically a combination of a liquid fuel and a liquid oxidizer—in a combustion chamber, where the chemical energy of the propellants is converted into thermal energy. This thermal energy is then expelled through a rocket nozzle to produce thrust. Liquid fuel rockets are known for their high efficiency and their ability to be throttled, making them suitable for various aerospace applications, including space exploration, satellite launches, and military uses. In the context of hypersonic flight, liquid fuel rockets play a crucial role in the initial boost phase, providing the powerful thrust necessary to propel hypersonic vehicles to speeds greater than Mach 5, while enabling precise control over acceleration and flight dynamics.

The core components of a liquid fuel rocket include the following:

- **Propellant Tanks**: Separate tanks store the liquid fuel and oxidizer. The fuel might be a substance like kerosene (RP-1) or liquid hydrogen (LH2), while the oxidizer is often liquid oxygen (LOX). These substances remain in liquid form at very low temperatures, especially for cryogenic fuels like liquid hydrogen.

- **Pumps and Plumbing**: Turbopumps are used to pressurize and deliver the fuel and oxidizer to the combustion chamber. These pumps are powered by a small amount of the propellants themselves, which are burned in a gas generator to create the power needed to drive the turbopumps.

- **Combustion Chamber**: The fuel and oxidizer are mixed and ignited in the combustion chamber. The high-temperature combustion of the propellants produces extremely hot, high-pressure gases, which are forced through the rocket's nozzle.

- **Nozzle**: The nozzle, usually a convergent-divergent (de Laval) type, expands and accelerates the exhaust gases as they exit the combustion chamber, converting the thermal energy into kinetic energy, thereby generating thrust. The nozzle design is optimized to ensure the gases are expelled at supersonic or hypersonic velocities, depending on the requirements of the flight.

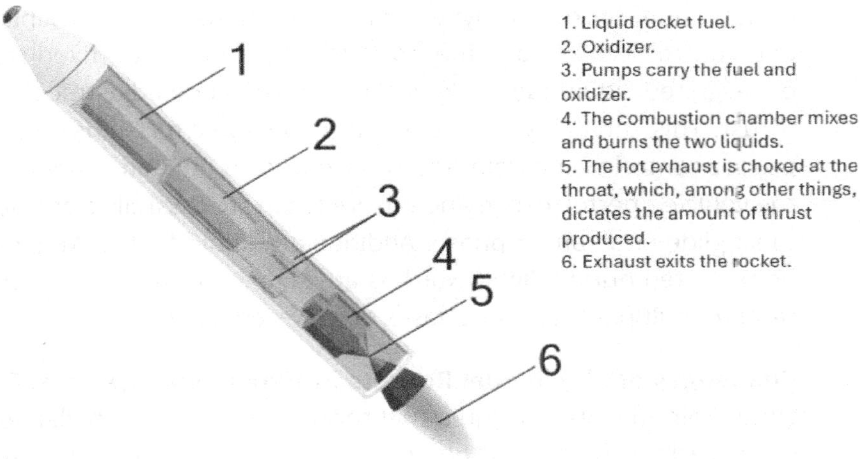

Figure 11: A simplified diagram of a liquid-fuel rocket. Pbroks13, CC BY-SA 4.0, via Wikimedia Commons.

In hypersonic flight, where vehicles travel at speeds exceeding Mach 5, liquid fuel rockets are often used in the boost phase to provide the initial thrust required to reach the desired altitude and velocity. This boost phase is crucial because hypersonic vehicles, especially those using scramjet engines or operating as boost-glide systems, require extremely high speeds before transitioning to their glide or scramjet-powered phases. The precise control over thrust provided by liquid fuel rockets allows for adjustments in acceleration and trajectory, which are critical for successfully reaching hypersonic speeds.

- **Thrust Generation and Acceleration**: During the boost phase, the liquid fuel rocket burns its propellants to rapidly accelerate the vehicle. The propellants are pumped into the combustion chamber, where they mix and ignite, creating an enormous amount of energy. The resulting high-pressure exhaust gases are expelled through the nozzle, generating the thrust needed to push the vehicle to hypersonic speeds. Because liquid fuel rockets can be throttled, the thrust can be increased or decreased as needed, allowing for more precise control over the vehicle's speed and altitude. This is particularly important when the vehicle needs to reach specific flight conditions for the next phase of its mission, such as entering the glide phase in a boost-glide system or engaging a scramjet engine.

- **Advantages in Hypersonic Systems**: One of the key advantages of liquid fuel rockets in hypersonic applications is their flexibility and control. Unlike solid fuel rockets, which burn at a fixed rate once ignited, liquid fuel rockets allow for real-time adjustments in thrust. This capability is essential in hypersonic missions, where precise control over acceleration and altitude is needed to optimize the vehicle's performance and ensure it can successfully transition to its glide or scramjet phase. Additionally, liquid fuel rockets can be restarted during flight, which is advantageous in missions that require multiple stages or phases of powered flight.

- **Challenges of Liquid Fuel Rockets in Hypersonic Systems**: Despite their advantages, liquid fuel rockets face several challenges, particularly in the context of hypersonic flight. One of the primary issues is thermal management. During hypersonic flight, the vehicle encounters extreme heat due to aerodynamic friction, especially when traveling through the denser parts of the atmosphere. This requires robust thermal protection systems to prevent overheating and potential damage to the rocket's components. Additionally, the cryogenic nature of many liquid propellants means that they need to be kept at extremely low temperatures, which complicates the design of the fuel tanks and plumbing systems. Any failure in maintaining the propellant's temperature could lead to a loss of performance or even engine failure.

- **Rocket Nozzle Design for Hypersonic Applications**: In hypersonic flight, the nozzle plays a crucial role in ensuring that the exhaust gases are expelled at the correct velocity to generate the necessary thrust. The design of the nozzle must account for the extreme temperatures and pressures encountered during the boost phase, as well as the need to minimize drag. In some cases, advanced materials, such as high-temperature alloys or ceramic composites, are used in the nozzle to withstand the thermal and mechanical stresses. Additionally, the nozzle must be shaped to ensure that the exhaust gases are efficiently expanded to hypersonic speeds, allowing the vehicle to achieve the necessary acceleration and altitude for its mission.

Once the liquid fuel rocket has accelerated the hypersonic vehicle to the required speed and altitude, the boost phase ends, and the vehicle transitions to either the glide phase (in a boost-glide system) or engages its scramjet engine (in an air-breathing hypersonic vehicle). At this point, the liquid fuel rocket's propulsion system shuts down, and the vehicle either glides through the atmosphere at hypersonic speeds or switches to its scramjet propulsion system for sustained high-speed flight.

In boost-glide systems, the vehicle separates from its rocket booster after reaching a high altitude and begins a controlled, unpowered descent through the atmosphere, using its aerodynamic shape to maintain hypersonic speeds. In scramjet-powered systems, the scramjet engine takes over once the vehicle reaches a speed where the incoming airflow is sufficient to sustain combustion in the engine. In both cases, the initial boost provided by the liquid fuel rocket is essential for achieving the necessary conditions for hypersonic flight.

Hybrid Rocket Systems

A hybrid rocket system is a propulsion technology that combines elements of both solid and liquid fuel rockets. In a hybrid rocket, a solid fuel is used alongside a liquid (or gaseous) oxidizer, which are stored separately and brought together in the combustion chamber to produce thrust. This setup allows for some of the advantages of both solid and liquid propulsion systems, including a simpler design compared to liquid rockets and better controllability compared to solid rockets. Hybrid rockets are being explored for various aerospace applications, including hypersonic flight, due to their potential for high performance, flexibility, and safety.

In a hybrid rocket system, the propulsion process involves the following key components and steps:

1. **Fuel and Oxidizer**: The core components of a hybrid rocket are its fuel and oxidizer. The fuel is in solid form and is stored inside the combustion chamber, while the oxidizer is stored separately in a tank and is injected into the combustion chamber when needed. Common solid fuels include materials such as rubber-based compounds (like hydroxyl-terminated polybutadiene, or HTPB) or paraffin wax. The oxidizer is typically a liquid or gaseous substance like liquid oxygen (LOX) or nitrous oxide (N_2O). These components are brought together in the combustion chamber, where they ignite and combust to produce thrust.

2. **Combustion Process**: The combustion process in a hybrid rocket system starts when the oxidizer is injected into the combustion chamber, where it comes into contact with the solid fuel. The oxi-

dizer flows over or through the surface of the solid fuel, where heat generated by the combustion causes the fuel to vaporize and mix with the oxidizer. The fuel-oxidizer mixture ignites, producing hot gases that are expelled through a nozzle to generate thrust. Unlike in a solid rocket, where the fuel and oxidizer are mixed together and burn uncontrollably once ignited, a hybrid rocket allows for more precise control over the combustion process. By regulating the flow of oxidizer into the combustion chamber, the thrust can be throttled, stopped, or even restarted mid-flight.

3. **Nozzle**: The hot gases produced by the combustion of the fuel and oxidizer are expelled through a nozzle, where they are accelerated to produce thrust. The design of the nozzle in a hybrid rocket is similar to that used in other rocket systems, typically a converging-diverging nozzle, which ensures efficient expansion and acceleration of the gases to maximize thrust. The nozzle design is critical for the rocket's performance, especially in hypersonic applications, where extremely high exhaust velocities are required to achieve the necessary speeds.

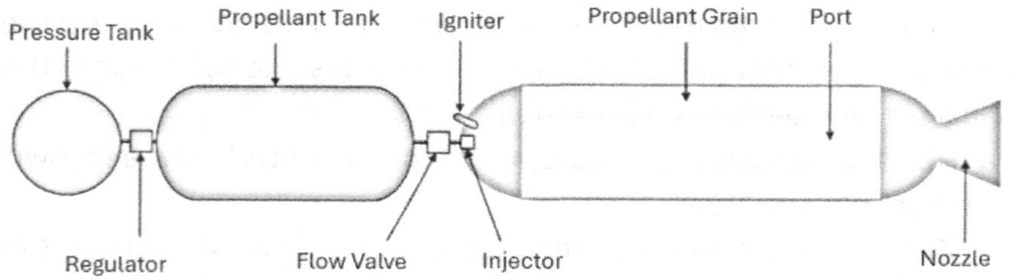

Figure 12: Hybrid Rocket Engine.

In hypersonic flight, where vehicles travel at speeds exceeding Mach 5 (five times the speed of sound), hybrid rocket systems can play a key role in the propulsion of certain types of missiles, spaceplanes, and other high-speed vehicles. Hypersonic systems require powerful, reliable propulsion to reach and sustain these high speeds, and hybrid rockets offer several potential benefits:

- **High Thrust-to-Weight Ratio**: Hybrid rocket systems can provide a high thrust-to-weight ratio, which is essential for the initial acceleration required to achieve hypersonic speeds. During the boost phase of a hypersonic flight, the hybrid rocket generates the necessary thrust to push the vehicle to high altitudes and velocities. The ability to throttle the engine allows for greater control over the acceleration profile, ensuring that the vehicle reaches its desired speed efficiently.

- **Throttling and Control**: One of the primary advantages of hybrid rockets over solid rockets is the ability to control the combustion process by adjusting the flow of the oxidizer. In hypersonic flight, this is particularly important because it allows for precise control over the vehicle's speed and trajectory. For example, the hybrid rocket can be throttled down to manage aerodynamic heating during certain phases of flight or throttled up to increase speed when needed. This flexibility is especially useful in missions that require varying thrust levels, such as manoeuvring during atmospheric flight or fine-tuning the vehicle's trajectory.

- **Safety and Simplicity**: Hybrid rocket systems are inherently safer than liquid rockets because the fuel and oxidizer are stored separately and only come into contact in the combustion chamber. This reduces the risk of accidental explosions or leaks during storage and handling. The simplicity of the hybrid system, with fewer moving parts than a liquid rocket, makes it more reliable and easier to maintain, which is crucial for hypersonic vehicles that operate under extreme conditions.

- **Thermal Management**: Hypersonic flight generates immense heat due to friction with the atmosphere, leading to temperatures that can exceed several thousand degrees Celsius. Hybrid rocket systems can incorporate advanced thermal protection systems (TPS) to manage the heat generated during both the propulsion and hypersonic flight phases. The solid fuel itself can be designed to contribute to thermal insulation, while the rocket's structure can include ablative or ceramic materials to protect it from the extreme heat encountered during hypersonic flight. In addition, since the combustion process in a hybrid rocket is more controlled than in a solid rocket, thermal loads can be better managed by adjusting the oxidizer flow.

- **Extended Burn Time**: In some hypersonic applications, such as space access vehicles or long-range missiles, hybrid rockets can provide an extended burn time compared to solid rockets. The ability to regulate the oxidizer flow allows the rocket to burn fuel more efficiently, leading to a longer duration of powered flight. This extended burn capability can be particularly advantageous in hypersonic boost-glide systems, where the vehicle needs to reach extremely high altitudes and velocities before transitioning into a glide phase through the atmosphere.

- **Use of Non-Cryogenic Propellants**: Hybrid rockets can use non-cryogenic propellants, such as nitrous oxide or other liquid oxidizers that do not require cryogenic storage conditions. This simplifies the design and operation of the rocket, especially in environments where managing cryogenic liquids, such as liquid oxygen, can be challenging. The use of non-cryogenic propellants also enhances the operational flexibility of hypersonic vehicles, as they can be stored and launched more easily in a variety of conditions.

Despite their advantages, hybrid rockets face several challenges when applied to hypersonic flight:

- **Combustion Efficiency**: One of the key challenges of hybrid rockets is achieving consistent and efficient combustion between the solid fuel and the liquid oxidizer. In hypersonic applications, where extremely high thrust and specific impulse are required, any inefficiencies in the combustion process can reduce the overall performance of the vehicle. Researchers are working on optimizing fuel-oxidizer combinations and combustion chamber designs to improve the performance of hybrid rockets in hypersonic systems.

- **Thrust-to-Weight Ratio Limitations**: Although hybrid rockets have a better thrust-to-weight ratio than solid rockets, they generally do not match the performance of liquid rockets in terms of specific impulse (the measure of fuel efficiency). For hypersonic applications that require extremely high accelerations and long-range flight, this can be a limiting factor.

- **Thermal Protection Requirements**: The extreme heat generated during hypersonic flight can put significant stress on the materials used in hybrid rockets, especially in the nozzle and combustion chamber. Advanced thermal protection systems, such as ablative materials or high-temperature ceramics, must be incorporated into the design to ensure the rocket can withstand the intense heat without compromising structural integrity.

Hybrid rocket systems offer a unique propulsion option for hypersonic flight, combining the simplicity and safety of solid rockets with the controllability of liquid rockets. Their ability to provide high thrust, control over the combustion process, and operational safety makes them a promising technology for future hypersonic applications. However, challenges related to combustion efficiency, thrust-to-weight ratios, and thermal management must be addressed to fully realize their potential in high-speed aerospace missions. As research continues, hybrid rocket systems may become a key component in the development of next-generation hypersonic weapons and vehicles.

Scramjet Engines

Ramjet and Scramjet Engines

A ramjet, short for "ram air jet engine," is an air-breathing engine that relies on the vehicle's high-speed forward motion to compress incoming air without the need for compressors or turbines. These components are typically found in traditional jet engines, but ramjets operate differently. The fundamental principle behind a ramjet involves slowing down the incoming supersonic air to subsonic speeds before it enters the combustion chamber. This is achieved through a process of compression via shock waves and a diffuser. Once the air is slowed and compressed, fuel is injected, mixed with the air, and ignited. The expanding exhaust gases provide the thrust necessary to propel the vehicle forward at high speeds.

One of the primary advantages of ramjets is their simplicity. With no moving parts, they are lighter and easier to build compared to more complex engines like turbojets or scramjets. Ramjets are also efficient in the supersonic speed range, typically operating between Mach 2 and Mach 4 or 5. This makes them particularly suitable for vehicles and weapons that need to sustain high speeds over long distances, such as supersonic cruise missiles. However, at speeds above Mach 5, ramjets become inefficient. The need to slow down the incoming air results in a significant increase in drag and heat, limiting their applicability in true hypersonic flight.

One example of a ramjet-powered system in use is the BrahMos missile, a supersonic cruise missile developed by India and Russia. The BrahMos operates primarily at Mach 2-3, but its advanced ramjet engine allows it to approach the lower end of hypersonic flight capabilities, though it remains primarily in the supersonic regime.

A scramjet, or "supersonic combustion ramjet," is a more advanced engine compared to the ramjet, designed specifically for sustained hypersonic speeds, typically Mach 5 and above. The key difference between a scramjet and a ramjet lies in how they handle airflow. Unlike ramjets, which slow down the incoming air to subsonic speeds, scramjets allow the air to remain supersonic as it passes through the entire engine, including the combustion chamber. This ability to maintain supersonic airflow enables scramjets to sustain extremely high speeds without the drag penalties associated with slowing down the air.

In a scramjet, the combustion process is far more complex. The fuel must be injected, mixed with the air, and ignited while the air is still moving at supersonic speeds, which requires precise timing and engineering. However, once operational, scramjets are highly efficient at Mach 5 and above, making them ideal for true hypersonic applications, such as high-speed missiles or space access vehicles. Like ramjets, scramjets are air-breathing engines, so they do not need to carry their own oxidizer for combustion, making them lighter and more efficient than rocket engines.

Despite these advantages, scramjets also have limitations. They only operate at extremely high speeds, generally Mach 5 and above. As a result, vehicles or weapons powered by scramjets require an initial boost from another engine or a rocket to accelerate to speeds where the scramjet can take over. Additionally, the combustion process in a scramjet is highly complex, requiring advanced materials and precise engineering to manage the intense heat and pressure of hypersonic flight.

The Hypersonic Air-breathing Weapon Concept (HAWC), developed by the United States in collaboration with DARPA, uses a scramjet engine to achieve sustained hypersonic flight at speeds above Mach 5. Another example is Russia's 3M22 Zircon, a scramjet-powered hypersonic missile capable of reaching speeds of Mach 8-9, designed to strike naval and land-based targets at incredible speeds, making it extremely difficult to intercept.

The primary difference between ramjets and scramjets is how they handle airflow. Ramjets slow down the incoming air to subsonic speeds before combustion, while scramjets maintain supersonic airflow through the engine, including during combustion. This fundamental difference dictates their operational speed ranges. Ramjets are effective at supersonic speeds, operating between Mach 2 and Mach 5, while scramjets are designed for hypersonic speeds, typically operating between Mach 5 and Mach 15 or higher.

In terms of efficiency, ramjets lose effectiveness at speeds above Mach 5 due to the increased drag associated with slowing down the airflow. Scramjets, on the other hand, are more efficient at these higher speeds because they maintain supersonic airflow, avoiding the drag penalties ramjets face. The combustion process is also different—ramjets perform combustion in subsonic airflow, while scramjets do so in supersonic airflow, making the scramjet combustion process more complex but enabling much higher speeds.

Both ramjets and scramjets have their roles in hypersonic weapon systems. Ramjets are more likely to be used in systems that operate in the lower hypersonic or supersonic regimes, such as cruise missiles designed for sustained high-speed flight over long distances. Scramjets, however, are better suited for higher hypersonic speeds, especially in weapons designed for long-range, ultra-fast strikes. These systems require sustained hypersonic flight to evade missile defence systems and reach distant targets quickly.

Both ramjets and scramjets are crucial to the development of high-speed weaponry. While ramjets provide valuable propulsion for lower-end hypersonic and supersonic systems, scramjets offer the most promising pathway for sustained hypersonic flight beyond Mach 5. The choice of engine depends on the specific mission requirements and the speed regime the weapon is designed to operate within.

Scramjet Design and Function

A scramjet engine, short for "supersonic combustion ramjet," is a type of air-breathing jet engine designed to operate at hypersonic speeds, typically Mach 5 and above. Unlike traditional jet engines, scramjets are designed to take advantage of the high velocity of incoming air and maintain supersonic airflow throughout the engine, including during combustion. This unique feature makes scramjets highly efficient for hypersonic flight, offering a promising propulsion technology for advanced aerospace applications, including hypersonic missiles, spaceplanes, and other high-speed vehicles.

Scramjet engines represent a significant advancement in jet propulsion technology, relying on the combustion of fuel and atmospheric oxygen to produce thrust. Unlike traditional rocket engines, which carry both fuel and an oxidizer, scramjet-powered aircraft carry only fuel and use atmospheric oxygen as the oxidizer. This reliance on atmospheric oxygen limits scramjets to suborbital atmospheric propulsion, where oxygen is available in sufficient quantities to maintain combustion.

The fundamental principle behind a scramjet is its ability to use the vehicle's own speed to compress incoming air, eliminating the need for complex mechanical components such as compressors and turbines that are common in conventional jet engines. The scramjet operates by maintaining supersonic airflow through the engine, even during combustion, allowing it to generate thrust efficiently at hypersonic speeds—where traditional engines become ineffective due to the limitations of airflow compression.

Air Intake and Compression

In a scramjet, the air intake system plays a critical role by capturing the supersonic airflow as the vehicle moves at hypersonic speeds. As the air enters the intake, it passes through a series of oblique shock waves, which serve to compress the air without decelerating it to subsonic speeds. This is a key distinction between ramjets and scramjets, as ramjets slow the airflow to subsonic levels before combustion. Scramjets maintain supersonic velocity throughout the engine, preventing the excessive drag and heat buildup that would otherwise make hypersonic flight inefficient.

As the air is compressed by the shock waves inside the intake, it is channelled into the combustion chamber. The compression is achieved without mechanical components, relying solely on the shock waves generated by the vehicle's forward motion. This process reduces the engine's complexity and weight, while also improving its ability to handle the extreme conditions of hypersonic flight.

Supersonic Combustion

Once the air is compressed, it enters the combustion chamber, where fuel is introduced and ignited. The combustion process in a scramjet is particularly challenging because the air is still moving at supersonic speeds. Unlike conventional engines, where combustion occurs at subsonic velocities, scramjets must ignite fuel in an environment where air flows rapidly through the combustion chamber. Typically, hydrogen is used as fuel because of its high energy content and its ability to ignite quickly.

The mixing of fuel and air, along with the ignition, must happen within milliseconds to ensure proper combustion while the air remains supersonic. This requires advanced fuel injection techniques and precise control of the combustion process. The combustion generates high-pressure exhaust gases, which expand rapidly and are ejected through the engine's nozzle, producing thrust.

Maintaining combustion in supersonic flow is one of the most significant technical challenges in scramjet design. The rapid speeds require careful management of airflow, fuel mixing, and heat, all of which must be controlled to ensure sustained propulsion. Despite these challenges, the ability of scramjets to maintain combustion at supersonic speeds allows them to achieve and sustain hypersonic velocities, typically above Mach 5.

Exhaust and Thrust Generation

After the fuel-air mixture ignites, the resulting hot gases expand and are expelled through a nozzle at supersonic speeds, generating thrust. The nozzle in a scramjet is typically a converging-diverging design, similar to those used in rocket engines. This nozzle accelerates the exhaust gases, maximizing the thrust produced and propelling the vehicle forward at hypersonic speeds.

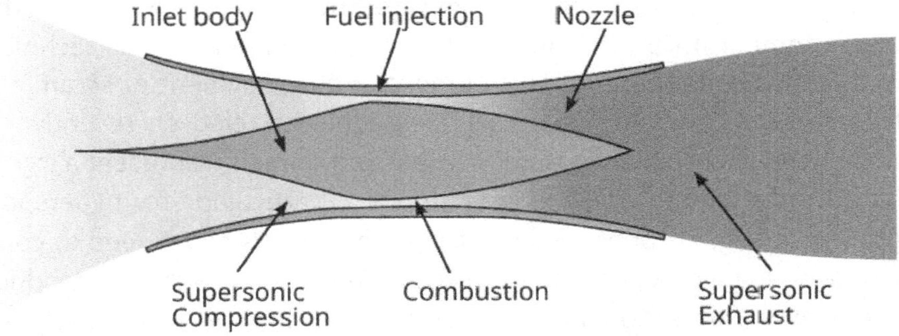

Figure 13: Diagram of principle of operation of a scramjet engine. Scramjet_ operation.png: en:User:Emoscopesderivative work: Luke490, CC BY-SA 3.0, via Wikimedia Commons.

One of the unique features of a scramjet engine is that its thrust increases with speed. As the vehicle accelerates, the scramjet becomes more efficient, allowing for sustained hypersonic flight over long distances. However, scramjets only function efficiently at speeds above Mach 5, meaning they require an initial boost from another propulsion system, such as a rocket, to reach the speeds necessary for scramjet operation.

Advantages of Scramjets

Scramjets are composed of three basic components: a converging inlet that compresses the incoming air, a combustion chamber where fuel is burned with atmospheric oxygen, and a diverging nozzle that accelerates the heated air to produce thrust. Unlike conventional turbojets or turbofans, scramjets have no moving parts, making them simpler and lighter, and eliminating many of the failure points associated with more complex engines. The absence of compressors and turbines allows scramjets to operate efficiently in the hypersonic regime, where traditional engines lose their effectiveness.

Challenges and Limitations

While scramjets are conceptually simple, their practical implementation is extremely difficult due to the harsh conditions of hypersonic flight. At these speeds, the vehicle experiences immense drag and heating, with temperatures in the engine and on the airframe far exceeding those encountered by conventional aircraft. These extreme conditions require the use of advanced materials and cooling technologies to protect the engine and vehicle from heat damage.

Moreover, maintaining combustion in supersonic flow presents significant challenges. The fuel must be injected, mixed, ignited, and burned within milliseconds, a process complicated by the need to manage the extreme temperatures and pressures inside the engine. These technical difficulties have limited the development and deployment of scramjets, with successful powered flights only being achieved relatively recently.

Scramjet engines, or supersonic combustion ramjets, represent a significant advancement in propulsion technology, particularly for hypersonic flight. These engines operate by utilizing the vehicle's high speed to compress incoming air, which is then mixed with fuel and ignited to produce thrust. Unlike conventional jet engines that rely on mechanical compressors and turbines, scramjets maintain supersonic airflow throughout the engine, including in the combustion chamber. This unique design eliminates the need for rotating parts, thereby reducing the engine's complexity and weight, which is crucial for high-speed applications [203-205].

The operational principle of scramjets hinges on the management of supersonic airflow. As air enters the engine at supersonic speeds, shockwaves are generated that compress the air without the need for mechanical compression. This process is essential for maintaining the necessary conditions for combustion while the airflow remains supersonic, a feature that distinguishes scramjets from ramjets, where the airflow is decelerated to subsonic speeds before combustion [203, 206]. The combustion process in scramjets is particularly challenging due to the short residence time of the reactants—typically around 1 millisecond—requiring precise fuel injection and rapid ignition to ensure efficient combustion [206, 207].

Fuel choice is critical in scramjet design, with hydrogen being a common option due to its high energy content and efficient combustion properties.

However, hydrocarbon fuels are also being explored for their practicality and safety in handling [205, 208]. The combustion of the fuel generates high-pressure exhaust gases that are expelled through a divergent nozzle, which accelerates the gases and maximizes thrust. The nozzle design is integral to the scramjet's performance, allowing for sustained hypersonic speeds by effectively managing the exhaust flow [205, 208].

Scramjets are effective only at speeds exceeding Mach 5, necessitating an initial boost from a rocket or another propulsion system to reach the required velocity for activation. Once operational, scramjets become increasingly efficient as speed increases, benefiting from enhanced air compression without the drag penalties associated with slower engines [203, 209]. However, the technology faces limitations, particularly regarding the sensitivity of scramjets to airflow speed. If the air slows below Mach 1 within the engine, it can lead to "choking," transitioning the engine to subsonic combustion and resulting in operational challenges [210].

Figure 14: The X-43A Hypersonic Experimental Vehicle, or "Hyper-X" is seen here undergoing ground testing at NASA's Dryden Flight Research Center, Edwards, California in December 1999. Defense Visual Information Distribution Service, Public Domain, NARA & DVIDS Public Domain Archive.

Research into scramjet technology has advanced significantly, with computational fluid dynamics (CFD) playing a crucial role in understanding the complex fluid dynamics involved in scramjet operation. Studies have shown how shockwaves and fuel flow interact to optimize combustion efficiency at high speeds [207, 211]. Despite the progress, challenges remain, particularly in achieving stable combustion in supersonic flows and developing materials that can withstand the extreme conditions of hypersonic flight [212]. Ground testing facilities and flight tests, such as those conducted by NASA, have validated various aspects of scramjet performance, yet the high costs and extreme conditions of hypersonic testing continue to pose significant hurdles for further development [80].

Scramjets in Hypersonic Flight

Scramjets are designed to operate in the hypersonic flight regime, which begins at speeds around Mach 5. They fill a crucial gap between turbojets, which are efficient at subsonic speeds, and rocket engines, which are ideal for speeds beyond the atmosphere. Turbojet engines, while efficient at lower speeds, become increasingly inefficient as vehicles approach transonic speeds. Ramjets, which use the forward motion of the vehicle to compress incoming air, are effective at supersonic speeds but lose efficiency as the vehicle approaches hypersonic velocities.

Scramjets are specifically designed for these high-speed regimes, where the incoming air cannot be slowed to subsonic speeds without significant losses in efficiency. By maintaining supersonic airflow throughout the engine, scramjets avoid the drag and heat buildup that limit the performance of ramjets at higher speeds.

However, because scramjets cannot operate at low speeds, vehicles equipped with scramjet engines require an initial boost from another propulsion system to reach the necessary operating speed. For example, in tests with the Boeing X-51A scramjet-powered vehicle, the test craft was carried to altitude by a B-52 aircraft and accelerated to Mach 4.5 using a detachable rocket booster before the scramjet engine was ignited.

Scramjets are a revolutionary technology that allows sustained hypersonic flight by using the vehicle's own speed to compress incoming air and maintain supersonic combustion. Their simple design, lack of moving parts, and ability to operate at extreme speeds make them an ideal propulsion system for high-speed aircraft and missiles. However, the technical challenges of maintaining combustion at hypersonic speeds and managing the extreme heat generated during flight mean that scramjet development is still in its early stages. Despite these challenges, scramjets hold great promise for the future of aerospace propulsion, particularly in the areas of hypersonic travel and space access.

Propulsion Challenges at Mach 5+

At speeds beyond Mach 5, vehicles enter the hypersonic flight regime, which introduces a host of significant propulsion challenges. These challenges stem from the extreme aerodynamic, thermal, and physical forces that occur at these high velocities. Hypersonic propulsion systems, such as ramjets, scramjets, and rocket engines, face unique difficulties in maintaining stable, efficient thrust while coping with the severe heat, pressure, and drag associated with hypersonic speeds.

Air Compression and Intake Challenges

The management of high-speed airflow entering hypersonic engines, particularly scramjets, presents significant challenges due to the complex interactions of shock waves, pressure, and temperature. At hypersonic speeds, the incoming air must be carefully compressed to maintain supersonic flow through the engine, which is critical for effective combustion. The design of the engine intake is paramount, as it must generate precise shock waves that decelerate and compress the air without allowing it to drop below supersonic speeds. Any miscalculation in this process can lead to choking, where the airflow slows excessively, resulting in potential engine failure [213, 214].

Research indicates that the flow quality in hypersonic intakes is significantly affected by shock wave interactions, which can lead to flow unsteadiness and increased aerodynamic drag [213]. For instance, the analysis of micro-ramps in hypersonic flows has shown that these configurations can mitigate total pressure loss and flow distortion, thereby enhancing the overall propulsive efficiency of hypersonic vehicles [213]. Furthermore, the performance of hypersonic inlets can be optimized through the application of heat sources and throttling effects, which help manage the complex flow dynamics at these extreme speeds [215].

The integration of airframe and propulsion systems is also crucial in addressing these challenges. Effective airframe-propulsion integration can reduce drag and improve thrust margins, which is essential for maintaining supersonic airflow through the engine ([216]. Additionally, the design of the scramjet intake must consider the ramp angles and configurations that influence shock wave strength, which directly impacts the efficiency of the intake [217].

Moreover, the phenomenon of thermal choking, where excessive heat release in the combustion chamber causes the flow to reach sonic conditions, poses a significant risk to scramjet operation. This can lead to flow separation and the formation of oblique shock trains, which disrupt the intended airflow and can cause unstart conditions [218, 219]. The sensitivity of scramjets to angle of attack (AOA) further complicates the intake design, as variations in AOA can lead to thermal choking if not properly managed [220].

Combustion and Fuel Injection

Combustion at hypersonic speeds presents significant challenges, particularly in the context of fuel injection and mixing. In traditional combustion engines, fuel and air are mixed at subsonic speeds, allowing sufficient time for combustion to occur. However, in hypersonic engines, such as scramjets, the airflow is supersonic, necessitating rapid mixing and ignition within milliseconds to achieve effective combustion [221, 222]. The high-speed airflow complicates the fuel-air mixing process, leading to issues such as poor mixing rates and reduced residence time for combustion, which are critical for maintaining flame stability [221, 223].

Hydrogen is often the fuel of choice for hypersonic applications due to its rapid combustion characteristics. The injection of hydrogen must be executed with precision to ensure thorough mixing with the high-speed airflow, which is essential for achieving stable combustion [224, 225]. Techniques such as direct injection (DI) and port fuel injection (PFI) are employed to optimize the fuel delivery process. Direct injection allows for better control over the timing and location of fuel introduction, which can mitigate issues related to pre-ignition and backfiring, common challenges in hydrogen combustion [226, 227]. Furthermore, the design of the injector plays a crucial role; parameters such as hole diameter and injection angle can significantly influence the operational conditions and performance of the combustion system [228, 229].

The thermal environment in hypersonic combustion chambers is another critical factor. At speeds exceeding Mach 5, the temperatures can become extreme, increasing the risk of uncontrolled combustion events, such as overheating or explosions [223]. Effective thermal management strategies are necessary to prevent structural damage to the engine components. Research has shown that optimizing the combustion chamber design and employing advanced cooling techniques can help maintain operational integrity under these harsh conditions [225, 230]. Additionally, the use of computational fluid dynamics (CFD) simulations has become essential in understanding and predicting the complex interactions between fuel injection, airflow, and combustion dynamics in scramjet engines [221, 230].

Thermal Management and Heat Loads

The management of thermal loads in hypersonic vehicles presents a significant engineering challenge due to the extreme temperatures generated by aerodynamic friction at speeds exceeding Mach 5. As these vehicles traverse the atmosphere, the friction between their surfaces and the surrounding air can lead to surface temperatures that may exceed several thousand degrees Celsius, particularly at critical areas such as the nose cone and engine inlets [125, 172]. This necessitates the implementation of robust thermal protection systems (TPS) to safeguard the vehicle's structural integrity and operational functionality.

Thermal protection systems are designed to withstand the intense heat generated during hypersonic flight. Common materials employed in TPS include ablative materials, ceramic composites, and advanced cooling systems. Ablative materials, which are engineered to erode and carry away heat, are particularly effective in high-temperature environments [231, 232]. Ceramic composites, known for their high thermal stability and low thermal expansion coefficients, are also utilized to manage heat loads effectively [233, 234]. For instance, the incorporation of materials such as silicon carbide (SiC) and aluminium-ceramic composites has been shown to enhance thermal durability and conductivity, making them suitable for high-power applications [234, 235].

Active cooling techniques are equally crucial in managing heat within the engine components of hypersonic vehicles. Regenerative cooling systems, which circulate fuel through the engine structure to absorb heat before combustion, are a prominent example of such techniques [236, 237]. This method not only helps to maintain optimal operating temperatures but also improves overall engine efficiency by utilizing the heat absorbed from the engine structure. Additionally, the use of advanced materials like graphene and boron nitride in thermal management systems has been explored for their exceptional thermal conductivity, which can significantly enhance heat dissipation capabilities [237].

The design of these thermal protection systems is inherently complex, as they must endure not only high temperatures but also the mechanical stresses associated with high-speed flight and thermal cycling [125, 236]. The interaction between aerodynamic forces and thermal loads can lead to significant structural challenges, necessitating a multidisciplinary approach to ensure the reliability and safety of hypersonic vehicles [238]. Furthermore, ongoing research into non-ablative TPS options is aimed at developing materials that can provide effective thermal management without the drawbacks associated with traditional ablative systems [115].

Drag and Pressure Challenges

At hypersonic speeds, particularly those exceeding Mach 5, drag becomes a critical challenge that significantly impacts vehicle performance and propulsion system efficiency. The interaction of high-speed airflow with the vehicle generates shockwaves, which can lead to substantial increases in drag. This phenomenon not only slows the vehicle but also diminishes its range and fuel efficiency. Studies indicate that the drag experienced by hypersonic vehicles can be reduced by employing specific aerodynamic configurations, such as aerospikes, which have been shown to decrease drag by as much as 60% under certain conditions ([239, 240]. Furthermore, the design of hypersonic vehicles must account for the balance between minimizing drag and ensuring sufficient thrust generation, a core engineering challenge in the development of efficient propulsion systems like scramjets [34].

The pressure build-up resulting from shockwaves can also induce instabilities in the airflow through the engine, complicating the combustion process. At hypersonic speeds, the interaction between shockwaves generated by the vehicle's nose and body and those formed within the engine's intake can disrupt airflow, leading to inefficient combustion and potentially catastrophic engine failures [241, 242]. Advanced computational fluid dynamics (CFD) modelling plays a crucial role in addressing these challenges by optimizing engine designs to minimize drag and ensure stable airflow under high-pressure conditions [243]. For example, the integration of the engine with the airframe can enhance aerodynamic performance and reduce overall drag, as demonstrated in studies focusing on waverider vehicles [120, 244].

Waverider vehicles are a type of hypersonic aircraft designed to ride or exploit the shockwave generated by their own flight at high speeds. This unique design allows waverider vehicles to achieve superior aerodynamic efficiency, especially at hypersonic speeds (Mach 5 and above). The concept of a waverider vehicle is rooted in the idea that it can "surf" on the shockwave it generates, creating a region of compressed air that enhances lift and reduces drag compared to conventional hypersonic vehicle designs.

At hypersonic speeds, vehicles create powerful shockwaves due to the compression of air in front of the vehicle. Waveriders are specifically designed to use these shockwaves to their advantage. Instead of having the shockwave form in front of the vehicle and dissipate, the waverider's shape ensures that the shockwave stays attached to the vehicle's leading edges, particularly underneath the body. This allows the vehicle to ride along the shockwave, leading to more efficient aerodynamic performance.

One of the key advantages of waverider vehicles is their ability to generate substantial lift by trapping high-pressure air beneath the vehicle's body. As the vehicle rides on its shockwave, the compressed air under the vehicle creates a lifting force. This helps to keep the vehicle aloft without the need for large, conventional wings. The shape of the waverider vehicle is typically slender and sleek, often resembling a wedge, to optimize this interaction with the shockwave.

Waveriders are designed to minimize drag, a critical factor in maintaining high-speed flight. At hypersonic speeds, aerodynamic drag increases dramatically, and conventional designs experience significant drag forces that limit speed and range. Waverider vehicles, by keeping the shockwave attached and using it to their benefit, reduce the overall drag on the vehicle. This makes waveriders more efficient for sustained hypersonic flight, as less energy is lost to drag, and the vehicle can maintain higher speeds for longer durations.

Waverider vehicles often have distinctively sharp, narrow shapes to maintain their interaction with the shockwave. The underside of the vehicle, which "rides" the shockwave, is usually flat or slightly contoured to capture the compressed air and generate lift. The vehicle's design is highly dependent on precise aerodynamic modelling, as even small changes in shape can dramatically affect the vehicle's ability to ride the shockwave efficiently. Computational fluid dynamics (CFD) and wind tunnel testing are commonly used to optimize waverider designs.

Waverider vehicles have several potential applications, particularly in the fields of aerospace and defence:

- **Hypersonic Missiles and Weapons**: Waverider designs are particularly well-suited for hypersonic missiles, which need to travel at very high speeds while minimizing drag and maximizing range. The ability of waveriders to sustain hypersonic speeds for long periods while remaining manoeuvrable makes them attractive for military applications.

- **Space Access Vehicles**: Waveriders are also being explored for use in spaceplanes or other reusable space access vehicles. Their efficiency at high speeds could make them ideal for reducing the cost of space travel, as they could potentially be used for atmospheric flight before transitioning to rocket-powered flight for space access.

- **Hypersonic Passenger Travel**: Although still largely conceptual, waverider designs could eventually play a role in future passenger aircraft capable of traveling at hypersonic speeds. Their ability to reduce drag and enhance fuel efficiency could make hypersonic commercial travel more viable in the future.

Moreover, the thermal protection system (TPS) of hypersonic vehicles is heavily influenced by the dissociation of gases in the shock layer surrounding the vehicle. Research has shown that under certain flight conditions, a significant portion of gas dissociation occurs in the transient excitation phase immediately behind the shockwave, which is critical for predicting TPS performance [245]. The need for effective drag reduction strategies is further underscored by the fact that increased wave drag not only elevates propellant consumption but also limits the operational range and payload capacity of hypersonic vehicles [241]. Thus, the development of innovative designs and materials, alongside advanced modelling techniques, is essential for overcoming the drag and pressure challenges faced in hypersonic flight.

Material Challenges

The development of materials capable of withstanding the extreme conditions encountered by vehicles operating at hypersonic speeds (Mach 5 and above) presents significant challenges. These vehicles must endure not only high temperatures but also intense mechanical stresses, which standard materials used in conventional aircraft and engines cannot withstand. The risk of melting, warping, or cracking under the thermal and pressure loads experienced during hypersonic flight necessitates the use of advanced materials specifically engineered for these conditions.

One of the primary material classes being investigated for hypersonic applications is ceramic matrix composites (CMCs). CMCs, particularly those reinforced with fibres, exhibit exceptional high-temperature resistance and mechanical strength, making them suitable for the harsh environments of hypersonic flight. For instance, Yang et al. [246] highlight that fibre-reinforced CMCs can overcome the brittleness and thermal shock resistance issues associated with traditional ceramics, thus making them ideal candidates for thermal protection systems in hypersonic and re-entry vehicles. Furthermore, Paul et al. [247] discuss the renewed interest in ultra-high temperature ceramics (UHTCs) for thermal protection systems, noting their ability to withstand extreme temperatures, which is critical for the leading edges of hypersonic vehicles. The development of UHTCs, such as those containing hafnium diboride (HfB2), is crucial as they possess high melting points and excellent thermal conductivity, which are essential for maintaining structural integrity under extreme conditions [248].

In addition to thermal resistance, the weight of materials is a critical factor in hypersonic vehicle design. Lightweight materials are necessary to ensure aerodynamic efficiency and minimize fuel consumption. CMCs are advantageous in this regard, as they typically offer a favourable strength-to-weight ratio compared to traditional metals. For example, Xue et al. [249]emphasize the importance of lightweight materials in the design of hypersonic aircraft powered by scramjet engines, where the configuration of engine components significantly influences overall vehicle performance. However, while CMCs provide many benefits, they are also susceptible to oxidation and surface recession, particularly in high-water vapor environments, which can limit their application in certain hypersonic conditions [250].

The ongoing research into the development of these materials is vital, as finding the right balance between heat resistance, weight, and durability remains a significant challenge in hypersonic propulsion systems. Recent advancements in the fabrication of CMCs, such as those utilizing reactive melt infiltration processes, have shown promise in enhancing the performance characteristics of these materials under hypersonic conditions [251]. Moreover, the integration of advanced materials into the design of hypersonic vehicles is essential for optimizing their aerodynamics and propulsion systems, as highlighted by Zheng et al. [216], who discuss the importance of airframe-propulsion integration in reducing drag and improving thrust margins.

Flight Control and Stability

Maintaining control and stability at hypersonic speeds poses significant challenges for propulsion systems due to the extreme aerodynamic forces encountered. These forces complicate vehicle steering and stabilization, particularly as the vehicle experiences varying pressure and temperature conditions throughout its flight path. Hypersonic engines must be designed to adapt to these conditions without compromising performance or stability. Research indicates that robust flight control systems are essential for managing these challenges, particularly through the use of advanced control strategies such as model predictive control and nonlinear control techniques [252-254].

Thrust vectoring is a common technique employed in hypersonic systems to adjust the vehicle's trajectory by controlling the direction of the engine's exhaust. This method is particularly effective in conjunction with sophisticated flight control systems that can respond rapidly to changes in aerodynamic forces. For instance, the integration of flight management units with electromechanical actuators has been shown to enhance control during hypersonic flight, as demonstrated in the X-43A vehicle [148]. Additionally, nonlinear control methods, such as feedback linearization and dynamic inversion, have been explored to improve the stability and performance of hypersonic vehicles under varying flight conditions [255, 256].

The design of flight control systems for hypersonic vehicles must also account for the nonlinear and coupling characteristics inherent in their dynamics. Research has highlighted the importance of adaptive control strategies that can accommodate uncertainties and variations in flight conditions, ensuring that the vehicle remains stable and controllable throughout its flight envelope [257, 258]. For example, studies have proposed using guardian maps to identify feasible ranges for static stability margins, which are critical for maintaining control during hypersonic flight [259]. Furthermore, advanced algorithms have been developed to manage actuator constraints and enhance fault tolerance, ensuring that the vehicle can maintain its trajectory even in the presence of system faults [260, 261].

Future Innovations in Propulsion

The future innovations in propulsion for hypersonic missiles are poised to significantly enhance their operational capabilities, leveraging advancements in scramjet technology, combined-cycle engines, and novel fuel injection strategies. As hypersonic flight becomes increasingly critical for both military and commercial applications, understanding these innovations is essential.

One of the most promising areas of development is the scramjet (supersonic combustion ramjet) engine, which has emerged as a cornerstone of hypersonic propulsion due to its ability to efficiently utilize atmospheric oxygen for combustion at high speeds. The performance of scramjets is heavily influenced by their design, particularly the fuel injection strategies employed. Recent studies have highlighted various fuel injection techniques that optimize mixing and combustion characteristics, which are crucial for achieving the high specific thrust required for hypersonic flight [203, 262]. Moreover, advancements in the understanding of heat transfer within scramjet systems have led to improved designs that enhance thermal management and overall efficiency [263].

In addition to scramjet technology, the integration of combined-cycle engines, such as the Turbine-Based Combined Cycle (TBCC), represents a significant innovation in hypersonic propulsion. These systems are designed to operate across a wide range of speeds, allowing for efficient transition from subsonic to hypersonic flight. The TBCC system utilizes a combination of turbojet, ramjet, and scramjet technologies to optimize performance throughout the flight profile, enabling sustained hypersonic speeds [264, 265]. This versatility is essential for future hypersonic missiles, which must be capable of rapid deployment and manoeuvrability in various operational scenarios.

Furthermore, the integration of advanced computational tools for modelling and simulation has revolutionized the design process for hypersonic propulsion systems. Tools such as the HYbrid PRopulsion Optimiser (HyPro) allow for extensive parametric studies and optimization of propulsion systems, facilitating the development of more efficient and reliable engines [80]. The ability to simulate complex interactions between the airframe and propulsion systems is critical for enhancing vehicle stability and performance at hypersonic speeds [216, 266].

Another innovative approach involves the exploration of stabilized detonation as a propulsion mechanism, which could provide higher thrust levels compared to traditional scramjet designs. This method leverages controlled detonation waves to enhance combustion efficiency, potentially leading to breakthroughs in hypersonic missile capabilities [267]. The ongoing research in this area indicates a promising future for propulsion technologies that can sustain extreme speeds and manoeuvrability.

The future of hypersonic missile propulsion is characterized by significant innovations in scramjet technology, combined-cycle engines, advanced modelling techniques, and novel combustion methods. These advancements not only promise to enhance the speed and efficiency of hypersonic missiles but also ensure their operational effectiveness in modern warfare scenarios.

Turbine-Based Combined Cycle (TBCC) Propulsion

Turbine-Based Combined Cycle (TBCC) propulsion is an advanced aerospace propulsion concept that integrates multiple engine types into a single system to optimize performance across a wide range of speeds, from low subsonic to hypersonic. The TBCC propulsion system typically combines a traditional turbine engine (such as a turbojet or turbofan) with a ramjet or scramjet engine, allowing the vehicle to efficiently transition between different propulsion modes depending on its speed. This system is designed to enable vehicles to achieve sustained hypersonic flight, making it highly relevant for next-generation hypersonic aircraft, missiles, and spaceplanes.

Figure 15: A Pratt & Whitney F100 engine next to a GE J85 at Hermeus headquarters in Atlanta. Hermeus has accepted its first F100 engine from Pratt & Whitney. Set to be integrated into the enhanced Chimera II turbine-based combined cycle (TBCC) propulsion system, this powerful engine component marks a critical milestone in Hermeus' development of Darkhorse which is set to unlock new possibilities in both military and commercial hypersonic applications. Hermeus, CC BY 2.0, via Flickr.

The primary goal of TBCC propulsion is to extend the operational speed range of a vehicle by using different engines for different speed regimes. The turbine engine handles propulsion at low to moderate speeds, while the ramjet or scramjet takes over at higher speeds, specifically for supersonic and hypersonic flight. This hybrid approach allows for greater efficiency and flexibility, as each engine type operates in its optimal performance window.

A TBCC system usually involves a seamless transition between the turbine and ramjet/scramjet engines, ensuring that the vehicle maintains continuous thrust as it accelerates from a standstill to hypersonic speeds. The combination of these technologies provides propulsion efficiency across a broad flight envelope, from take-off to hypersonic cruising.

A typical TBCC propulsion system operates in several phases, with different engines taking over at various stages of the flight, depending on the speed of the vehicle.

1. Subsonic to Low Supersonic (Turbine Engine Phase)

In the initial phase of flight, such as during take-off and low-speed flight, the vehicle relies on a conventional turbine engine, such as a turbojet or turbofan. These engines are highly efficient at subsonic and low supersonic speeds and can provide the necessary thrust for acceleration and initial climb.

- **Turbojet or Turbofan Engine**: At low speeds, the turbine engine draws in air, compresses it using multiple compressor stages, mixes it with fuel, and ignites it in the combustion chamber. The high-pressure exhaust gases are then expelled through a nozzle, generating thrust. Turbofan engines also include a bypass fan, which provides additional thrust and improves fuel efficiency.

- **Low-Speed Efficiency**: The turbine engine is optimized for low to moderate speeds, making it ideal for take-off, climb, and acceleration to supersonic speeds (up to Mach 2-3). During this phase, the ramjet or scramjet is not yet active, as they require higher speeds to function effectively.

2. Supersonic to Hypersonic Transition (Ramjet/Scramjet Takeover)

As the vehicle reaches higher speeds, typically above Mach 2 or 3, the turbine engine begins to lose efficiency due to the high drag and heating associated with compressing air at supersonic velocities. At this point, the ramjet or scramjet engine starts to take over. In many TBCC systems, there is a seamless transition between the turbine and the ramjet/scramjet engines to ensure continuous thrust.

- **Ramjet Engine**: At speeds between Mach 2 and Mach 5, a ramjet engine becomes efficient. In a ramjet, the incoming air is compressed by the forward motion of the vehicle without the need for rotating compressor blades. The compressed air is mixed with fuel, ignited, and the high-energy exhaust gases provide thrust. Ramjets are highly efficient at supersonic speeds but become inefficient at speeds above Mach 5.

- **Scramjet Engine**: At speeds above Mach 5, a scramjet (supersonic combustion ramjet) takes over. Unlike a ramjet, a scramjet keeps the incoming air moving at supersonic speeds through the combustion chamber. This enables the scramjet to sustain extremely high velocities, making it ideal for hypersonic flight. Scramjets are most efficient at speeds between Mach 5 and Mach 15 or more.

3. Hypersonic Flight (Scramjet Phase)

Once the vehicle reaches hypersonic speeds (Mach 5 and above), the scramjet becomes the primary propulsion system. At these speeds, the scramjet can take full advantage of the vehicle's velocity to compress the incoming air and maintain supersonic combustion, producing continuous thrust without the drag penalties associated with slowing the air down to subsonic speeds.

- **Sustained Hypersonic Flight**: In this phase, the vehicle achieves sustained hypersonic flight, using the scramjet for efficient propulsion over long distances. The scramjet's design allows for extended range and high-speed manoeuvrability, making it ideal for hypersonic missile systems, space access vehicles, or military aircraft that need to cover vast distances in short periods.

Figure 16: Chimera is a proprietary turbine-based combined cycle (TBCC) engine that will take Quarterhorse to 90,000+ ft with speeds exceeding 3,000 mph. Hermeus, CC BY 2.0, via Flickr.

Turbine-Based Combined Cycle (TBCC) propulsion offers several key advantages that make it a powerful solution for high-speed flight. One of its primary benefits is the ability to provide efficient propulsion across a wide speed range, from subsonic to hypersonic velocities. This versatility is achieved by seamlessly transitioning between turbine engines, which are effective at lower speeds, and ramjet or scramjet engines, which excel at hypersonic speeds. A single engine type cannot cover such a broad range of speeds, but the TBCC system combines the strengths of each, allowing for sustained performance throughout the entire flight envelope.

Another advantage of TBCC propulsion is its efficiency across multiple flight regimes. The system capitalizes on the fuel efficiency of turbine engines at lower speeds and the high-speed capabilities of ramjets or scramjets during hypersonic flight. This approach optimizes performance during every phase of flight, whether the vehicle is taking off, accelerating, or cruising at hypersonic speeds. By using the appropriate engine for each speed regime, TBCC propulsion not only maximizes fuel efficiency but also minimizes drag, making it an ideal solution for extended, high-speed missions.

TBCC systems also contribute to increased range and speed, which are critical in military applications. Integrating multiple propulsion systems into a single vehicle enables it to achieve long-range, high-speed flight. This is particularly beneficial for missions involving strike weapons, reconnaissance, or rapid global reach, where speed and range are essential for mission success.

However, there are several challenges associated with TBCC propulsion. One of the primary technical hurdles is the complexity of transitioning between the turbine engine and the ramjet/scramjet. Achieving a smooth and seamless transition without a significant loss of thrust or stability is critical for the vehicle's performance. This requires advanced control systems and precise timing. Any delays or inefficiencies in the transition process could affect the overall effectiveness of the propulsion system.

Thermal management is another significant challenge in hypersonic flight, especially during the scramjet phase, where friction with the atmosphere generates extreme heat. Properly managing thermal loads on both the engine and the vehicle's structure is crucial to prevent engine failure or structural damage. This requires advanced materials and thermal protection systems (TPS) that can withstand the high temperatures encountered during hypersonic flight.

Lastly, integrating multiple propulsion systems into a single airframe adds weight and design complexity. The vehicle must accommodate both the turbine engine and the ramjet/scramjet, which can increase the overall size and mass of the system. Ensuring that both propulsion systems function optimally within the same airframe presents significant design challenges, particularly in terms of balancing weight distribution, aerodynamics, and structural integrity.

TBCC propulsion is seen as a key technology for future hypersonic vehicles, particularly in military applications. Hypersonic missiles, reconnaissance aircraft, and space access vehicles are some of the areas where TBCC systems are expected to play a major role. The ability to transition between propulsion systems enables these vehicles to cover large distances at extreme speeds, providing significant strategic advantages in terms of rapid response and strike capabilities.

Figure 17: PSL control room test engineers and operators, GE J 85 TBCC engine test. Defense Visual Information Distribution Service, Public Domain, NARA & DVIDS Public Domain Archive.

Turbine-Based Combined Cycle propulsion offers a versatile and efficient solution for vehicles operating across a broad range of speeds, from subsonic to hypersonic. By integrating turbine engines with ramjet or scramjet technologies, TBCC systems can provide continuous propulsion across the entire flight envelope, making them ideal for next-generation aerospace applications, especially in hypersonic flight.

PART II - Development and Testing

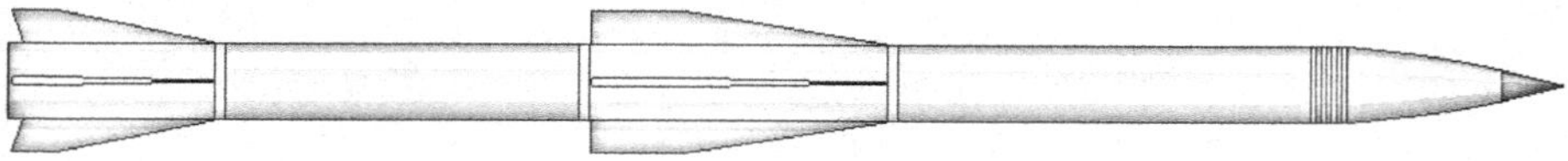

EARLY DEVELOPMENT OF HYPERSONIC WEAPONS

Historical Context: From Ballistic Missiles to Hypersonics

Historical Perspective of United States Missile Development

Missile technology has undergone steady and dramatic development since the 1950s, profoundly transforming the way combat is conducted. One of the early breakthroughs came in 1956 with the AIM-9 Sidewinder. Its proportional guidance accuracy significantly increased its lethality and improved the air-to-air combat exchange ratio. This was demonstrated in September 1958 during the Sidewinder's first combat application when Republic of China F-86 aircraft shot down four People's Republic of China MIG-17 aircraft without any losses. The following year, in 1957, the development of the Russian R-7 rocket motor marked the advent of the world's first intercontinental ballistic missile (ICBM), enabling a global reach for missile targets [268].

In 1960, the SA-2 Guideline (V-77) rocket motor showcased its ability to intercept high-altitude targets. This was exemplified during the famous incident when a Soviet SA-2 shot down an American U-2 reconnaissance aircraft piloted by Gary Powers. The same year, solid aluminized propellant technology enabled the creation of a safer, high-performance rocket motor for the Polaris submarine-launched ballistic missile (SLBM). This marked the beginning of a new era in missile propulsion [268].

Advancements continued in the late 1960s with the development of radar seekers, leading to the first combat demonstration of a beyond-visual-range (BVR) surface-to-air missile. In 1968, the U.S. Navy's Talos missiles were launched from the USS Long Beach and successfully downed two North Vietnamese MIG aircraft from a distance of over 50 nautical miles. Precision-guided munitions like the Guided Bomb Unit-10 (GBU-10) further revolutionized air combat by drastically reducing the number of sorties required to destroy targets. For example, in May 1972, laser-guided bombs successfully destroyed the Thanh Hoa Bridge in Vietnam, a target that had withstood over 800 unguided bomb attacks over several years [268].

In 1972, the development of the Short Range Attack Missile (SRAM) enhanced missile survivability and the ability of bombers to destroy multiple targets during a single mission. This gave the U.S. Air Force's B-52 and B-1 bombers an edge in standoff attacks against heavily defended targets. By 1979, the Tomahawk missile introduced lightweight turbine propulsion, enabling long-range standoff strikes. During Operation Desert Storm in 1991, 297 Tomahawk missiles were launched with a success rate of over 90% in destroying their intended targets.

Figure 18: AGM-69 short range attack missiles (SRAMS) and Mark 28 thermo-nuclear bombs, background, in the bomb bay of a B-52H Stratofortress aircraft, as a downloading operation takes place during Exercise GLOBAL SHIELD '84. The U.S. National Archives, Public Domain, via Picryl.

In 1982, the introduction of ramjet propulsion in the SS-N-22 Sunburn missile provided a rapid, time-critical capability to strike ship targets. This system increased missile survivability by enhancing flight speed, altitude, and range. In 1985, the operational capability of the ground-to-air Stinger missile system, featuring an infrared/ultraviolet seeker, improved target acquisition and countermeasure resistance. During the Soviet-Afghan War, Stinger missiles shot down over 200 Soviet aircraft and helicopters [268].

In the late 1980s, the AA-11 Archer missile introduced thrust vector control (TVC), enhancing lethality in short-range air-to-air combat by allowing large off-boresight engagements. Around the same time, the Hellfire missile, equipped with a digital processor, became a multi-mission missile with enhanced reliability. In 1991, Apache helicopters launched Hellfire missiles to destroy Iraqi radar sites during Operation Desert Storm, clearing the way for F-117 aircraft [268].

Technological advancements continued into the 2000s with the introduction of GPS/INS guidance for the Joint Direct Attack Munition (JDAM), providing a low-cost, all-weather precision strike capability. By 2013, more than 250,000 JDAMs had been produced, with over 20,000 dropped in combat. The Patriot Advanced Capability-3 (PAC-3) missile's kinetic hit-to-kill accuracy demonstrated its effectiveness in ballistic missile defence during Operation Iraqi Freedom in 2003, while the Standard Missile-3 (SM-3) showcased long-range missile defence by successfully destroying a satellite in 2008 [268].

Figure 19: Joint Direct Attack Munition. Vestman, CC BY 2.0, via Flickr.

In 2010, air-breathing propulsion technology, such as meteor-ducted rockets, enabled standoff air-to-air capabilities with twice the range of conventional missiles like the AMRAAM. Finally, in 2012, the Advanced Precision Kill Weapon System (APKWS) was introduced, offering the accuracy and range of the Hellfire missile at a significantly reduced weight and cost. These continuous advancements in missile technology have played a crucial role in shaping modern military capabilities, making missiles more accurate, survivable, and capable of addressing a wide array of combat scenarios [268].

Ballistic Missiles to Hypersonics

The development of ballistic missiles and their evolution into hypersonic weapons represents a history marked by significant technological advancements, shifts in military strategy, and intense international competition. This progression began during World War II, with the German V-2 rocket in 1944, recognized as the first long-range guided ballistic missile. It employed liquid-fuelled engines and followed a suborbital trajectory, laying the foundation for modern ballistic missile technology. The V-2's success demonstrated the destructive potential of missile systems, influencing post-war missile development by nations like the U.S. and the Soviet Union.

During the Cold War, ballistic missile technology advanced significantly. In the 1950s, both the U.S. and Soviet Union developed Intercontinental

Ballistic Missiles (ICBMs) capable of delivering nuclear warheads over vast intercontinental distances. The Soviet Union's R-7 Semyorka, launched in 1957, and the U.S. Atlas missile in 1959 are notable examples. Furthermore, the same ICBM technology was applied to space exploration, with the Soviet Union's launch of Sputnik in 1957, driving further missile advancements. In the 1960s and 1970s, technological advancements like Multiple Reentry Vehicles (MRVs) and Multiple Independently Targetable Reentry Vehicles (MIRVs) allowed a single missile to carry and deliver multiple nuclear warheads to different targets. Around the same time, Submarine-Launched Ballistic Missiles (SLBMs), such as the U.S. Polaris missile, introduced the concept of launching ballistic missiles from submarines, enhancing second-strike capabilities and nuclear deterrence strategies.

Figure 20: Russian Topol-M ICBM tanks during the Victory parade 2012. Stefan Un, CC BY 2.0, via Wikimedia Commons.

The Titan missile series represents a significant advancement in both military defense and space exploration, developed during the Cold War era. Initially designed as intercontinental ballistic missiles (ICBMs), Titan missiles were later adapted for space launch purposes, playing a critical role in both nuclear deterrence and NASA missions. Their development reflected the technological and strategic goals of the United States during a period marked by global competition.

The Titan I missile was the first version in the Titan ICBM series, developed by the Glenn L. Martin Company (which later became Martin Marietta) in the late 1950s. It became operational in the early 1960s and formed part of the U.S. Air Force's strategic missile force. The Titan I was designed as a nuclear deterrent during the height of the Cold War, with the capability of delivering nuclear warheads over intercontinental distances. The missile was housed in underground silos, offering a level of protection from pre-emptive strikes. However, its liquid-fuelled engines required fuelling before launch, making it less responsive compared to later solid-fuelled ICBMs. Titan I had an impressive range of approximately 6,300 miles (10,100 km) and was capable of carrying a single nuclear warhead.

Figure 21: A vintage Titan Nuclear Missile preserved in original launch silo, now the Titan Missile Museum, Arizona. The titan was the United States' first multistage intercontinental ballistic missile (ICBM), in use from 1963 until 1987. Zcobb99, CC0, via Wikimedia Commons.

The Titan II, developed in the early 1960s, was an improved version of Titan I, with more advanced technology, greater range, and a higher payload capacity. It was the first ICBM to carry a re-entry vehicle equipped with a thermonuclear warhead, enhancing its destructive potential. Unlike its predecessor, Titan II used hypergolic propellants, which could be stored for extended periods, enabling rapid launch. Titan II not only played a crucial role as a part of the U.S. nuclear deterrent but also became a key asset in NASA's space exploration efforts, particularly for launching astronauts into space during the Gemini missions of the mid-1960s. Titan II remained a vital component of U.S. defence until its retirement in the mid-1980s and continued to be used as a space launch vehicle until 2003.

Following its military use, the Titan II was repurposed as a space launch vehicle, supporting both NASA and U.S. Air Force missions. It was instrumental in launching satellites, probes, and manned missions, particularly during the space race. The series evolved with the development of Titan III and Titan IV, which were specifically designed for heavy payloads. Titan IV, the largest and most powerful in the series, was responsible for launching military satellites and NASA's Cassini probe to Saturn.

The Titan missile series stands as one of the most versatile and pivotal developments in the U.S. missile and space programs. It served a dual purpose, acting as a formidable nuclear deterrent and as a reliable vehicle for space exploration. Titan II, in particular, played a significant role in NASA's Gemini program, enabling human spaceflight and laying the foundation for the Apollo missions to the Moon. The Titan missile's legacy continues to influence modern missile and space launch technologies.

The post-Cold War era saw the maturation of missile technologies, with improvements in precision and miniaturization during the 1990s. Advancements in guidance systems and the miniaturization of warheads made ballistic missiles more accurate and effective. Concurrently, the development of missile defence systems like the U.S.'s Patriot and Aegis systems spurred further innovations in missile technology to evade interception.

Figure 22: An MIM-104 Patriot missile is fired by members of Battery B, 8th Battalion, 43rd Air Defense Artillery. The U.S. National Archives, Public Domain, via NARA & DVIDS Public Domain Archive.

The transition to hypersonic weapons began during Cold War research into hypersonic flight. Hypersonic speeds, defined as greater than Mach 5, offer strategic advantages in speed, range, and manoeuvrability, though early technological progress was constrained by materials and propulsion challenges. Experimental hypersonic programs from the 1960s to the 1980s, such as the X-15 (a hypersonic rocket-powered aircraft) and scramjet engine development, laid the theoretical foundations for future hypersonic weapons.

In the 21st century, the development of hypersonic weapons intensified, with the U.S., Russia, and China pursuing various systems. Boost-glide vehicles emerged in the 2000s, combining a rocket-propelled boost phase with a gliding phase, allowing hypersonic travel through the atmosphere while evading missile defences. Russia's Avangard, operational in 2019, is a notable nuclear-capable boost-glide vehicle that can achieve speeds over Mach 20. China has also developed the DF-ZF hypersonic glide vehicle, tested and incorporated into its arsenal. The U.S. has focused on systems like the Hypersonic Air-breathing Weapon Concept (HAWC) and the Conventional Prompt Strike (CPS) program, aiming for speed and precision in non-nuclear missions. Meanwhile, scramjet-powered hypersonic cruise missiles, such as the Boeing X-51 Waverider and Russia's 3M22 Zircon, represent another technological path to hypersonic speeds using atmospheric air for sustained high-speed travel.

The rise of hypersonic weapons has significant global implications. Their speed, manoeuvrability, and ability to evade existing missile defence systems offer a unique advantage in modern warfare. Hypersonic weapons can strike high-value targets with minimal warning, creating concerns about a new arms race, particularly among the U.S., Russia, and China. These weapons challenge current defence infrastructures and could destabilize the global strategic balance.

Despite these advancements, hypersonic weapons development still faces significant challenges. These include managing heat loads during hypersonic flight, sustaining propulsion beyond Mach 5, and accurately guiding weapons at such high speeds. Solving these technical issues is critical for making hypersonic weapon systems operational. Additionally, developing defences against hypersonic threats is now a priority for many nations, with new radar systems, directed energy weapons, and counter-hypersonic technologies in development.

Looking ahead, hypersonic technology could expand beyond military applications. Future uses could include space exploration or high-speed commercial travel, as governments and private companies continue research and development in hypersonic propulsion systems. The evolution of ballistic missiles into hypersonic weapons represents a critical advancement in military technology, with far-reaching implications for global security and future innovation.

Cold War Hypersonic Research

During the Cold War, hypersonic research emerged as a pivotal area of interest, primarily due to its potential to transform military and aerospace capabilities. Hypersonic speeds, defined as those exceeding Mach 5, promised significant advancements in missile systems, aircraft, and space exploration. Both the United States and the Soviet Union heavily invested in developing hypersonic technologies, motivated by the strategic competition of the era. This research aimed at achieving rapid strike capabilities, enhancing missile performance, and advancing space travel, reflecting the intense military rivalry that characterized the period [269, 270].

One of the earliest and most notable programs was the U.S. X-15 project, which began in the late 1950s and continued through the 1960s. The X-15 was a rocket-powered aircraft designed to explore the challenges of hypersonic flight, achieving speeds over Mach 6.7 and altitudes exceeding 100 kilometres, thus crossing the threshold of space [270]. The data collected from the X-15 program provided crucial insights into high-speed flight dynamics, aerodynamic heating, and control mechanisms at extreme velocities, which would later inform the development of future hypersonic and space vehicles [269, 270]. This program exemplified the innovative spirit of the era and laid the groundwork for subsequent advancements in aerospace technology.

In parallel, both the U.S. and Soviet Union explored hypersonic missile technologies to enhance their ballistic missile systems. Hypersonic weapons were seen as a means to deliver warheads to targets much faster than traditional ballistic missiles, thereby reducing reaction times and complicating interception by missile defence systems [270]. The development of these technologies was driven by advancements in materials science, propulsion systems, and aerodynamics, which aimed to address the extreme conditions experienced at hypersonic speeds, such as high thermal loads and aerodynamic pressures [270]. The strategic implications of hypersonic weapons were profound, as they promised to alter the landscape of military engagements.

Research into ramjets and scramjets also gained momentum during this period. These engines were considered essential for achieving sustained hypersonic flight within the atmosphere. Ramjets operate efficiently at speeds between Mach 2 and 4, compressing incoming air to generate thrust without moving parts, while scramjets are designed to function at speeds greater than Mach 5 by allowing air to pass through the engine at supersonic speeds [270]. Both propulsion systems were critical to the vision of developing air-breathing hypersonic vehicles capable of covering vast distances at extreme speeds, thus enhancing military and exploratory capabilities [270].

The Soviet Union similarly pursued hypersonic technologies, although much of their work remained classified. They experimented with hypersonic missiles and aircraft designs, including hypersonic glide vehicles (HGVs) that could manoeuvre unpredictably at extreme speeds, thereby complicating interception efforts by missile defence systems [270]. Despite the technical challenges and high costs associated with hypersonic research during the Cold War, the era laid a robust foundation for modern hypersonic technologies. The principles developed during this time—such as thermal protection, advanced materials, and high-speed propulsion—continue to influence contemporary hypersonic weapons and aerospace vehicles [270].

Cold War hypersonic research represented a fusion of ambitious scientific exploration and intense military competition. It sought to push the boundaries of speed, with applications ranging from space exploration to rapid-strike weaponry. While many projects from this era were experimental and faced significant technical hurdles, the lessons learned have had a lasting impact on modern aerospace engineering and military technology [270].

Early Military Experiments and Lessons Learned

Early military hypersonic experiments, conducted primarily during the Cold War, were driven by the desire to achieve rapid strike capabilities and explore advanced aerospace technologies. These early efforts aimed to push the boundaries of speed, manoeuvrability, and performance at Mach 5 and above, often leading to significant technical challenges but also laying the groundwork for future advancements in hypersonic technologies. The lessons learned from these early experiments shaped the trajectory of hypersonic research and continue to influence modern military and aerospace developments.

One of the most prominent early military hypersonic experiments was the U.S. X-15 program, which began in the late 1950s and continued through the 1960s. The X-15 was a rocket-powered aircraft designed to reach hypersonic speeds and altitudes at the edge of space. During its 199 missions, the X-15 achieved speeds of over Mach 6.7 and altitudes exceeding 100 kilometres (62 miles). It provided invaluable data on aerodynamic heating, structural integrity, and flight control in the hypersonic regime. One key lesson learned from the X-15 was the critical importance of thermal protection systems. At hypersonic speeds, vehicles encounter extreme heat due to air friction, requiring advanced materials and heat shields to prevent structural failure. The X-15's research on heat management directly influenced the development of future spacecraft and hypersonic weapons.

In addition to aerodynamic and thermal challenges, early experiments revealed the difficulty of controlling vehicles at hypersonic speeds. The X-15 program demonstrated that traditional aerodynamic control surfaces, such as fins and rudders, became less effective at high velocities due to shockwave interactions and reduced air density. As a result, military researchers explored alternative control mechanisms, such as reaction control systems (RCS) that used small rocket thrusters to adjust a vehicle's orientation. This lesson on control at extreme speeds became a vital component of subsequent hypersonic vehicle designs, particularly for re-entry vehicles and missiles that had to navigate through the atmosphere at hypersonic speeds.

Figure 23: EC88-0180-1The X-15 #2 56-6671 launches away from the B-52 mothership with its rocket engine ignited. The white patches near the middle of the ship are frost from the liquid oxygen used in the propulsion system, although very cold liquid nitrogen was also used to cool the payload bay, cockpit, windshields, and nose.early 1960s. Defense Visual Information Distribution Service, Public Domain, via Picryl.

Another important early experiment was the development of the Dyna-Soar project (Dynamic Soaring), a U.S. Air Force program initiated in the 1950s. The Dyna-Soar was designed as a hypersonic spaceplane capable of both suborbital and orbital flight. The concept involved using a boost-glide approach, where a rocket would propel the vehicle to high altitudes, and then the craft would glide at hypersonic speeds through the atmosphere toward its target. Although the project was cancelled before reaching operational status, it contributed key insights into the challenges of hypersonic glide vehicles. One of the major lessons learned from the Dyna-Soar project was the difficulty of sustaining hypersonic glide while maintaining manoeuvrability. The project showed that managing aerodynamic forces and stability at hypersonic speeds was far more complex than anticipated, particularly due to shockwave-boundary layer interactions and high drag forces.

The Soviet Union also pursued early military hypersonic experiments, although much of their research remained classified during the Cold War. One notable project was the development of hypersonic glide vehicles, which could be launched atop ballistic missiles and glide through the atmosphere at high speeds. These early experiments focused on evading missile defence systems by making the vehicle manoeuvrable at hypersonic speeds. The lessons learned from these efforts influenced the later development of modern hypersonic glide weapons, such as Russia's Avangard hypersonic missile system.

Another critical lesson learned during these early hypersonic experiments was the importance of materials science. Hypersonic flight introduces extreme heat and pressure, requiring vehicles to be constructed from advanced materials capable of withstanding temperatures exceeding 1,500°C (2,732°F). Early experiments showed that conventional materials were inadequate for prolonged hypersonic flight. This led to the development of advanced heat-resistant materials, such as ablative coatings, ceramic composites, and alloys designed to protect critical vehicle components from heat damage.

Propulsion challenges were another significant aspect of early hypersonic experiments. Ramjet and scramjet engines were seen as promising technologies for sustained hypersonic flight. However, early experiments with these engines revealed difficulties in achieving stable combustion at high speeds. The extreme airflow and high temperatures in scramjet engines made it difficult to ignite and sustain fuel combustion. These lessons underscored the need for further research into air-breathing propulsion systems and their integration into hypersonic vehicles.

Early military hypersonic experiments provided critical insights into the challenges of achieving and sustaining hypersonic flight. These experiments highlighted the importance of thermal protection, vehicle control, materials science, and propulsion technology. Although many early hypersonic projects were experimental or cancelled before reaching operational status, the lessons learned from these efforts laid the foundation for modern hypersonic weapon systems and aerospace technologies. Today's hypersonic research continues to build on the knowledge gained from these early military experiments, advancing the field toward operational hypersonic capabilities in both military and civilian applications.

Modern Hypersonic Weapon Programs

Key Programs: U.S., Russia, China, and Others

The development of hypersonic weapons is a rapidly advancing area in military technology, with multiple countries investing in research programs to create high-speed, nearly impossible-to-intercept systems. These weapons are characterized by their ability to travel at speeds greater than Mach 5 (five times the speed of sound), and their extreme velocity and manoeuvrability pose significant challenges for current defence systems. Several nations are pursuing advanced hypersonic weapon programs to ensure strategic advantages in global military power.

Several countries around the world are actively pursuing hypersonic missile programs, but due to the secretive nature of military research, unclassified information about the true capabilities of these weapons is limited. This lack of transparency makes it difficult to fully evaluate the effectiveness and readiness of these missile systems. While countries may make claims about the progress of their hypersonic programs, there is often a gap between reported capabilities and real-world performance. This discrepancy can be attributed to the technical challenges of developing hypersonic weapons, the extensive testing required, and the need to overcome significant hurdles in materials science, propulsion, and guidance technologies [1].

Despite these challenges, China, Russia, and the United States are widely considered to have the most advanced hypersonic missile programs.

These nations have reportedly already deployed hypersonic missiles or are expected to do so within the next few years. Both China and Russia have claimed to possess operational hypersonic weapons capable of delivering either conventional or nuclear warheads, with Russia's Avangard hypersonic glide vehicle and China's DF-ZF being the most notable examples. The United States, while not yet fully deploying its hypersonic missiles, is conducting extensive testing and development to catch up with its geopolitical rivals. The US has several ongoing hypersonic programs, including the Hypersonic Glide Vehicle (HGV) and Hypersonic Cruise Missile (HCM) projects, which aim to create a range of weapons capable of both conventional and nuclear strikes [1].

Beyond these leading nations, multiple other countries, such as France, India, Japan, North Korea, and Iran, have also initiated their own hypersonic missile development programs. However, these efforts are generally considered to be in earlier stages of development compared to the more advanced work being done in the US, China, and Russia. For example, India is developing the BrahMos-II hypersonic missile in collaboration with Russia, while France is focusing on a hypersonic glide vehicle under the ASN4G program. Japan has announced plans to develop its own hypersonic cruise missile as part of its broader effort to modernize its defence capabilities. North Korea and Iran have also made claims about hypersonic weapons development, but due to the secretive nature of their military programs, little reliable information is available about their actual capabilities [1].

One key feature of the global race to develop hypersonic missiles is the increasing tendency for countries to collaborate with one another. Many nations have established international partnerships to accelerate innovation, pool resources, and share the high costs associated with developing these advanced weapons. For example, India and Russia are jointly developing the BrahMos-II, while France has partnered with Germany on a variety of hypersonic technology research initiatives. Another notable collaboration is the AUKUS security partnership between Australia, the United Kingdom, and the United States, which includes the development of hypersonic and counter-hypersonic technologies. Such collaborations allow countries to share knowledge and reduce the financial burden of hypersonic missile development while ensuring they remain competitive in the global arms race [1].

However, even with collaboration, developing effective hypersonic weapons is extremely challenging, as these systems must overcome technical barriers related to speed, manoeuvrability, thermal management, and guidance systems. The need for precise targeting at such high speeds makes navigation systems a key area of focus, while the intense heat generated during hypersonic flight necessitates advanced materials capable of withstanding extreme temperatures. Additionally, propulsion systems, particularly for hypersonic cruise missiles that require sustained flight at Mach 5 or higher, represent a major technological challenge that few countries have fully mastered [1].

In Australia, the SCIFiRE (Southern Cross Integrated Flight Research Experiment) project is being developed in collaboration with the United States. This joint initiative aims to produce operational hypersonic weapons that are expected to enter service between 2026 and 2031. The project integrates Australia's strategic defence requirements with the U.S.'s technological advancements in hypersonics. SCIFiRE symbolizes a critical leap forward in Australia's military capabilities, aligning its defence strategy with cutting-edge aerospace technologies [271].

Brazil is also working on its hypersonic program, with its primary project being the 14-X hypersonic vehicle. While specific details about the 14-X are not extensively publicized, the program reflects Brazil's commitment to advancing in the hypersonic technology field as part of its broader aerospace research efforts. Brazil's 14-X adds to the growing global interest in developing hypersonic systems for both military and civilian applications.

China has made significant strides in developing hypersonic weapons, with several systems already operational or in advanced stages of testing. The DF-ZF is one of China's key hypersonic glide vehicles, launched from the DF-17 ballistic missile. This system enables precision strikes at extremely high speeds, with unpredictable flight paths that make interception challenging. Another significant hypersonic weapon developed by China is the YJ-21, a hypersonic anti-ship missile that, according to reports, successfully avoided detection in a December 2023 computer-simulated strike on U.S. warships. Additionally, China is working on various other systems like the KF-21, KD-21, and the DF-27, each designed for long-range, high-speed strikes. The Starry Sky 2 and MF-1 hypersonic vehicles further demonstrate China's robust investment in both military and civilian applications of hypersonic technology.

France has a deep history in hypersonic research, with its efforts dating back to the VERAS hypersonic glide vehicle project in 1965, although this was cancelled in 1971. Currently, France has several active hypersonic programs, including the ASN4G, an air-launched cruise missile being developed to replace the ASMP missile for pre-strategic nuclear deterrence by 2035. Additionally, France's LEA Hypersonic Cruise Demonstrator, launched in 2003, aims to validate technologies for the ASN4G program. Another significant project, Prométhée, focuses on a scramjet missile, with successful testing around 2021-2022. France's VMaX (Véhicule Manœuvrant Expérimental) program recently saw its first successful flight test in June 2023, with plans for a follow-up VMaX-2 project expected in 2024 or 2025. France is also exploring the Espadon hypersonic combat aircraft, reflecting its forward-thinking approach to integrating hypersonic technology into both missile systems and aircraft.

Germany has also contributed to hypersonic research through its SHEFEX II project, which is overseen by the German Aerospace Center (DLR). The project focuses on high-speed atmospheric flight and the development of heat-resistant materials essential for hypersonic vehicles. SHEFEX II represents Germany's efforts to push the boundaries of aerospace engineering and hypersonic flight capabilities [272].

India is actively advancing its hypersonic weapons development with key projects like the BrahMos-II and the Hypersonic Technology Demonstrator Vehicle. The BrahMos-II is a joint venture with Russia, aiming to produce a hypersonic missile capable of both anti-ship and land-attack missions [273]. India's HGV-202F project is also part of its broader hypersonic development, positioning the nation as a key player in the hypersonic arms race.

Iran recently unveiled two hypersonic missiles, marking its entry into this advanced technology field. In June 2023, Iran introduced the Fattah-1, a hypersonic ballistic missile. Later in November 2023, Iran revealed the Fattah-2, a hypersonic cruise missile designed for increased manoeuvrability and enhanced striking capabilities. These developments highlight Iran's growing ambitions to modernize its military arsenal with cutting-edge hypersonic systems.

Japan is also focusing on developing hypersonic technologies to strengthen its defence capabilities. The Hyper Velocity Gliding Projectile is designed to provide enhanced protection for Japan's islands, while the Hypersonic Cruise Missile (HCM) is expected to boost Japan's ability to conduct long-range precision strikes. Japan's hypersonic projects reflect its strategic goal to reinforce its defence against regional threats.

North Korea's contribution to the hypersonic race comes in the form of the Hwasong-8, part of its ongoing ballistic missile development program. North Korea aims to leverage hypersonic technology to strengthen its deterrence capabilities and extend its long-range strike potential.

Russia is widely considered a leader in hypersonic weaponry, with numerous systems already operational. The 3M22 Zircon hypersonic cruise missile, capable of reaching speeds up to Mach 9, is designed to target naval ships with extreme precision. Russia's Avangard hypersonic glide vehicle, mounted on intercontinental ballistic missiles (ICBMs), is renowned for its ability to evade missile defence systems due to its unpredictable flight paths. The Kh-47M2 Kinzhal, an air-launched hypersonic missile, has already been deployed, while other systems like the Kh-95 and Vympel R-37M continue to undergo development. Additionally, Russia's Zmeevik anti-ship missile and the Gremlin hypersonic guided rocket extend its hypersonic capabilities across various platforms, from air to sea and ground.

South Korea is modernizing its defence capabilities with the Hycore missile, a scramjet-based two-stage missile designed for hypersonic speeds. This project is part of South Korea's broader strategy to enhance its military's technological edge and counter regional threats.

The United Kingdom has also launched several initiatives to develop hypersonic systems. The Hypersonic Air Vehicle Experimental (HVX) program, launched in 2022, aims to create operational hypersonic technology by the early 2030s. As part of the HVX program, the UK unveiled Concept V, a single-engine hypersonic aircraft that showcases the UK's vision for future air-based hypersonic platforms.

The United States is leading the charge with a wide array of hypersonic weapon programs. The Boeing X-51 Waverider is a scramjet-powered test vehicle designed to demonstrate sustained hypersonic flight. The DARPA Falcon Project and Hypersonic Air-breathing Weapon Concept (HAWC) are part of DARPA's initiatives to develop hypersonic systems for the military. The U.S. Air Force is also working on the Hypersonic Attack Cruise Missile (HACM), a precision-strike weapon, while the Army and Navy are collaborating on the Long-Range Hypersonic Weapon (LRHW) and Conventional Prompt Strike (CPS) programs, both of which centre on a shared hypersonic glide body. The OpFires project is another DARPA-led initiative aimed at creating a ground-launched hypersonic missile, and the HALO anti-ship missile under the U.S. Navy's OASuW Inc 2 program extends U.S. hypersonic capabilities into maritime warfare. The AIM-260 JATM and Lockheed Martin Mako are additional air-launched hypersonic systems expected to bolster U.S. tactical and strategic options.

The global race to develop hypersonic weapons is reshaping the future of warfare, with many nations vying to secure technological superiority in this domain. Hypersonic weapons offer unprecedented speed, manoeuvrability, and offensive capabilities, driving countries to explore diverse avenues of research and development, from air-launched systems to glide vehicles and anti-ship missiles. As these technologies mature, they are expected to fundamentally alter military strategies and global defence dynamics.

United States Hypersonic Air-breathing Weapon Concept (HAWC)

The United States Hypersonic Air-breathing Weapon Concept (HAWC) is an advanced military program aimed at developing and testing hypersonic weapons powered by air-breathing scramjet engines. Managed by the Defense Advanced Research Projects Agency (DARPA) in partnership with the U.S. Air Force, HAWC seeks to create a missile system capable of flying at hypersonic speeds—greater than Mach 5—with high levels of manoeuvrability and sustained flight within the Earth's atmosphere. As part of the U.S. military's broader strategy to advance hypersonic technologies, HAWC is designed to ensure the U.S. maintains technological parity with global competitors such as Russia and China, who are also making significant progress in hypersonic weapons development.

The Hypersonic Air-breathing Weapon Concept (HAWC) is one of several U.S. initiatives aimed at developing an operational hypersonic cruise missile that uses scramjet technology. Scramjets, or supersonic combustion ramjets, enable the missile to achieve speeds exceeding Mach 5 (approximately 3,800 mph or 6,174 km/h), giving it the ability to strike long-range targets quickly while evading modern missile defense systems. The primary goal of HAWC is to provide the U.S. with a new high-speed, precision strike weapon that can operate effectively in contested environments.

Key features and capabilities include:

- **Air-Breathing Scramjet Propulsion:** At the core of HAWC's technological advantage is its scramjet engine, which uses air-breathing propulsion to sustain hypersonic speeds. Scramjets differ from traditional rocket engines in that they compress incoming air at supersonic speeds, mix it with fuel, and ignite the mixture to produce thrust. This method eliminates the need for onboard oxidizers, making the missile lighter and more efficient. Scramjet propulsion allows HAWC to maintain sustained hypersonic speeds over long distances while remaining within the atmosphere, which is crucial for both manoeuvrability and operational flexibility.

- **Hypersonic Speed and Manoeuvrability:** HAWC is engineered to travel at speeds greater than Mach 5, with the potential to reach Mach 7 or higher. At these velocities, the missile can cover vast distances in just a few minutes, drastically reducing the time available for enemies to detect, track, and intercept it. Unlike traditional ballistic missiles that follow a predictable arc, HAWC's hypersonic speed is paired with high manoeuvrability, enabling it to evade advanced missile defence systems. This capability is especially valuable in targeting high-value assets in heavily defended areas.

- **Long-Range Strike Capability:** HAWC is intended to function as a long-range strike weapon capable of hitting targets deep within enemy territory. Its combination of speed and manoeuvrability makes it an ideal weapon for prompt precision strikes against key military targets, such as radar installations, air defence systems, command centres, and even mobile targets like aircraft carriers. With the ability to strike from long distances, HAWC can launch attacks without requiring aircraft to enter enemy airspace, thus minimizing risk to personnel and equipment.

- **Air Launch Platform:** HAWC is designed to be launched from a range of aircraft, including fighter jets such as the F-15 and F-35, as well as bombers like the B-52. Launching from an aircraft provides the missile with a significant performance boost in both altitude and speed, enhancing its range and effectiveness. An air-launched hypersonic weapon like HAWC also enables the U.S. military to maintain standoff capabilities, allowing aircraft to remain outside of enemy air defences while still delivering an impactful strike.

The HAWC program is being developed by two major U.S. defence contractors, Lockheed Martin and Raytheon Technologies, both of which were awarded contracts by DARPA to develop competing versions of the missile system. These versions underwent extensive wind tunnel testing and flight trials to validate the scramjet propulsion technology and overall missile performance.

Notable milestones:

- In **September 2021**, the U.S. Air Force and DARPA conducted the first successful **flight test** of the HAWC missile. During this test, the missile was air-launched and reached **sustained hypersonic speeds**, validating the scramjet engine's functionality and showcasing the missile's potential for future operational use.

- Another successful flight test occurred in **March 2022**, where the missile demonstrated its long-range hypersonic capabilities. These tests are crucial for refining the technology and preparing it for eventual deployment.

The development of HAWC has far-reaching implications for U.S. military strategy and the broader global hypersonic arms race. HAWC is part of the U.S.'s strategy to maintain a technological edge over global competitors such as Russia and China, both of whom are advancing their own hypersonic weapons programs. Russia, for example, has deployed the 3M22 Zircon, and China has developed the DF-ZF hypersonic glide vehicle. The successful development and deployment of HAWC will ensure that the U.S. remains competitive in this critical area of defence technology.

Additionally, HAWC is particularly valuable in Anti-Access/Area Denial (A2/AD) environments, where adversaries such as China and Russia might use long-range missile systems and advanced air defence networks to prevent U.S. forces from accessing strategic regions. HAWC's speed and manoeuvrability make it a powerful tool for penetrating defended areas and neutralizing key targets, giving the U.S. military greater operational flexibility.

HAWC also contributes to first-strike capabilities, enabling the U.S. to conduct rapid, precise attacks on time-sensitive or heavily defended targets before the enemy can react. This makes HAWC an essential asset for pre-emptive military actions.

Finally, HAWC fits into the broader U.S. military strategy of prompt global strike, which seeks to give the U.S. the ability to strike anywhere in the world quickly using conventional weapons. HAWC provides a non-nuclear hypersonic weapon option for delivering devastating strikes, expanding the U.S.'s strategic options in global military engagements.

Despite the success of HAWC's development so far, several challenges remain before the missile becomes fully operational.

- **Thermal management** is a significant challenge, as flying at hypersonic speeds generates immense heat that can damage the missile's structure and electronics. Effective **thermal protection systems** must be perfected to ensure the missile can withstand these extreme conditions.

- **Guidance and control** are also areas that require refinement. At hypersonic speeds, plasma clouds generated by the missile's flight can interfere with traditional guidance systems like GPS, so advanced inertial navigation systems (INS) and sensors are needed to maintain accuracy.

- Finally, integration and deployment with existing military aircraft and systems will be essential for HAWC's operational success. Ensuring seamless integration with current military platforms will require close coordination and testing.

The Hypersonic Air-breathing Weapon Concept (HAWC) marks a major leap forward in U.S. hypersonic weapons development. With its scramjet-powered hypersonic speeds, manoeuvrability, and long-range strike capabilities, HAWC offers a potent tool for the U.S. military to engage high-value targets in contested environments. As testing continues, HAWC will play a critical role in shaping the future of modern warfare, ensuring that the U.S. maintains its competitive edge in hypersonic technology and its ability to project military power globally.

Development Timelines and Milestones

The development of hypersonic weapons has been a complex and multi-decade process, shaped by advancements in aerospace technology, materials science, and propulsion systems. Below is an outline of the key timelines and milestones in the evolution of hypersonic weapons from their inception to the present day:

1. **Early Foundations (1940s-1950s): Hypersonic Flight Concept**
   - **1940s:** The idea of hypersonic flight began to take shape during the latter stages of World War II, as scientists explored the limits of aircraft speed and missile technology. The development of the V-2 rocket by Nazi Germany reached supersonic speeds, laying a foundation for later work on higher-speed flight.

   - **1950s:** With the start of the Cold War, the U.S. and the Soviet Union ramped up research into hypersonic technology as part of broader missile and aerospace programs. Rocket-powered experimental vehicles like the X-1 (the first plane to break the sound barrier) and the X-15 helped provide early data on high-speed flight conditions, although they were not true hypersonic weapons.

2. **X-15 Program and Hypersonic Research (1960s)**
   - **1960-1968:** The X-15, a U.S. experimental hypersonic aircraft, con-

ducted nearly 200 test flights and achieved speeds of up to Mach 6.7. The X-15 provided valuable insights into the thermal and aerodynamic challenges of high-speed flight, particularly in managing heat generated by air friction at hypersonic speeds.

- **1960s-1970s:** Various hypersonic glide and re-entry vehicle experiments, such as the Dyna-Soar project (cancelled before full development), demonstrated that controlled hypersonic flight was technologically feasible, but still in the experimental phase.

3. Cold War Research and Early Milestones (1970s-1980s)

- **1970s:** The U.S. and the Soviet Union continued hypersonic research focused on re-entry vehicles for intercontinental ballistic missiles (ICBMs) and hypersonic glide vehicles. The goal was to enhance manoeuvrability and speed in order to evade missile defences.

- **1980s:** The Soviet Union developed concepts for hypersonic weapons as part of its ICBM program, including designs for manoeuvrable warheads capable of hypersonic speeds during the re-entry phase. Research in the U.S. continued with projects such as NASA's scramjet engine experiments, which explored air-breathing propulsion at hypersonic velocities.

4. Breakthrough Experiments and Hypersonic Weapon Concepts (1990s)

- **1990s:** The collapse of the Soviet Union shifted focus from large-scale military programs to more targeted research into hypersonic technologies. U.S. projects such as the National Aerospace Plane (NASP) aimed to develop a single-stage-to-orbit hypersonic aircraft but were ultimately cancelled due to technical challenges.

- **1990s:** Experimental programs like the X-43 and X-51 Waverider further developed the scramjet (supersonic combustion ramjet) technology necessary for sustained hypersonic flight. These programs tested the potential of air-breathing hypersonic engines, achieving speeds of over Mach 9 in some cases.

5. 21st Century Hypersonic Weapons Race (2000s)

- **2000s:** Hypersonic research re-emerged as a high priority for military development. The U.S., Russia, and China began focusing on hypersonic weapons capable of delivering precision strikes at unprecedented speeds, sparking a new arms race.

- **2001-2010:** The U.S. began testing hypersonic vehicles like the X-43A, which achieved a record speed of Mach 9.8 in 2004. The X-51 Waverider, developed in the late 2000s, successfully flew in 2010, marking an important milestone in scramjet-powered flight.

6. **Boost-Glide Systems and Military Applications (2010s)**
    - **2010-2020:** Russia, China, and the U.S. intensified development of boost-glide hypersonic vehicles, which use rocket boosters to reach high altitudes and speeds before gliding at hypersonic velocities. Russia's Avangard boost-glide system, capable of Mach 20 speeds, became operational in 2019. China's DF-ZF hypersonic glide vehicle was tested successfully multiple times in this period.
    - **2018:** The U.S. Hypersonic Air-breathing Weapon Concept (HAWC) and the Conventional Prompt Strike (CPS) programs were launched to develop both boost-glide systems and scramjet-powered cruise missiles.
7. **Hypersonic Weapon Systems Operational (2020s)**
    - **2020s:** Hypersonic missile systems began entering operational use. Russia declared its **3M22 Zircon**, a scramjet-powered hypersonic cruise missile, operational in 2020, and China integrated its DF-ZF into its missile arsenal. The U.S. continued to advance several hypersonic programs, aiming to field operational weapons by mid-decade.
    - **2021:** The U.S. conducted successful tests of HAWC, while Russia's Avangard system continued to be a focal point of global hypersonic discussions.
8. **Global Arms Race and Strategic Implications**
    - **Ongoing:** The development of hypersonic weapons has triggered an arms race, as countries like the U.S., Russia, and China seek to gain strategic advantages by fielding weapons capable of evading missile defence systems and delivering rapid, precise strikes.
    - **Challenges:** Developing effective missile defences against hypersonic threats remains a top priority. Advanced radar, directed-en-

ergy weapons, and new technologies are being pursued to counter hypersonic systems.

9. **Future Developments and Expansion**
   - **Future:** Ongoing research is focused on improving materials, propulsion technologies (e.g., scramjets), and the accuracy of hypersonic systems. Additionally, hypersonic technology is being considered for space exploration and high-speed commercial travel applications.

This timeline underscores the steady advancement of hypersonic weapons technology, moving from experimental research to operational systems capable of transforming modern warfare. Each phase of development has built on lessons learned in the previous era, culminating in today's hypersonic arms race.

Testing and Prototyping: Challenges and Achievements

The development of hypersonic weapons presents a series of formidable challenges and notable achievements across various domains, including thermal protection, aerodynamic stability, propulsion systems, testing methodologies, and guidance systems. Each of these areas has seen significant advancements that are critical for the successful deployment of hypersonic technologies.

Thermal Protection and Materials Science

The extreme heat generated by atmospheric friction at hypersonic speeds, often exceeding 2,000°C (3,600°F), poses a significant challenge for hypersonic weapon systems. Traditional materials are inadequate, as they tend to melt or degrade under such conditions. This necessitates the development of advanced thermal protection systems (TPS) that can withstand these extreme temperatures while maintaining structural integrity [138, 274]. Recent advancements in materials science have led to the creation of ceramic composites, ablative coatings, and carbon-carbon composites, which have shown promise in protecting hypersonic vehicles during flight. Notable examples include the U.S. X-51 Waverider and Russia's Avangard boost-glide vehicle, both of which successfully integrated these advanced materials, allowing for sustained hypersonic flight without disintegration.

Aerodynamic Stability and Control

Achieving stable flight at hypersonic speeds is inherently challenging due to unique aerodynamic forces and shock waves. The design of hypersonic vehicles must optimize their shape to manage these forces effectively, ensuring that they can manoeuvre while maintaining control [275, 276]. Successful prototypes like the X-43A and X-51A Waverider have demonstrated stable hypersonic flight using air-breathing scramjet engines, effectively managing their trajectories and sustaining high speeds through optimized aerodynamic designs. Furthermore, China's DF-ZF hypersonic glide vehicle has also shown significant advancements in manoeuvrability at hypersonic speeds, marking a critical achievement in this domain.

Propulsion Systems: Scramjets and Boost-Glide Technology

The development of reliable propulsion systems, particularly scramjets and boost-glide technologies, remains a major hurdle in hypersonic weaponry. Scramjets must ignite fuel in a high-speed airflow, which requires precise timing and fuel mixing [203]. The X-51A Waverider's successful demonstration of sustained hypersonic flight at Mach 5.1 in 2013 exemplifies the feasibility of scramjet technology [277]. Additionally, Russia's Zircon missile, operational since 2020, utilizes a scramjet engine to achieve speeds of Mach 8-9, showcasing the viability of this propulsion technology. The integration of boost-glide systems, such as the Avangard and HAWC (Hypersonic Air-breathing Weapon Concept), further confirms the advancements made in achieving long-range hypersonic flight with high manoeuvrability.

Testing in Wind Tunnels and Real-Flight Conditions

Testing hypersonic vehicles presents unique challenges, as standard wind tunnels cannot accurately replicate the extreme conditions encountered during hypersonic flight. To address this, specialized shock tunnels and arc-heated wind tunnels have been developed, capable of simulating hypersonic conditions up to Mach 10 [278, 279]. These facilities allow for the testing of materials and aerodynamic designs under realistic conditions, providing critical data for engineers. Notable contributions from institutions like the Langley Research Center and the Australian T4 shock tunnel have been instrumental in advancing hypersonic research. However, full-scale flight tests, such as those conducted with the X-43 and X-51, remain the ultimate benchmark for performance validation.

Figure 24: The AEDC Aerodynamic and Propulsion Test Unit at Arnold Air Force Base supports recent testing for the Air Force Research Laboratory Medium Scale Critical Components Scramjet program. The Northrop Grumman-produced engine was successfully operated at at conditions above Mach 4 and has set the record for highest thrust produced by an air-breathing hypersonic engine in Air Force history. Defense Visual Information Distribution Service, Public Domain, via NARA & DVIDS Public Domain Archive.

Guidance, Navigation, and Control Systems

Guidance systems for hypersonic vehicles face unique challenges due to rapid atmospheric traversal, high temperatures, and the need for instantaneous adjustments. Traditional navigation methods struggle under the intense conditions of hypersonic flight [280]. Modern hypersonic vehicles now incorporate advanced inertial navigation systems (INS) combined with AI-based predictive algorithms, enabling them to maintain accurate trajectories and make real-time course corrections [189, 281]. The U.S. Conventional Prompt Strike (CPS) program has made significant strides in refining these guidance systems, ensuring precision in targeting and enhancing the evasion capabilities of hypersonic weapons [282, 283].

The field of hypersonic weaponry is marked by significant challenges and achievements across multiple domains. The integration of advanced materials, innovative aerodynamic designs, reliable propulsion systems, specialized testing methodologies, and sophisticated guidance technologies collectively contribute to the progress in hypersonic capabilities.

HYPERSONIC WEAPON TESTING FACILITIES

Wind Tunnels and Specialized Testing Equipment

The hypersonic testing process is a comprehensive and complex series of steps aimed at validating the design, performance, and durability of hypersonic vehicles under extreme conditions. The goal is to simulate and analyse the extreme environments hypersonic vehicles encounter, such as high temperatures, pressures, and speeds, to ensure that the vehicles can perform effectively in real-world situations.

Wind tunnel testing is an essential phase in the development of hypersonic vehicles, as it allows engineers to simulate the extreme conditions these vehicles face during high-speed flight. Hypersonic wind tunnels, particularly shock tunnels and arc-heated wind tunnels, are specialized facilities designed to recreate the intense heat, pressure, and aerodynamic forces experienced at hypersonic speeds. These tunnels utilize high-pressure gas bursts to generate supersonic or hypersonic airflow, providing a controlled environment for engineers to examine the behaviour of a vehicle's design, materials, and overall structure under conditions that mimic real flight scenarios [284, 285].

In these simulations, engineers focus on how a vehicle interacts with high-speed airflow, emphasizing aerodynamic shape, structural integrity, and thermal protection systems (TPS). The unique challenges of hypersonic flight, such as extreme atmospheric friction that generates significant heat, necessitate robust thermal protection to safeguard both the vehicle's surface and its internal components [109, 124]. While wind tunnels cannot perfectly replicate all aspects of hypersonic flight, such as sustained high speeds or precise temperature gradients, they provide invaluable data for understanding how various designs and materials respond to these conditions [189].

One of the most significant contributions of wind tunnel testing is the refinement of aerodynamic properties of hypersonic vehicles. By simulating airflow in a controlled setting, engineers can observe airflow patterns around the vehicle at extreme speeds, optimizing its shape to reduce drag and enhance stability [286, 287]. Additionally, wind tunnel tests facilitate a detailed examination of thermal protection systems, which are crucial for managing the intense heat experienced during hypersonic flight. Testing these systems in a simulated environment allows engineers to identify vulnerabilities in materials or designs and make necessary adjustments to ensure the vehicle can withstand thermal stresses during actual flight [288, 289].

Moreover, wind tunnel testing is critical for assessing the structural integrity of hypersonic vehicles under high-speed conditions. The aerodynamic forces exerted on the vehicle can create substantial pressure on its frame, necessitating thorough evaluations to ensure that the design can endure these forces without compromising performance [290, 291]. The data gathered from wind tunnel experiments help identify potential structural weaknesses and inform adjustments before advancing to more complex testing stages, such as flight tests [214, 292]. Although wind tunnel testing cannot fully replicate the dynamic nature of actual hypersonic flight, it lays a vital foundation for early-stage design validation, guiding engineers in refining their designs to meet the challenges of hypersonic travel [172, 231].

Figure : MARHy wind tunnel. Sandra c, CC BY-SA 4.0, via Wikimedia Commons.

As an example of a wind tunnel, the MARHy wind tunnel, which stands for Mach Adaptable Reflective Hypersonic wind tunnel, is specifically designed for hypersonic flow testing. It is used to simulate the extreme conditions encountered during hypersonic flight, such as high-speed airflow, extreme temperatures, and pressure. This type of wind tunnel allows engineers and researchers to test various aspects of hypersonic vehicle design, including aerodynamics, thermal protection systems, materials, and propulsion technologies, in a controlled environment before proceeding to full-scale flight tests.

The MARHy wind tunnel is particularly valuable for testing vehicles or components that will experience speeds greater than Mach 5. It provides high-quality data on how different designs perform under the intense conditions of hypersonic travel, such as aerodynamic stability, thermal loads, and structural integrity. Researchers use it to improve the design and performance of missiles, re-entry vehicles, spacecraft, and other hypersonic systems by experimenting with different materials, shapes, and configurations.

In addition to its application for military and aerospace projects, the MARHy wind tunnel also plays a role in advancing the development of scramjet engines and other propulsion technologies aimed at sustaining hypersonic speeds, as well as studying shock wave behaviours that occur at these high velocities. The tunnel helps in generating critical data for these systems to better understand and overcome the challenges posed by hypersonic flight.

Using a wind tunnel for hypersonic testing involves several carefully planned steps to simulate the extreme conditions a vehicle or component would encounter at speeds greater than Mach 5. Here's a detailed explanation of how hypersonic wind tunnel testing is conducted:

1. Selection of Test Model and Objectives

The first step in using a wind tunnel for hypersonic testing is to define the test objectives and select a suitable model of the vehicle or component. The test model can be a scaled-down version of a hypersonic vehicle, such as a missile, spacecraft, or aircraft, or it could be a specific component, like a thermal protection system (TPS), control surface, or engine part.

Engineers determine what aspects of the design need to be tested, such as:

- Aerodynamic performance (e.g., lift, drag, stability)
- Heat resistance and thermal behaviour
- Structural integrity
- Shock wave interactions
- Control surface efficiency

2. Wind Tunnel Setup

Hypersonic wind tunnels, such as shock tunnels or arc-heated wind tunnels, are specially designed to generate high-speed airflow that simulates hypersonic conditions. Depending on the wind tunnel type, the setup process varies:

- **Shock Tunnels**: Use high-pressure gas to create shock waves that accelerate the airflow to hypersonic speeds. The air is compressed and heated as it passes through the shock waves, simulating the conditions experienced during hypersonic flight.
- **Arc-Heated Wind Tunnels**: Use electrically heated gases to simulate the high thermal loads of hypersonic travel. These tunnels are used to study the thermal protection systems and material response to extreme heat.

The tunnel is calibrated to generate the desired flow conditions, such as Mach number (speed), temperature, and pressure.

3. Mounting the Model

The model is securely mounted inside the wind tunnel's test section. This section is typically designed to minimize interference with the airflow and the data collection process. The mounting system must be strong enough to withstand the intense forces exerted by the hypersonic airflow while keeping the model in the correct orientation for testing.

Depending on the test objectives, the model might be mounted at different angles of attack or orientations to study various aerodynamic behaviours or to test how control surfaces perform under different flight conditions.

4. Instrumentation and Sensors

Before running the test, the model is fitted with various sensors and instrumentation to collect critical data during the experiment. These sensors are used to measure:

- Pressure distribution
- Heat flux (heat transfer to the surface)
- Aerodynamic forces (lift, drag, moment)
- Shock wave formation and behaviour
- Structural responses (vibration, deformation)
- Temperature of the surface and key components

This data is essential for analysing how the model performs under hypersonic conditions.

5. Running the Test

Once the wind tunnel is calibrated and the model is mounted, the test is ready to begin. The wind tunnel is activated, creating hypersonic airflow around the model. The airflow is generated using high-pressure gases (in shock tunnels) or heated gases (in arc-heated tunnels) to achieve speeds above Mach 5.

During the test, engineers observe how the model interacts with the high-speed air:

- **Shock waves**: Engineers study the formation of shock waves around the model, including bow shocks and oblique shocks that form due to the high-speed flow. Shock waves are crucial for understanding drag, heat transfer, and aerodynamic stability.

- **Thermal behaviour**: The model is exposed to high thermal loads, and the sensors record how heat is transferred to and absorbed by the material. This helps in evaluating the effectiveness of thermal protection systems.

- **Aerodynamics**: Engineers measure the forces acting on the model (lift, drag, side forces) and assess how the vehicle will perform in flight. The stability and control of the vehicle at different angles and speeds are also evaluated.

6. Data Collection and Analysis

As the test progresses, real-time data is collected from the sensors and transmitted to a computer system for analysis. High-speed cameras, Schlieren imaging (for visualizing shock waves), and other diagnostic tools may also be used to capture visual data on the airflow and shock wave formation.

After the test, engineers conduct a thorough analysis of the data to assess:

- **Aerodynamic performance**: Lift, drag, and stability characteristics.

- **Thermal protection**: How well the thermal protection system performed under extreme heat loads.

- **Structural integrity**: Whether the model experienced any deformation, stress, or vibration that could affect its performance.

This data is compared to predictions made using computational fluid dynamics (CFD) simulations and previous test results.

7. Iterative Testing

Hypersonic wind tunnel testing is often part of an **iterative design process**. Based on the results from each test, engineers refine the design of the vehicle or component. They may adjust the shape, materials, or other features to improve performance. Further tests are then conducted to validate these changes until the design meets the required performance criteria.

8. Limitations and Complementary Testing

Although wind tunnels provide invaluable insights into hypersonic behaviour, they cannot fully replicate every aspect of hypersonic flight, such as sustained flight times and the dynamic environment over long ranges. Therefore, prototype flight tests are eventually required to confirm the performance of the design in real-world conditions.

Wind tunnel testing is complemented by CFD simulations and flight testing to fully understand how a hypersonic vehicle or component will perform during actual missions.

Simulation Tools and Computational Models

After wind tunnel testing, the next step is to test individual components of the vehicle under laboratory conditions. Component-level testing is a crucial phase in the development of hypersonic vehicles, particularly after initial wind tunnel evaluations. This stage focuses on assessing individual components such as thermal protection systems (TPS), control surfaces, and propulsion systems under controlled laboratory conditions. The objective is to ensure that these components can endure the extreme thermal and pressure loads characteristic of hypersonic flight, which can reach speeds exceeding Mach 5.

One of the primary components tested is the thermal protection system, which is essential for safeguarding the vehicle from the intense heat generated during flight. Research indicates that ablative materials, which serve as sacrificial layers, are vital in protecting vehicles and probes during hypersonic re-entry through planetary atmospheres. These materials are designed to absorb and dissipate heat, thereby preventing structural damage to the vehicle [293]. The performance of TPS can be critically evaluated through various thermal testing methodologies, including arc jet testing, which simulates the thermal environment experienced during flight [294]. Such tests help in validating the design and material selection for TPS, ensuring they meet the rigorous demands of hypersonic conditions [295].

In addition to TPS, propulsion systems, particularly scramjet combustion chambers, undergo extensive testing to evaluate their performance under hypersonic conditions. The HIFiRE program, which focuses on hypersonic flight technologies, emphasizes the importance of ground tests to validate propulsion and airframe integration [130]. These tests are designed to isolate specific parameters that affect performance, allowing engineers to fine-tune designs before full-system evaluations [296]. The data gathered from these tests are crucial for understanding the interactions between different components and the overall system behaviour during flight.

The HIFiRE (Hypersonic International Flight Research Experimentation) program was a collaborative effort between the United States and Australia aimed at advancing research in hypersonic flight. Initiated in the mid-2000s, the program focused on investigating and demonstrating key technologies required for sustained hypersonic flight, including air-breathing scramjet engines, high-temperature materials, and vehicle design capable of withstanding extreme conditions.

The HIFiRE program was created to achieve several critical objectives. First, it aimed to develop and demonstrate a range of essential hypersonic technologies. These included scramjet propulsion, advanced aerodynamics, thermal protection systems, and guidance and control technologies, all of which are crucial for successful hypersonic flight. Another key focus was on exploring air-breathing propulsion, particularly scramjets, which could enable sustained flight at speeds greater than Mach 5. Additionally, the program sought to understand aerothermal effects—how the extreme heat generated during hypersonic flight affects materials and vehicle integrity. This research helped advance knowledge in thermal protection systems and high-temperature materials. The program also involved a series of flight tests to validate the technologies and designs being developed, providing valuable data on how hypersonic vehicles perform in real-world conditions.

HIFiRE was a joint initiative between the Defense Science and Technology Group (DSTG) in Australia and the U.S. Air Force Research Laboratory (AFRL). Collaboration also included industry and academic partners, such as Boeing, along with several universities. This partnership contributed significantly to the program's design, testing, and analysis efforts.

Figure 26: Technicians mount the Hypersonic International Flight. Defense Visual Information Distribution Service, Public Domain, via NARA & DVIDS Public Domain Archive Public Domain Archive.

The HIFiRE program comprised multiple flight experiments designed to test specific hypersonic technologies and gather data on vehicle performance at high speeds. One of the initial phases, HIFiRE-0, involved ground-based tests and wind tunnel experiments to simulate hypersonic conditions. These tests validated key design concepts and propulsion systems before flight testing began.

The first major flight test, HIFiRE-1, was conducted in 2009. This experiment tested a specially designed vehicle at Mach 7, providing valuable insights into aerodynamic performance and shock-wave interactions at hypersonic speeds. Further tests, such as HIFiRE-2 and HIFiRE-3, focused on scramjet propulsion and demonstrated supersonic combustion, while later phases (HIFiRE-4 through HIFiRE-8) explored more advanced concepts such as sustained scramjet flight, thermal protection, and materials performance. These tests helped refine scramjet technology and demonstrated its potential for future hypersonic systems.

The HIFiRE program made significant contributions to the field of hypersonic research. Among its most notable achievements was the successful validation of scramjet propulsion, proving that these engines could operate efficiently at hypersonic speeds. Additionally, the program provided critical insights into the aerodynamic and thermal behavior of vehicles traveling at these extreme velocities, including how materials respond to the intense heat generated during flight. The international collaboration between the U.S. and Australia showcased the importance of combined efforts in advancing cutting-edge aerospace technologies.

The HIFiRE program laid the groundwork for future hypersonic research and development. The knowledge and data generated from the program have influenced ongoing projects related to hypersonic weapons, high-speed reconnaissance systems, and next-generation aerospace vehicles. The technological advancements made through HIFiRE continue to shape current research in both military and civilian sectors, potentially contributing to innovations in hypersonic missiles and even commercial hypersonic transport systems in the future.

Control surfaces are another critical area of focus during component-level testing. The stability and control of hypersonic vehicles are influenced by the design and performance of these surfaces. Flight-test experiments, such as those conducted under the HIFiRE program, provide valuable insights into how control surfaces respond to various flight conditions, enabling engineers to optimize their designs for better manoeuvrability and stability [297]. The integration of advanced flight control methods further enhances the safety and efficiency of hypersonic vehicles by ensuring that actuator performance is rigorously tested and validated [148].

By isolating each component during testing, engineers can identify potential weaknesses and address them before progressing to full-system tests. This meticulous approach is essential for developing the resilience and durability of hypersonic vehicles, ensuring they can perform reliably in real-world flight conditions. The iterative nature of component-level testing allows for continuous improvement and innovation in the design and materials used in hypersonic vehicle development.

Computational Fluid Dynamics (CFD) is an essential tool in the design and analysis of hypersonic vehicles, providing critical insights into their aerodynamic performance under extreme conditions. The ability of CFD to simulate airflow, heat transfer, and material interactions at hypersonic speeds allows engineers to predict vehicle behaviour before physical testing, thus enhancing design efficiency and safety.

CFD simulations are particularly valuable in understanding the complex interactions that occur in hypersonic flows, such as shock wave formation, boundary layer transition, and thermal effects. For instance, Pehrson [298] highlights that CFD is a primary design tool for predicting hypersonic boundary layers, which significantly influence heating, entropy production, and drag generation in hypersonic vehicles. Additionally, Chen and Zhao [299] emphasize the importance of aerothermoelastic analysis, which integrates thermal and structural dynamics to assess how heating affects the vehicle's aerodynamic properties. This integration is crucial, as the aeroelastic behaviour of hypersonic vehicles is sensitive to temperature-induced structural variations, which can be effectively modelled using CFD [299].

Aerothermoelastic analysis is a multidisciplinary approach used to understand the interaction between aerodynamic forces, thermal effects, and structural elasticity, particularly in high-speed flight regimes such as hypersonic flight. It plays a crucial role in the design and analysis of high-speed aerospace vehicles, where aerodynamic loads, extreme temperatures, and the structural flexibility of materials come together in complex ways. This analysis ensures that the vehicle's structure can withstand the combined effects of these forces without undergoing failure, excessive deformation, or instability. Here is how to conduct an aerothermoelastic analysis, explained in detail:

Aerothermoelastic analysis involves three interconnected domains: aerodynamics, thermal effects, and structural elasticity.

1. **Aerodynamics**: At high speeds, particularly in hypersonic regimes (Mach 5+), aerodynamic forces exert significant pressure and shear stresses on the surface of the vehicle. The shock waves generated by hypersonic motion can induce localized high-pressure zones that must be accounted for in the analysis. These forces influence

the structural deformation of the vehicle and need to be accurately modelled using computational fluid dynamics (CFD) or aerodynamic theories.

2. **Thermal Effects**: The friction between the air and the surface of the vehicle generates intense heat, which causes thermal expansion and stresses in the structure. In aerothermoelastic analysis, the distribution of heat across the vehicle's surface (often referred to as the thermal load) is calculated using heat transfer methods. Materials can experience high temperatures that degrade their strength and stiffness, and this must be considered when assessing the vehicle's overall stability and performance.

3. **Structural Elasticity**: This refers to the flexibility of the vehicle's structure, which can deform under aerodynamic and thermal loads. Understanding how the structure reacts to these forces is essential for ensuring that the vehicle maintains its integrity without excessive deformation. Finite element analysis (FEA) is typically used to model the structural behaviour under combined aerodynamic and thermal loads.

Key steps in aerothermoelastic analysis include:

1. **Modelling the Aerodynamic Forces**: The first step is to model the aerodynamic forces acting on the vehicle during flight. This is usually done using CFD simulations or experimental data from wind tunnel tests. The aerodynamic model should capture the variations in pressure and shear forces across the surface of the vehicle as it travels at high speeds. In hypersonic flight, the interaction of shock waves with the structure must also be accounted for, as they can lead to significant localized pressure spikes.

2. **Thermal Analysis**: The next step is to conduct a thermal analysis to determine how heat is distributed across the vehicle's surface. This involves solving heat conduction equations to find out how the temperature varies throughout the structure. It's critical to model the effects of thermal expansion, as materials exposed to high temperatures will expand differently, leading to thermal stresses. The thermal analysis will provide information about the thermal gradients and temperature distributions across the structure, which are inputs for the subsequent structural analysis.

3. **Structural Analysis Under Aerothermal Loads**: Once the aerodynamic and thermal forces are known, they are applied to the finite element model of the structure. This model assesses how the structure deforms in response to combined aerothermal loading. The structural analysis uses elasticity theory and material properties (which may change with temperature) to determine whether the vehicle will experience excessive deformation or failure under the expected flight conditions. In particular, engineers are concerned with buckling, flutter, and thermal expansion. Buckling occurs when compressive aerodynamic or thermal stresses cause parts of the structure to collapse. Flutter is an aeroelastic phenomenon where aerodynamic forces interact with structural flexibility, potentially leading to catastrophic failure. Thermal expansion adds an extra layer of complexity, as the material stretches or shrinks in response to heat.

4. **Coupling the Aerodynamic, Thermal, and Structural Domains**: Aerothermoelastic analysis is an iterative process, where the interactions between aerodynamic forces, thermal effects, and structural responses are continually refined. The fluid-structure interaction (FSI) approach is used to simulate how changes in structural shape due to aerodynamic or thermal effects influence the flow of air over the vehicle, and vice versa. This iterative process is computationally intensive, as it involves running simulations that account for the back-and-forth interactions between airflow, temperature changes, and structural deformation.

5. **Assessing Stability and Performance**: After solving the coupled aerodynamic, thermal, and structural equations, the vehicle's stability and performance are assessed. The primary goal is to ensure that the vehicle remains structurally sound and performs as intended without failure under real-world flight conditions. If the analysis shows any **unstable modes** (e.g., flutter) or **excessive deformation** (due to thermal expansion or aerodynamic loads), the design must be revised by optimizing materials, geometry, or control systems.

Several numerical tools and methods are used for aerothermoelastic analysis. For structural and thermal analysis, finite element methods (FEM) are the standard. For aerodynamics, CFD simulations are essential to model the flow of air around the structure. Multidisciplinary optimization (MDO) techniques are often used to find the best balance between aerodynamic efficiency, structural strength, and thermal resilience.

For high-speed applications such as hypersonic vehicles, coupled solvers that integrate CFD, thermal, and structural solvers are used to handle the highly nonlinear interactions between the domains. For experimental validation, wind tunnel tests at high temperatures and pressures are often conducted to verify the computational results.

Aerothermoelastic analysis is crucial for designing high-speed vehicles such as hypersonic aircraft, missiles, or spacecraft. By examining how aerodynamic forces, thermal loads, and structural elasticity interact, engineers can ensure that these vehicles remain stable, perform efficiently, and avoid failure under extreme conditions. This complex analysis requires advanced computational tools and simulations to assess how different forces influence one another, ultimately leading to safer and more effective aerospace designs.

Moreover, the limitations of physical testing in hypersonic conditions make CFD an attractive alternative for design validation. Seager and Agarwal [300] point out that the high costs and technical challenges associated with hypersonic wind tunnel testing often render it impractical, thereby necessitating the use of CFD to optimize vehicle shapes for reduced drag and heat transfer. This sentiment is echoed by Lamorte and Friedmann [161], who discuss the necessity of reliable computational results for the performance and stability analyses of hypersonic vehicles, given the challenges in validating complex aerodynamic models through traditional testing methods.

CFD also allows for the exploration of scenarios that are difficult to replicate in wind tunnels, such as varying atmospheric conditions and flight profiles. For example, the work of Dharavath et al. [301] demonstrates the use of CFD to simulate both external and internal flows in a hypersonic vehicle, providing insights into scramjet combustor performance and overall vehicle thrust. This capability to simulate diverse conditions enhances the predictive power of CFD, enabling engineers to identify potential design flaws early in the development process.

Prototype Flight Testing, Telemetry and Data Collection, and Post-Flight Analysis

Prototype flight testing is a critical phase in the development of hypersonic vehicles, which follows preliminary assessments such as wind tunnel tests and component-level evaluations. During this phase, full-scale flight tests are conducted under controlled conditions, often in restricted airspace, to assess the vehicle's performance in real atmospheric conditions. For instance, vehicles like the X-43 and X-51A Waverider utilize a launch method where they are carried to altitude by rockets or aircraft before igniting their scramjet engines, allowing them to reach the necessary speeds for testing [302, 303]. This method is essential for evaluating various performance metrics, including speed, maneuverability, and thermal performance, which are critical for the success of hypersonic flight missions [109, 304].

Telemetry systems play a pivotal role during these prototype flight tests, providing real-time data on crucial flight parameters such as speed, temperature, altitude, vibration, and pressure. This data is vital for engineers to monitor the vehicle's adherence to design specifications and to identify any anomalies that may occur during flight [296, 305]. In the event of a critical failure, self-destruct systems are often employed to mitigate risks associated with loss of control, ensuring that the vehicle does not pose a threat to populated areas [306]. The telemetry data collected during these tests is invaluable for understanding the vehicle's behaviour across different atmospheric layers and under varying thermal conditions, thereby informing future design improvements [307, 308].

Post-flight analysis is another essential component of the prototype testing process. After each flight, engineers conduct a thorough review of the telemetry data alongside video footage, often utilizing techniques such as Schlieren imaging to visualize shock waves and airflow patterns around the vehicle [302, 309]. This analysis allows for a detailed comparison between the actual flight performance and the predictions made from earlier wind tunnel tests and computational fluid dynamics (CFD) simulations [310, 311]. Discrepancies identified during this phase are critically examined to refine the vehicle's design and technology, particularly focusing on aspects such as guidance systems, thermal protection, and propulsion efficiency [101, 312]. The insights gained from post-flight evaluations are crucial for advancing the technology towards operational readiness and ensuring the success of future missions [313].

Case Studies of Successful and Failed Tests

Successful Hypersonic Tests:

- **NASA X-15 Program (1960s)**
  - **Overview:** The X-15 was a rocket-powered aircraft designed to explore high-speed flight at altitudes up to the edge of space. It set numerous speed and altitude records, reaching speeds of Mach 6.72 (4,520 mph) and altitudes of over 100 kilometres.
  - **Key Achievements:** The X-15 was one of the first successful hypersonic aircraft, providing critical data on aerodynamics, thermodynamics, and materials at hypersonic speeds. Its flights helped develop control systems and thermal protection technologies that informed future space missions.
  - **Impact:** The program contributed invaluable knowledge to NASA's manned spaceflight programs, particularly Apollo, and laid the groundwork for modern hypersonic research.

Figure 27: X-15 test pilots - Thompson, Dana, and McKay. Defense Visual Information Distribution Service , Public Domain, NARA & DVIDS Public Domain Archive Public Domain Archive.

- **Boeing X-51A Waverider (2010-2013)**
  - **Overview:** The X-51A Waverider was a scramjet-powered experimental vehicle designed to demonstrate sustained hypersonic flight using air-breathing propulsion. It was launched from a B-52 bomber and accelerated by a rocket booster before igniting its scramjet engine.
  - **Key Achievements:** The final test in 2013 reached Mach 5.1, sustaining hypersonic speeds for over three minutes. This was one of the most successful tests of a scramjet-powered vehicle and proved the feasibility of air-breathing hypersonic propulsion.
  - **Impact:** The X-51A program was a significant milestone in demonstrating the viability of scramjet technology, contributing to future development in hypersonic weapons and aerospace systems.

Figure 28: X-51A Waverider. Mark Jones Jr., CC BY 2.0, via Flickr.

- **Avangard Hypersonic Boost-Glide Vehicle (Russia, 2019)**

- **Overview:** The Avangard is a Russian hypersonic boost-glide vehicle capable of reaching speeds exceeding Mach 20. It is launched by an intercontinental ballistic missile (ICBM) and can glide through the atmosphere while manoeuvring to evade missile defences.

- **Key Achievements:** Successfully tested in December 2018, the Avangard became operational in 2019, marking one of the first deployable hypersonic weapons. It demonstrated precision targeting and the ability to perform complex evasive manoeuvres at hypersonic speeds.

- **Impact:** The Avangard's success has accelerated global interest in hypersonic weapons, particularly in the U.S. and China, and raised concerns about a new arms race in hypersonic technologies.

- **Hypersonic Air-breathing Weapon Concept (HAWC)**

  - **Overview:** The HAWC program, developed by the U.S. Air Force and DARPA, aims to demonstrate an operational hypersonic missile using scramjet propulsion. Tests conducted in 2021 successfully achieved hypersonic speeds exceeding Mach 5.

  - **Key Achievements:** The successful HAWC flight demonstrated sustained scramjet-powered hypersonic flight, collecting critical data on propulsion, thermal management, and aerodynamics.

  - **Impact:** HAWC is poised to play a major role in developing next-generation hypersonic cruise missiles for military use.

Failed Hypersonic Tests:

- **NASA X-43A Scramjet (2001)**

  - **Overview:** The X-43A was an experimental hypersonic aircraft designed to test scramjet propulsion. The first flight test in 2001 aimed to reach Mach 7.

  - **Failure:** During the first test, the booster rocket that was supposed to accelerate the X-43A lost control and broke apart mid-flight, causing the mission to fail before the scramjet could be tested.

- **Lessons Learned:** NASA identified the need for more precise control of the booster phase in hypersonic test flights. Subsequent tests in 2004 were successful, with the X-43A achieving Mach 9.6.
- **Impact:** Despite the initial failure, the X-43A's eventual success contributed significantly to scramjet research.

- **HTV-2 Falcon (DARPA, 2010-2011)**
  - **Overview:** The HTV-2 (Hypersonic Technology Vehicle 2) was part of DARPA's effort to develop a hypersonic glide vehicle capable of traveling at Mach 20. It was designed to test the feasibility of global strike capabilities using hypersonic vehicles.
  - **Failure:** Both test flights (in 2010 and 2011) ended prematurely. In the first flight, telemetry was lost after nine minutes of flight, and in the second, the vehicle self-destructed due to aerodynamic instability after a few minutes of successful hypersonic flight.
  - **Lessons Learned:** The HTV-2 tests revealed significant challenges related to vehicle stability and thermal management at extreme hypersonic speeds. DARPA acknowledged that controlling the vehicle under sustained hypersonic conditions was more difficult than anticipated.
  - **Impact:** While the HTV-2 did not achieve all of its objectives, the program provided critical data for the continued development of hypersonic glide vehicles.
- **Yu-71 (Russia, 2015)**
  - **Overview:** The Yu-71, part of Russia's hypersonic glide vehicle research, was tested as a potential nuclear delivery system capable of reaching Mach 10. The system was launched atop an ICBM and entered the hypersonic glide phase.
  - **Failure:** Several early test flights were reported as unsuccessful, with the vehicle failing to reach its designated target or achieve controlled flight. Issues with navigation and thermal management were cited as reasons for the failures.

- **Lessons Learned:** The failures highlighted the difficulty in transitioning from the boost phase to controlled hypersonic glide, as well as challenges in material durability under extreme heat and aerodynamic forces.
- **Impact:** Despite the setbacks, Russia continued to improve its hypersonic programs, eventually leading to the successful deployment of the Avangard system.

Part III - Military and Strategic Implications

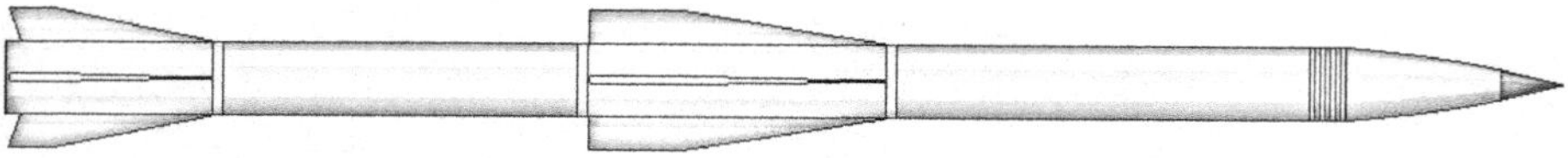

Modern Hypersonic Weapon Programs

Offensive Capabilities and Strike Precision

Hypersonic missiles, with the ability to deliver either conventional or nuclear warheads, represent a transformative leap in military technology. These weapons are capable of long-range strikes against highly valuable military targets, such as headquarters, airfields, or even large naval assets like aircraft carriers. Additionally, they can be used as terror weapons against civilian population centres due to their sheer speed and destructiveness. The development of hypersonic missiles has introduced significant new challenges for defence systems worldwide, especially given their speed, manoeuvrability, and ability to evade conventional missile defences.

One of the key challenges posed by hypersonic missiles is their extreme speed. Traveling at speeds greater than Mach 5 (five times the speed of sound), these weapons significantly reduce the time available for detection, identification, tracking, and engagement. This speed compresses the decision-making window for defensive systems, making it difficult for current sensor and missile defence systems to react in time. Traditional missile defence systems, which were designed to handle slower, more predictable threats, struggle to keep pace with the rapid movement and manoeuvrability of hypersonic missiles.

Furthermore, hypersonic missiles are extremely difficult to intercept due to their unpredictable flight trajectories. Unlike ballistic missiles, which follow a more or less predictable parabolic arc, hypersonic missiles can change direction mid-flight, making it much harder to predict their impact points. This unpredictability requires defensive systems to have 360-degree coverage to ensure comprehensive protection. Current missile defence architectures, which often focus on specific high-value areas, may need to be reimagined to provide persistent and omnidirectional surveillance capabilities. Once a hypersonic missile begins to manoeuvre, tracking it becomes significantly more challenging, further complicating interception efforts.

Because hypersonic missiles can travel vast distances across multiple domains—through air, space, and across oceans—defensive measures against them must also be multi-national and multi-domain in nature. This means that protecting against hypersonic threats cannot be the responsibility of a single nation or system. Instead, it requires a collaborative approach involving multiple countries and layers of defence. Defence networks would need to integrate air, space, and sea-based sensors and interceptors, sharing real-time intelligence to maintain tracking on these fast-moving targets across their entire trajectory.

Another major challenge is that today's missile interceptors are ill-suited for engaging hypersonic threats, particularly during the glide phase. Hypersonic missiles often cruise at altitudes between 30 and 60 kilometres (in the stratosphere), which falls outside the effective range of both exo-atmospheric and terminal-phase endo-atmospheric interceptors. While terminal-phase intercept is technically possible when the missile descends toward its target, this method provides only limited territorial protection. The terminal-phase interception, by its nature, only defends a small area, leaving vast areas vulnerable. Additionally, the number of interceptors currently available is insufficient to provide adequate protection for all potential targets, particularly in light of the high speeds and manoeuvrability of hypersonic missiles.

In summary, the emergence of hypersonic missile technology has introduced several critical challenges to global defence systems. The extreme speed and manoeuvrability of these weapons reduce response times and make tracking and interception far more difficult. Furthermore, existing missile defence systems are not equipped to handle the altitudes and trajectories at which hypersonic missiles operate, requiring new and innovative approaches to ensure protection. The defence against these weapons will require multinational collaboration, comprehensive 360-degree coverage, and the development of new technologies capable of addressing this growing threat.

Stealth and Low Observability

Stealth and low observability are critical factors in the effectiveness of hypersonic weapons. While their sheer speed allows them to evade or outrun many radar and detection systems, hypersonic weapons must also contend with radar and heat signatures that could potentially compromise their stealth. The unique design features and materials used in hypersonic vehicles help minimize these detectable signatures, enhancing their survivability and making them much harder to intercept by conventional defence systems. This combination of speed and stealth is a significant contributor to the overall strategic value of hypersonic weapons in modern warfare.

One of the main ways hypersonic vehicles achieve radar cross-section reduction is through the use of radar-absorbing materials (RAM) and specialized aerodynamic shapes. Radar cross-section (RCS) refers to how detectable an object is by radar; the larger the RCS, the easier it is for radar systems to detect and track the object. Hypersonic vehicles are designed to minimize their RCS, making them appear smaller and harder to detect by enemy radars. Radar-absorbing materials, such as specialized coatings or composites, are commonly used to absorb or deflect radar waves, reducing the amount of radar energy that bounces back to the detection system. This technique, often referred to as stealth technology, is similar to the methods used in stealth aircraft, but hypersonic weapons require even more advanced materials due to the extreme heat and stress generated by high-speed flight.

In addition to radar-absorbing materials, the aerodynamic shape of hypersonic vehicles plays a key role in reducing radar detection. Hypersonic weapons often feature sleek, angular designs with minimal protrusions, which help scatter radar waves in multiple directions rather than reflecting them directly back to the radar source. This minimizes the chances of the radar detecting the vehicle. For example, the smooth, tapered design of hypersonic glide vehicles (HGVs) allows them to travel at high altitudes while presenting a low radar signature, making them particularly difficult for traditional radar systems to track. The ability to reduce radar cross-section through both material and design is essential for maintaining the stealth of hypersonic weapons, especially when operating in hostile environments with advanced detection capabilities.

However, despite efforts to reduce radar signatures, the heat generated by hypersonic vehicles during flight poses a significant challenge to their stealth capabilities. As a vehicle moves at hypersonic speeds, it experiences extreme friction with the air, which generates enormous amounts of heat. This heat creates an infrared (IR) signature that can be detected by heat-seeking missiles or sensors. Managing this heat is crucial for maintaining low observability, as a high IR signature can expose the vehicle's location to enemy tracking systems.

To address this issue, hypersonic weapons incorporate advanced heat management techniques to reduce their infrared signature. One approach is to distribute the heat across the vehicle's surface more evenly, preventing localized hot spots that could increase detectability. By using aerodynamic designs that spread the heat load over larger surfaces, hypersonic vehicles can avoid concentrated areas of high temperature that would be easily detected by infrared sensors. Additionally, specialized coatings can be applied to the exterior of the vehicle to reflect or absorb heat, reducing the amount of infrared radiation emitted. These coatings are designed to withstand the extreme temperatures generated during hypersonic flight while minimizing the vehicle's thermal signature.

In some cases, hypersonic vehicles use insulated materials or cooling systems to further manage heat. Insulated materials, such as ceramic composites, can help shield sensitive components from extreme heat while also preventing the vehicle from radiating excessive heat. Some hypersonic designs may incorporate active cooling systems that use fuel or other coolants to manage heat buildup, although these systems are more complex and less common due to the constraints of weight and space. Regardless of the specific method, the goal is to reduce the vehicle's infrared signature, making it harder for heat-seeking systems to track and target the weapon.

In combination, the radar cross-section reduction and heat management techniques employed by hypersonic weapons make them highly stealthy and difficult to detect. Traditional missile defence systems, which rely on radar and infrared sensors to detect incoming threats, struggle to track hypersonic vehicles due to their minimized signatures and incredible speed. By the time a radar or infrared sensor detects the weapon, it may already be too late to intercept it due to the rapid pace at which it approaches its target.

The ability to maintain low observability is a defining feature of hypersonic weapons, contributing to their strategic superiority on the battlefield. Their stealth capabilities, combined with their speed and manoeuvrability, allow them to evade detection systems and strike targets with little to no warning. This makes hypersonic weapons highly effective tools for modern military operations, where avoiding detection and maintaining the element of surprise are critical for success. As technology continues to evolve, further advancements in radar-absorbing materials and heat management systems will likely enhance the stealth capabilities of future hypersonic weapons, solidifying their role as a key element of next-generation warfare.

Payload Capacity

Payload capacity is a crucial aspect of hypersonic weapons, determining their versatility and the types of missions they can execute. Hypersonic weapons can be equipped with a wide variety of payloads, ranging from conventional explosives for precision strikes to nuclear warheads for strategic deterrence, and even penetrating warheads for destroying fortified targets. The type of payload carried by a hypersonic weapon is closely tied to the mission objectives and the specific design of the weapon system. Hypersonic vehicles, by virtue of their extreme speed and manoeuvrability, are capable of delivering these payloads with precision and efficacy, making them formidable tools in modern military arsenals.

One of the most common payload types for hypersonic weapons is conventional warheads. These are designed to carry high-explosive charges that can be used for precision strikes against critical military assets or infrastructure. Conventional warheads are particularly effective when deployed on hypersonic weapons due to the extreme speed and impact force with which the weapon hits its target. This combination of speed and precision allows hypersonic missiles to strike high-value targets, such as aircraft carriers, missile defence systems, or critical infrastructure like power plants or communication networks, before the enemy has time to respond or reposition their assets.

The speed of hypersonic weapons enhances the lethality of conventional warheads by increasing the kinetic energy upon impact, allowing for more destructive power without the need for large quantities of explosives. Even with relatively smaller explosive charges, the velocity at which the weapon strikes the target amplifies the force of the explosion, making hypersonic missiles highly effective against time-sensitive or mobile targets. Furthermore, the ability of hypersonic weapons to evade traditional radar and missile defence systems ensures that conventional strikes can be carried out with minimal risk of interception, increasing their reliability and effectiveness in modern combat scenarios.

In addition to conventional warheads, some hypersonic weapons, particularly those developed by nations like Russia and China, are equipped with nuclear warheads. The integration of nuclear payloads into hypersonic systems significantly alters the strategic calculus, as the combination of hypersonic speed and nuclear capability creates a formidable deterrent. Hypersonic weapons armed with nuclear warheads have the potential to bypass existing missile defence systems due to their speed and manoeuvrability, making them extremely difficult to track and intercept. This ability to penetrate even the most advanced defence shields enhances their effectiveness as a deterrent, as it reduces the likelihood that an adversary could successfully defend against a nuclear strike.

The deployment of nuclear-armed hypersonic weapons also raises the stakes in terms of global security and arms control. The speed with which these weapons can be launched, travel, and reach their target creates an environment where reaction times are dramatically shortened, potentially escalating conflicts before diplomatic or defensive measures can be implemented. Countries with hypersonic nuclear capabilities, such as Russia's Avangard system, are able to project significant nuclear deterrence over vast distances, as these systems can carry nuclear warheads across continents while evading detection. This capability fundamentally alters the balance of power in global military strategy, as it complicates efforts to maintain stability through mutual defence systems and missile shields.

Another type of payload designed for hypersonic weapons is the penetrating warhead, which is specifically engineered to destroy hardened or fortified targets, such as underground bunkers, command centres, or deeply buried military installations. Penetrating warheads take advantage of the high velocity of hypersonic weapons to achieve deep penetration before detonating. As the hypersonic missile strikes its target, the sheer kinetic energy allows the warhead to burrow into the target's structure, bypassing layers of fortification that would typically protect against traditional munitions.

These penetrating warheads are often designed with reinforced casings to withstand the initial impact and penetrate deeply before triggering the explosive charge. This makes them particularly effective against hardened bunkers or underground facilities that are otherwise impervious to standard bombs or missiles. The ability of hypersonic weapons to reach such targets at high speed, combined with the precision offered by modern guidance systems, makes them ideal for neutralizing command and control centres, nuclear storage facilities, or other critical underground infrastructure that are essential to an adversary's operational capability.

In addition to their utility against hardened targets, penetrating warheads also have applications in anti-ship missions, where they can be used to strike large naval vessels with precision. When equipped on hypersonic anti-ship missiles, penetrating warheads can pierce through the outer armour of ships, reaching vital internal structures before detonating. This allows hypersonic missiles to effectively target heavily armoured vessels, such as aircraft carriers or destroyers, providing a powerful tool for naval dominance and anti-access/area denial (A2/AD) strategies.

The payload capacity of hypersonic weapons adds to their strategic flexibility and makes them invaluable in a wide range of military operations. Whether armed with conventional explosives for high-precision strikes on valuable targets, nuclear warheads for strategic deterrence and global power projection, or penetrating warheads for destroying fortified or armoured structures, hypersonic weapons offer unmatched versatility and lethality. The combination of their extreme speed, precision, and ability to evade detection makes them a revolutionary force in modern warfare, capable of altering the balance of power and providing significant tactical and strategic advantages to the nations that possess them.

Offensive Capabilities

Hypersonic missiles represent a significant advancement in military technology, characterized by their ability to travel at speeds exceeding Mach 5, which allows for rapid strikes against time-sensitive targets. The exceptional speed of these missiles drastically reduces the reaction time available to adversaries, making them particularly effective for neutralizing high-value or mobile targets before they can relocate or activate defensive measures. This capability is critical in military operations where swift elimination of strategic assets, such as command centres and missile launchers, is essential for gaining a tactical advantage [2, 314].

In addition to their speed, hypersonic missiles possess extended range and global reach, particularly those utilizing a boost-glide mechanism. This method involves launching the missile to high altitudes with a rocket before it glides through the atmosphere at hypersonic speeds. Such capabilities enable these missiles to strike targets across intercontinental distances without the necessity for forward-deployed assets, thereby enhancing their strategic utility in long-range precision strike operations [15, 21]. The combination of speed and range allows for effective engagement of targets that may be geographically constrained or heavily defended, further solidifying the operational advantages of hypersonic systems [2, 31].

Manoeuvrability is another defining feature of hypersonic missiles, setting them apart from traditional ballistic missiles, which follow predictable trajectories. Hypersonic vehicles, including boost-glide and scramjet-powered missiles, can alter their flight paths mid-course, making them exceedingly difficult to track and intercept by existing missile defence systems [315]. This enhanced manoeuvrability allows hypersonic missiles to evade detection and countermeasures, significantly increasing their likelihood of successful engagement against well-defended targets [2, 314]. The unpredictable flight paths of these missiles challenge current defence strategies, necessitating a re-evaluation of how military forces protect critical infrastructure from such advanced threats [2, 314].

The precision of hypersonic missiles is further augmented by advanced guidance systems, including GPS and inertial navigation systems, which enable accurate targeting even at extreme speeds. Some systems incorporate real-time data from sensors or satellites, allowing for mid-flight course adjustments to ensure strikes are executed within meters of intended targets [314, 315]. This level of precision not only enhances operational effectiveness but also minimizes collateral damage, making hypersonic missiles particularly valuable for precision strikes against strategic military infrastructure [314].

Moreover, the kinetic energy generated by hypersonic missiles upon impact is substantial, providing a unique advantage in terms of destructive capability. The high velocity of these missiles means that their kinetic energy alone can inflict significant damage, even without explosive payloads. This characteristic is particularly effective against hardened targets such as bunkers and missile silos, allowing hypersonic missiles to penetrate deeper and deliver more substantial damage compared to conventional munitions of similar size [2, 314].

Hypersonic missiles also exhibit versatility in warhead deployment, capable of carrying various types of warheads, including conventional explosives, thermobaric warheads, or nuclear payloads. This flexibility enables military planners to tailor their strike options according to mission objectives, ranging from conventional precision strikes to nuclear deterrence capabilities [2, 314]. The potential to deliver nuclear payloads with such speed and precision introduces a new dimension to nuclear strategy, as these missiles could potentially bypass existing anti-ballistic missile systems, complicating deterrence dynamics [2, 314].

Finally, the ability of hypersonic missiles to overcome existing missile defence systems is one of their most significant advantages. Traditional missile defences are primarily designed to counter predictable threats, whereas hypersonic missiles travel at unprecedented speeds and can manoeuvre during flight, making them exceedingly difficult to detect and intercept [2, 314]. This capability compels military strategists to reconsider defence postures and develop new countermeasures to address the evolving threat posed by hypersonic weapons [2, 314].

Defensive Countermeasures Against Hypersonics

The manoeuvrability, altitude, and speed of hypersonic missiles present significant challenges for existing missile defence systems, which were primarily designed to counter ballistic missiles. Unlike ballistic missiles that follow predictable, parabolic flight paths, hypersonic missiles can manoeuvre unpredictably during their flight and travel at lower altitudes, making them harder to detect and intercept. These challenges lead to later detection, giving defenders less time to make decisions and respond. Traditional missile defence systems may struggle to keep up with the complexity and speed of hypersonic threats, requiring significant upgrades and adaptations to remain effective [1].

Missile defence systems, like those used by the UK, US, and other nations, are designed with the primary goal of deterring attacks and protecting critical assets from missile threats. These systems are made up of three key components: detection, threat validation, and interception [1].

1. Detection: This involves a network of sensors that identify and track incoming missiles. Detection systems typically include satellites equipped with infrared scanning capabilities to detect the heat signatures from a missile's rocket exhaust and surface-based radar systems located on land or naval vessels. These sensors work together to provide early warnings of a missile launch, allowing defensive systems to begin tracking the missile's flight.

2. Threat validation: Once a missile is detected, data from the sensors are sent to a command and control centre, where analysts determine whether the threat is real and decide on an appropriate response. During this phase, the missile's trajectory is calculated, and the information is relayed to the appropriate interceptor system to prepare for engagement.

3. Interception: The final stage involves interceptor missiles, which are launched to collide with the incoming missile and destroy it before it can reach its target. Interceptors can target different phases of a missile's flight path, either in its boost, midcourse, or terminal phases. However, this process is technically challenging and expensive, especially when dealing with long-range missiles like intercontinental ballistic missiles (ICBMs), which require multiple interceptors and highly accurate targeting systems.

For example, the US Ground-based Midcourse Defense (GMD) system, designed to counter ICBMs, has cost approximately $53 billion over 25 years, with plans for an additional $10 billion in spending from 2020 to 2025. Despite this investment, its interception success rate has been around 55%, and some analysts believe that GMD would be inadequate to counter a large-scale attack involving multiple missiles launched simultaneously [1].

The speed, altitude, and manoeuvrability of hypersonic missiles introduce new challenges to existing missile defence systems. These systems are designed for intercepting ballistic missiles, which are easier to predict due to their fixed trajectories and higher altitudes. Hypersonic missiles, however, fly at lower altitudes within the atmosphere and have the ability to manoeuvre throughout their flight, making them more difficult to detect and track. Surface-based radars, which many missile defence systems rely on, may not detect a hypersonic missile until late in its flight, giving defenders much less time to respond [1].

One of the biggest challenges is the compressed decision-making timeline. When hypersonic missiles are detected late, defenders have limited time to analyse the threat, calculate the missile's trajectory, and launch interceptors. Additionally, hypersonic missiles can perform evasive manoeuvres in the final moments of their flight, potentially evading interception entirely. Even if an interceptor is able to destroy the missile, if the interception happens too close to the target, the resulting debris could still cause significant damage [1].

To effectively counter hypersonic missiles, existing defence systems will need to be upgraded to increase warning times and improve interception capabilities. One potential solution is to enhance detection systems by integrating space-based sensors, such as infrared satellites. These satellites could detect the surface heating of hypersonic missiles as they travel through the atmosphere, providing earlier warnings compared to surface-based radar alone. Increasing the detection range would give defenders more time to prepare and launch interceptors, reducing the likelihood of late interception [1].

Additionally, hypersonic missiles travel primarily within the atmosphere, which limits the types of interceptors that can be used. Traditional interceptors designed to engage ballistic missiles in space are ineffective against hypersonic missiles that operate at lower altitudes. Therefore, interceptors designed to engage within the atmosphere must be selectively located near high-value targets, ensuring sufficient defensive coverage. However, because atmospheric interceptors have shorter ranges, ensuring adequate coverage of all potential targets could require significant investments in additional interceptor sites and infrastructure [1].

Defence contractors are already working with the US and EU to develop novel interceptors capable of flying at hypersonic speeds and engaging both ballistic and hypersonic missiles. These advanced interceptors would need to be able to track and match the speed of a hypersonic missile, while also having the ability to predict and counter the missile's manoeuvres. Such systems would represent a significant leap in missile defence technology, but their development and deployment would likely take years and require substantial financial resources [1].

The development of effective defence systems against hypersonic missiles will be technologically challenging and costly. Nations will need to invest in upgraded sensors, new interceptors, and more sophisticated tracking systems. As defence systems evolve to counter hypersonic threats, they will also need to adapt to the possibility of multiple, simultaneous attacks, as hypersonic technology continues to advance [1].

In conclusion, hypersonic missiles present a unique challenge to traditional missile defence systems due to their unpredictable flight paths, lower altitudes, and high speeds. To address these threats, nations will need to make significant investments in new detection technologies, faster interceptors, and enhanced command-and-control systems. While advancements are being made, the development and deployment of reliable counter-hypersonic systems will take time, and in the interim, hypersonic missiles could remain a disruptive force in global military strategy [1].

Countering Hypersonic Weapons

Hypersonic weapons, including cruise missiles, glide vehicles, and boost-glide systems, travel at speeds of Mach 5 or higher, making them extraordinarily difficult to intercept. Their speed, coupled with in-flight manoeuvrability and low-altitude flight paths, complicates tracking and interception, making them more elusive than traditional ballistic missiles. As a result, defence research has shifted toward developing countermeasures capable of neutralizing these advanced weapons. Counter-hypersonic strategies include the development of high-power lasers, microwave weapons, railguns, hypersonic interceptors, and cyber-based "soft-kill" weapons that can target critical subsystems. Each of these technologies presents unique advantages in defending against the new threat posed by hypersonic vehicles.

High-power lasers are being researched as one of the most promising counter-hypersonic measures. These lasers could potentially disable a hypersonic weapon by targeting and disrupting the boundary layer of high-speed air that forms around the vehicle. By heating and destabilizing this air layer, the hypersonic missile's aerodynamics could be compromised, causing a loss of control or outright destruction. In addition to physically damaging the vehicle, lasers could also be used to blind or "dazzle" its navigation and targeting sensors, making it more difficult for the missile to stay on course or locate its intended target.

Tracking and navigating these weapons is another significant challenge due to their high speed and the lower altitudes they typically travel at, often below the radar horizons of traditional defence systems. Satellite imaging, using optical, microwave, or radar sensors, becomes crucial for early detection and ongoing tracking of hypersonic weapons in flight. Navigation systems, particularly Global Navigation Satellite Systems (GNSS), are also critical, and this reliance opens up another countermeasure option: "spoofing." By sending false GNSS signals, defence systems can mislead a hypersonic weapon, tricking it into thinking it is on course while leading it off-target.

Artificial intelligence (AI) plays an essential role in emerging hypersonic defence systems. AI can assist in a range of tasks, including the detection of hypersonic weapon launches, simulating and recommending optimal countermeasures, calculating the best intercept paths, and guiding the countermeasures to their targets. The speed at which hypersonic weapons travel leaves little time for human decision-making, and AI-enabled systems help to automate threat assessment and interception decisions based on real-time data, improving response times.

Cyber-based countermeasures offer another avenue of defence. Targeted cyberattacks, such as Distributed Denial-of-Service (DDoS) attacks, can exploit vulnerabilities in the software controlling the hypersonic vehicle's flight or navigation systems. By confusing, overloading, or shutting down these critical systems, cyberattacks can disable or divert hypersonic weapons mid-flight. Cyberattacks can also be used to disable the infrastructure supporting hypersonic weapon operations, such as land-, sea-, or air-based launch platforms.

Decision support systems are integral to counter-hypersonic defence strategies. Given the limited time available to respond to hypersonic threats, these systems help centralize and automate decision-making processes. They combine pre-existing information with real-time data to assess the threat and rapidly choose the most appropriate countermeasure. This capability is critical, as hypersonic threats require quick and precise decisions to deploy effective defences.

In summary, countering hypersonic weapons requires a multifaceted approach that leverages cutting-edge technology across various domains. High-power lasers, satellite tracking, AI, cyber-based attacks, and advanced decision support systems all contribute to a comprehensive defence strategy. These technologies, combined with international collaboration and real-time surveillance, are crucial in defending against the evolving hypersonic threat.

Early Warning Systems

The integration of space- and stratospheric-based early warning systems into missile defence architectures is critical for addressing the evolving threats posed by hypersonic missiles and other advanced weaponry. Such systems enhance the capability to detect, track, and characterize missile launches in real-time, thereby providing essential intelligence for national and allied defence strategies.

Space-based early warning radars (SBEWR) are particularly advantageous due to their ability to cover extensive areas and provide early detection of hypersonic threats. Research indicates that SBEWR in low Earth orbit can achieve greater detection ranges and longer warning times, which is crucial for intercepting fast-moving targets with small radar cross sections [316]. This capability allows for the monitoring of missile launches from various platforms, including ground, sea, and air, thereby enhancing situational awareness and response times for defence forces.

Moreover, the interoperability of these systems with NATO and other allied networks is essential for a cohesive defence strategy. The integration of space and stratospheric sensors with ground-based systems can facilitate a comprehensive approach to missile defence, allowing for the sharing of data and coordinated responses to threats [317]. This interoperability is vital for ensuring that early warning systems can effectively communicate and operate within a broader missile defence framework, thereby enhancing deterrence and defence capabilities against potential adversaries.

In addition to space-based systems, surface-based early warning radars play a crucial role in a comprehensive missile defence architecture. These radars are capable of detecting and tracking threats at long ranges and can be integrated into a command and control (C2) network that includes both national and allied systems [317]. The combination of surface-based and space-based sensors allows for persistent tracking of missile trajectories, even as hypersonic missiles manoeuvre, thereby improving the accuracy of threat assessments and impact predictions.

The challenges posed by hypersonic missiles necessitate a multi-layered defence approach that incorporates both space and surface-based early warning systems. As hypersonic threats continue to evolve, the integration of advanced detection technologies and collaborative defence strategies will be paramount in ensuring national and allied security [316]. Continuous advancements in radar technology and sensor capabilities will further enhance the effectiveness of early warning systems, enabling timely and informed responses to emerging threats.

Command and Control

The existing Ballistic Missile Defense (BMD) systems, such as those deployed by Thales Raytheon Systems at the NATO Ballistic Missile Defence Operations Center (BMDOC) in Ramstein since 2016, face significant challenges in countering the emerging threat of hypersonic missiles. These hypersonic weapons are characterized by their high speeds, manoeuvrability, and ability to operate at lower altitudes, which allows them to evade traditional BMD systems that rely on established detection and tracking methodologies [5, 50]. The limitations of current BMD systems stem from their reliance on conventional radar and infrared sensors, which are often inadequate for tracking the unpredictable flight paths of hypersonic missiles [56]. As hypersonic missiles can fly below the radar horizon, the architecture of sensor networks must evolve to incorporate denser and more interconnected systems capable of detecting and tracking these threats effectively [5, 50].

The complexity of hypersonic missile threats is further compounded by their potential launch platforms, which include submarines, ships, and aircraft positioned closer to targets. This proximity reduces the time available for interception, often to mere minutes, and complicates the attribution of responsibility for attacks, thereby increasing strategic uncertainty [5]. The rapid flight of hypersonic missiles necessitates a rethinking of command

and control (C2) systems, which must adapt to manage launch platforms effectively and ensure they maintain a safe distance from high-value assets (HVAs) [5]. The tactical response to hypersonic threats requires minimizing human intervention in decision-making processes, leading to an increased reliance on artificial intelligence (AI) and network-based capabilities to enhance situational awareness and expedite responses [5].

An effective architecture for countering hypersonic threats would involve a multi-layered defence system, integrating various sensor types, including radars and space-based infrared sensors, with a robust C2 system [56]. This integration is crucial for maintaining seamless communication across all levels of defence, from strategic to tactical, particularly as reaction times decrease from minutes to seconds [56]. The adaptation of the C2 Integrated Air and Missile Defense (IAMD) system is essential, requiring enhancements in surveillance systems to include multi-sensor trackers capable of handling the high speeds and manoeuvrability of hypersonic targets [56]. Moreover, missile engagement and fire control systems must facilitate collaborative engagement functions to improve response coordination [56].

The role of the system orchestrator becomes critical in managing the allocation of capabilities and authority across the distributed C2 architecture, ensuring effective operation of all nodes within the system [56]. By integrating real-time sensor data and incorporating new sensors specifically designed for high-speed and highly manoeuvrable targets, the missile defence system can enhance its ability to monitor hypersonic threats from launch to impact [56]. Enhanced detection capabilities will allow for the characterization of hypersonic missile trajectories and the discrimination between different types of targets, which is vital for assessing adversarial intent and determining appropriate defensive or offensive actions [5].

Detection, Identification and Tracking

The role of sensors in hypersonic missile defence is pivotal for early warning, persistent tracking, classification, and identification of threats. These sensors are engineered to detect incoming missiles, classify their types, and ascertain their hostility. The primary objective of these sensors is to facilitate subsequent defensive actions, such as engaging the missile with available weaponry and activating the fire-control system for interception. Depending on the area that needs protection and the capabilities of both the hypersonic missile and the defence interceptors, these tasks can be executed by a single sensor or a network of multiple sensors, which may vary in type. The accuracy of these sensors is crucial, as the interception windows are typically narrow, necessitating rapid and precise responses to threats [39, 42].

Interception of hypersonic missiles presents significant challenges due to their high speeds and manoeuvrability. While interception may be feasible during the missile's boost phase, this often exceeds the range of interceptors. The exo-atmospheric phase may also prove too distant for effective engagement. The glide phase, which occurs post-re-entry for glide vehicles or during the terminal phase for certain hypersonic cruise missiles, is often identified as the most viable opportunity for interception. During this phase, the missile's speed typically decreases just before impact, providing another potential interception opportunity. Successful interception hinges on accurately estimating the hit point, which requires precise trajectory tracking of the threat. The manoeuvrability of hypersonic missiles complicates this process, as premature launches of interceptor missiles can lead to failure due to the unpredictable flight paths of the missiles [13, 73, 318].

The time available for early detection and identification of a missile is critically short, often limited to mere seconds. Identification relies on a comprehensive assessment matrix that evaluates various characteristics, including the missile's origin, radar signature, shape, velocity, and flight trajectory. At the command and control (C2) level, these features are analysed against pre-established threat models to initiate a pre-planned response. The main challenges in this process include the need for early detection to allow sufficient time for identification and reaction, overcoming detection limitations posed by the missile's low altitude, and ensuring accurate data for classification. Furthermore, tracking missiles over vast distances, often crossing international borders, necessitates collaboration among nations to maintain a consistent and precise track on the threat [319, 320].

Detection, tracking, classification, and identification of hypersonic threats depend on a diverse network of sensor types, each fulfilling specific roles. Radar and infrared sensors are typically employed to detect missiles as they emerge above the horizon, generally at distances of 600 to 1,200 kilometres for targets at cruise altitudes between 20 to 80 kilometres. Dedicated or volunteered radars are then utilized for persistent tracking across the operational theatre, providing the necessary range, velocity, and elevation coverage to manage the aggressive manoeuvres of hypersonic missiles. These sensors also classify missiles into categories such as hypersonic glide vehicles (HGVs) or hypersonic cruise missiles (HCMs), and determine whether the object poses a threat based on sensor data and intelligence from a threat library [5, 26].

Once a threat is identified, sensors contribute valuable information for intelligence gathering, passive defence measures (such as alerting populations and preparing for impact), counteraction (targeting the launch site), and active defence (engaging the threat). The sensors facilitate the designation of the missile for fire-control radar systems, a process known as Threat Evaluation and Weapon Assignment (TEWA). Networking heterogeneous sensors into a smart grid enhances the operational capabilities of missile defence systems. This sensor grid allows for persistent tracking across the battlespace, overcoming horizon limitations. Dynamic sensor tasking, data fusion, and real-time information distribution across borders provide a comprehensive view of the air, ballistic, and hypersonic threat environment. Multiple sensors with overlapping coverage collaborate to ensure continuous tracking, even if one sensor loses track, and the grid can adapt to sensor jamming by adjusting multi-frequency coverage and utilizing unjammed sensors to maintain tracking and classification [5, 39, 321].

Interception Strategies

Intercepting hypersonic missiles presents significant challenges compared to traditional ballistic missiles due to their extraordinary speed, manoeuvrability, and flight profiles. Various interception strategies are currently under development or deployment to counter these advanced threats. These strategies can be categorized into three primary phases: boost-phase interception, midcourse interception, and terminal-phase interception.

Boost-Phase Interception occurs immediately after a missile's launch when it is still ascending. This phase is characterized by relatively lower speeds and a more predictable trajectory, making it an ideal time for interception. However, the logistical challenge arises from the distance between the missile launch point and interceptor systems, which may be far apart, complicating timely engagement [2, 322]. The effectiveness of this strategy relies on rapid detection and response capabilities, as well as the ability to deploy interceptors quickly to the launch area [322].

Midcourse Interception, or glide phase interception, occurs after the boost phase when hypersonic missiles can travel vast distances at high speeds. During this phase, the missiles often exhibit advanced manoeuvrability, making interception more complex. Interceptor systems must be equipped with high-speed tracking capabilities and advanced guidance systems to predict the missile's trajectory accurately [278, 323]. The development of hypersonic interceptors, which can match the speed and manoeuvrability of the incoming threat, is crucial for effective midcourse interception [324, 325].

Terminal-Phase Interception takes place just before impact when the missile is typically slowing down. While this phase offers a final opportunity for interception, the time available is extremely limited, and the missile may be too close to its target for effective engagement. Interceptor systems must possess fast reaction times and high precision to succeed in this critical phase [19, 280]. The challenges of terminal-phase interception are compounded by the need for advanced guidance laws that can adapt to the rapid changes in the missile's trajectory as it approaches its target [73].

In terms of interceptor technologies, several systems are being developed to counter hypersonic threats effectively. Hypersonic interceptors are designed to chase and intercept incoming hypersonic missiles, requiring advanced guidance systems capable of predicting the missile's trajectory while matching its speed and manoeuvrability [69, 318]. Directed energy weapons, such as high-power lasers, represent another promising approach, as they can disable hypersonic missiles by disrupting their aerodynamic flow or blinding onboard sensors [2, 60]. However, the challenge remains in maintaining a focused energy beam on a fast-moving target for sufficient duration [2].

Railguns utilize electromagnetic forces to launch projectiles at extremely high speeds, theoretically allowing them to intercept hypersonic missiles effectively. These systems offer the advantage of low-cost, high-speed projectiles but face challenges related to power requirements and accuracy at high speeds [2, 60]. Additionally, traditional missile defence systems, such as THAAD and Aegis, are being adapted to engage hypersonic threats, although they may require significant upgrades to be effective against these new types of missiles [2, 322].

Lastly, electronic and cyber warfare strategies, including jamming and spoofing, provide a soft-kill approach to counter hypersonic missiles. By disrupting the missile's guidance and communication systems, these methods can mislead the missile or render it inoperable [2, 326]. Cyberattacks targeting the missile's control systems or the infrastructure supporting its launch can further enhance defence capabilities against hypersonic threats [2, 326].

Strategic Advantages in Modern Warfare

The strategic impact of hypersonic weapons in modern warfare is profound, driven by their unique technical characteristics of extreme speed, manoeuvrability, and ability to evade traditional defence systems. These attributes give hypersonic weapons a distinct role in shaping military strategy, transforming how nations plan for both offensive and defensive operations. The deployment of hypersonic technologies presents significant advantages in terms of first-strike capabilities, anti-access/area denial (A2/AD) strategies, and survivability, marking them as key tools in the arsenals of technologically advanced military forces.

One of the most critical roles that hypersonic weapons play is in first-strike capability. Their speed—traveling at Mach 5 or greater—gives them the ability to reach targets in minutes, reducing the time available for an adversary to detect, track, and respond to an incoming threat. This makes hypersonic weapons ideal for pre-emptive strikes against high-value military assets, such as command centres, missile defence installations, air bases, or strategic infrastructure like communication hubs and power grids. The rapid nature of a hypersonic strike means that it can neutralize enemy defences before they can react, creating a window of strategic advantage. The ability to strike quickly and with precision could enable a military to cripple an adversary's capabilities in the early stages of a conflict, potentially altering the course of a war before it fully escalates.

In addition to their first-strike capabilities, hypersonic weapons are also vital components of Anti-Access/Area Denial (A2/AD) strategies. A2/AD refers to the military strategy of using long-range weapons to prevent or restrict an enemy's ability to operate in specific geographic areas, particularly contested zones such as the South China Sea, the Baltic Sea, or areas near NATO countries. Hypersonic anti-ship missiles, for example, are particularly effective in this role. They can travel at extremely high speeds and execute evasive manoeuvres, making them difficult for naval forces to detect and intercept. As a result, hypersonic weapons can be deployed to deter or destroy enemy naval assets, such as aircraft carriers or destroyers, preventing them from operating freely in contested waters. This form of deterrence is highly effective because it forces adversaries to reconsider entering or maintaining a presence in a particular region, thus creating a form of strategic dominance without necessarily engaging in full-scale conflict.

Hypersonic weapons also offer unparalleled survivability and evasion compared to traditional missile systems. One of the core features of hypersonic weapons, particularly Hypersonic Glide Vehicles (HGVs) and Hypersonic Cruise Missiles (HCMs), is their ability to manoeuvre unpredictably during flight. Unlike ballistic missiles, which follow a predictable arc, hypersonic weapons can change direction and altitude mid-flight. This makes them extremely difficult to intercept using current missile defence technologies, which rely on predicting the trajectory of incoming threats to engage and destroy them. Hypersonic weapons can evade these systems through rapid, unpredictable movements, leaving little to no time for interception. Additionally, their speed makes it difficult for radar and other tracking systems to maintain continuous observation. Even if detected, their flight path can change so quickly that defence systems often cannot respond in time, making defensive countermeasures largely ineffective against hypersonic threats.

The survivability of hypersonic weapons is further enhanced by their ability to operate at both high and low altitudes. Hypersonic Glide Vehicles travel at high altitudes, often in the upper atmosphere, where they can avoid much of the Earth's curvature and remain out of range of many ground-based radar systems. At these altitudes, the weapons can cover long distances, which expands their operational reach and makes them harder to detect by standard tracking systems. Hypersonic Cruise Missiles, on the other hand, fly at much lower altitudes, where they can exploit terrain features and the curvature of the Earth to evade radar. This low-altitude flight path also reduces the time available for adversaries to detect, track, and intercept the missile, enhancing the weapon's survivability during flight. These features make hypersonic weapons particularly useful for penetrating modern missile defence systems, which were primarily designed to intercept slower, less manoeuvrable ballistic missiles.

Potential Uses in Military Operations

The potential uses of hypersonic missiles in military operations are varied and depend on factors such as the missile type (either Hypersonic Glide Vehicles, HGVs, or Hypersonic Cruise Missiles, HCMs), range, speed, warhead type, and launch platform. These variables determine the missile's suitability for different types of missions, and in many cases, the high cost of developing and manufacturing hypersonic missiles might limit the number available for deployment. Therefore, selective use of hypersonic weapons is likely, focusing on specific high-value or time-critical missions where their capabilities provide the greatest advantage [1].

One of the key applications for hypersonic missiles is the ability to rapidly strike mobile assets. Hypersonic missiles, especially short-range versions, have very short flight times, drastically reducing the time it takes to reach their target. This compressed timeline leaves adversaries with a much shorter response window, limiting their ability to move assets or deploy countermeasures in time. For example, hypersonic missiles could be used to target high-value, time-sensitive, or mobile targets such as aircraft carriers, transport vehicles, or command centres. These are often critical assets that play pivotal roles in an adversary's military operations. The speed and manoeuvrability of hypersonic missiles would make it exceedingly difficult for defensive systems to react quickly enough to intercept the missile or for the target to relocate. In this context, hypersonic missiles could provide a decisive advantage, allowing military forces to neutralize key assets before they can be effectively defended [1].

Another important use of hypersonic missiles is for long-range precision strikes. Conventionally armed hypersonic missiles could be deployed to strike distant, well-defended targets, such as military bases, radar stations, or parts of missile defence systems that were previously difficult to reach using traditional missile technology. These precision strikes could be used to disable critical infrastructure, thereby weakening an adversary's overall defensive capabilities. For instance, targeting radar installations or missile defence systems could hinder the enemy's ability to defend against subsequent attacks, opening the way for larger-scale operations involving other types of missiles or airstrikes. The ability to strike from a distance also helps protect the launch platform—whether it's a ship, aircraft, or ground installation—from an adversary's defences, as the hypersonic missile's range allows it to attack without requiring the launch platform to come within range of enemy forces [1].

Hypersonic missiles could also play a significant role in enhancing nuclear deterrence. Nuclear-armed HGVs, in particular, could be designed to bypass an adversary's missile defence systems, strengthening the deterrent effect of possessing such weapons. The high speed and unpredictable flight path of a hypersonic glide vehicle would make it difficult for missile defence systems to intercept, ensuring that the missile reaches its target. In the context of nuclear deterrence, this capability could provide an additional layer of assurance that a retaliatory strike would be successful, potentially deterring an adversary from launching a nuclear attack in the first place. However, the effectiveness of nuclear-armed HGVs in this role, as well as their potential impact on global stability, is a subject of ongoing debate. Some analysts argue that the deployment of hypersonic nuclear weapons could lead to arms race escalation and increase the risk of conflict, while others maintain that their presence would merely reinforce existing deterrent strategies [1].

Despite their potential, some experts suggest that physical limitations on hypersonic missile performance could reduce their advantages. One key issue is that Hypersonic Glide Vehicles (HGVs), in particular, experience air resistance as they glide through the atmosphere, which can cause them to slow down as they approach their target. This reduction in speed could make them more vulnerable to interception by missile defence systems, especially if those systems are equipped to detect and track incoming threats at longer ranges. The fact that HGVs rely on manoeuvrability to evade defences also presents challenges. While the ability to change direction makes them more difficult to intercept, the act of manoeuvring causes the missile to lose speed, further increasing the risk that it might be intercepted before reaching its target [1].

Hypersonic missiles offer a range of military applications, including the ability to rapidly strike high-value, mobile assets, conduct long-range precision strikes on well-defended targets, and potentially enhance nuclear deterrence by bypassing missile defence systems. However, there are still physical and technical challenges to overcome, such as the slowing effects of air resistance on HGVs and the balance between speed and manoeuvrability. These challenges must be addressed for hypersonic missiles to fully realize their potential as game-changing weapons in modern military operations.

Global Hypersonic Arms Race

Key Players: U.S., Russia, China, India, and Europe and National Strategies and Investments

The Hypersonic Weapons Market is experiencing rapid growth and significant investment due to the increasing need for advanced defence systems. In 2023, the market size was valued at USD 7.12 billion, and it is projected to grow at a Compound Annual Growth Rate (CAGR) of over 10% from 2024 to 2032. This growth is driven by strategic military objectives, including the need to deter adversaries, enhance national security, and modernize military arsenals [327].

Hypersonic weapons are becoming a critical asset for nations seeking to maintain a strategic edge, as they offer unparalleled speed (Mach 5 or higher) and manoeuvrability. This makes them difficult to track and intercept, a feature that is particularly important in countering asymmetric warfare tactics and emerging global threats. Countries such as the United States, China, and Russia have led hypersonic development efforts, with programs like the Conventional Prompt Strike (CPS) in the U.S. aimed at delivering rapid strikes anywhere in the world. For instance, in 2023, the U.S. Air Force awarded Raytheon Technologies a USD 985 million contract for the development of hypersonic cruise missiles [327].

Despite the immense potential of hypersonic weapons, their development comes with challenges. These systems require significant financial investment due to the high costs associated with research, development, and testing. The development of propulsion systems, thermal protection, and advanced materials are particularly expensive, creating budget constraints for countries with limited resources. Additionally, ensuring the reliability and effectiveness of these weapons demands extensive testing, further adding to development costs [327].

Key trends in the hypersonic weapons market include the integration of Artificial Intelligence (AI), Machine Learning (ML), and advanced materials. These technologies improve the precision, efficiency, and reliability of hypersonic weapons by enhancing guidance systems, predictive maintenance, and structural durability. For example, AI and ML are being used to optimize flight paths and improve target acquisition in dynamic environments. Meanwhile, materials like carbon composites are crucial for withstanding the extreme conditions of hypersonic flight [327].

International collaboration is another growing trend, with defence contractors, governments, and research institutions partnering to share resources and expertise. This collaboration is crucial for accelerating hypersonic weapons development. Notable examples include the Tactical Boost Glide (TBG) program, which involves partnerships between Lockheed Martin and DARPA [327].

In terms of market segments, the airborne platform is expected to witness significant growth, with a projected CAGR of 10% during the forecast period. Airborne hypersonic weapons, launched from aircraft like bombers or specialized jets, offer greater range, flexibility, and tactical advantages. The Air-Launched Rapid Response Weapon (ARRW) is one such example, designed to strike targets at hypersonic speeds while evading missile defence systems [327].

As the hypersonic weapons market continues to expand, key players like Lockheed Martin, Raytheon Technologies, Northrop Grumman, and Thales Group are at the forefront, driving innovation and securing significant contracts from global defence agencies [327].

The global hypersonic arms race is significantly influenced by the strategic initiatives of the United States, Russia, and China, as these nations heavily invest in the development of hypersonic weapons. Hypersonic technology, characterized by speeds exceeding Mach 5, is perceived as a transformative capability in modern warfare due to its speed, manoeuvrability, and potential to bypass conventional missile defence systems.

The United States is at the forefront of this hypersonic arms race, having allocated nearly $15 billion for hypersonic systems development between fiscal years 2015 and 2024. This investment encompasses both offensive hypersonic missiles and defensive technologies aimed at countering threats posed by such weapons [4]. The U.S. Department of Defense (DoD) collaborates with NASA and the Department of Energy to enhance hypersonic technology further, reflecting the strategic importance attributed to these systems [5]. The U.S. military views hypersonic weapons as essential for maintaining a competitive edge over adversaries, particularly in the context of evolving global threats [35].

The United States has been engaged in the development and testing of hypersonic weapons since the 1950s. However, over the past few decades, U.S. investment in these systems has not kept pace, especially when compared to the efforts of other global powers such as China and Russia. Both countries have made significant advancements in their hypersonic programs. Russia, for instance, deployed six hypersonic Kinzhal missiles in Ukraine earlier this year, marking the largest single strike using hypersonic missiles during the conflict. In addition to China and Russia, several other nations including Australia, Iran, North and South Korea, Brazil, Germany, Israel, India, and Japan are also pursuing the development of hypersonic technology [328].

The recent uptick in U.S. funding for hypersonic weapons development comes amid heightened tensions between the U.S. and China, with bilateral relations reaching their lowest point in decades. According to a January report from the Congressional Budget Office, the Department of Defense has proposed a budget of $15 billion for Army and Air Force hypersonic programs for the years 2023 to 2027. This figure does not include the Navy's hypersonic program, which secured a $1.1 billion contract with defence manufacturer Lockheed Martin in February to integrate a hypersonic system into Zumwalt-class destroyers [328].

For years, U.S. defence officials have expressed concern that the U.S. is lagging behind China in hypersonic weapons development. This sentiment gained further credence in 2021 when China conducted a test of a hypersonic, nuclear-capable weapon. The test caught many within the defence community off guard and highlighted China's rapid progress in the field of hypersonics [328].

Russia has also made substantial advancements in hypersonic weaponry, integrating these systems into its military modernization efforts. Notable developments include the Avangard hypersonic glide vehicle and the Zircon hypersonic cruise missile, both capable of reaching speeds up to Mach 9 [2]. These systems are designed to enhance Russia's strategic deterrence capabilities, encompassing both nuclear and conventional roles. The Russian military's focus on hypersonic technology is indicative of a broader strategy to counter NATO and U.S. military capabilities, thereby reshaping the global security landscape [5].

China, similarly, has been proactive in hypersonic weapons development, conducting multiple tests of systems like the DF-ZF hypersonic glide vehicle, which is engineered to evade missile defence systems [5]. China's hypersonic advancements are part of a larger military strategy aimed at enhancing its global power projection and addressing perceived threats from the U.S. and its allies [2]. The rapid development of hypersonic capabilities by China underscores the urgency with which these nations are pursuing technological superiority in this domain [5].

While the primary focus remains on the U.S., Russia, and China, other nations such as India, France, Japan, and Australia are also exploring hypersonic technologies, albeit at a slower pace [5]. Collaborative initiatives, such as the U.S.-Australian Hypersonic International Flight Research Experimentation (HIFiRE) program, are indicative of a growing international interest in hypersonic research and development [5]. The escalating investments in hypersonic technology reflect its perceived strategic importance, with billions allocated not only for the development of offensive systems but also for the necessary infrastructure to support testing and production [35].

United States

The U.S. Department of Defense (DOD) is actively pursuing the development of hypersonic weapons through multiple programs, including the Navy's Conventional Prompt Strike (CPS) program, as well as initiatives led by the Air Force, Army, and DARPA. These efforts aim to enable the U.S. military to strike high-value, hardened, or time-sensitive targets rapidly with conventional warheads. Hypersonic weapons are seen as a strategic tool to counter adversaries' advanced missile defence systems, many of which are designed to deny U.S. military access to contested areas. In recognition of their importance, hypersonic weapons are highlighted in the 2018 National Defense Strategy as essential to ensuring the U.S. military's ability to "fight and win" future wars [329].

U.S. hypersonic weapons are conventionally armed, in contrast to the nuclear-armed systems being developed by Russia and China. This difference means U.S. hypersonic weapons will need a higher level of accuracy to achieve their objectives. While nuclear-armed hypersonic systems can afford to be less precise due to the devastating blast effects of nuclear warheads, conventionally armed systems must be more technically sophisticated to ensure they hit their intended targets with precision [329].

The U.S. Navy is a leading force in developing hypersonic weapons. It has undertaken the development of a Common Hypersonic Glide Body, which will be used across multiple military services. The Navy's CPS program pairs this glide body with a booster system to form an All Up Round (AUR), which both the Navy and Army will use. Initial flight tests of the AUR, such as the one in June 2022, were unsuccessful, but subsequent tests have shown more promise. By 2025, the Navy plans to deploy hypersonic weapons on Zumwalt-class destroyers and potentially on Ohio-class submarines. In addition to the CPS program, the Navy is developing the Offensive Anti-Surface Warfare Increment 2 (OASuW Inc 2), also known as Hypersonic Air-Launched OASuW (HALO), for air-launched capabilities [329].

The U.S. Army is developing the Long-Range Hypersonic Weapon (LRHW), also known as Dark Eagle, which uses the same glide vehicle as the Navy's CPS program but with a different launch platform. The Army plans to field two additional batteries of LRHW by 2027, with a range of over 1,725 miles, capable of hitting high-value targets and breaking through anti-access/area denial (A2/AD) defences [329].

The U.S. Air Force has been developing the Air-Launched Rapid Response Weapon (ARRW), a hypersonic glide vehicle designed to be launched from bombers, but recent tests have seen mixed success, leading the Air Force to shift its focus to the Hypersonic Attack Cruise Missile (HACM) program. HACM is expected to have wider applications and could be carried by both bombers and fighter jets [329].

DARPA has also played a critical role in advancing hypersonic technology, with programs like Tactical Boost Glide (TBG) and Hypersonic Air-breathing Weapon Concept (HAWC). These efforts focus on developing air-launched hypersonic systems and integrating them with platforms like the Navy's Vertical Launch System [329].

The U.S. is also investing heavily in hypersonic missile defence. The Missile Defense Agency (MDA) is developing counter-hypersonic systems such as the Glide Phase Interceptor (GPI) and the Hypersonic and Ballistic Tracking Space Sensor (HBTSS), with plans to field these systems by the mid-2030s. In parallel, DARPA is working on Glide Breaker, a system designed to intercept hypersonic missiles at long ranges [329].

As hypersonic weapons development advances, the U.S. has faced challenges related to testing infrastructure. With limited hypersonic flight test facilities and ranges, the DOD has recognized the need to expand its testing infrastructure. Universities such as Purdue and Texas A&M have contributed by developing high-speed wind tunnels, and DOD has announced plans to modernize missile test ranges. Congress continues to allocate funding to enhance hypersonic testing capabilities, with recent appropriations aimed at improving test ranges and hypersonic research facilities.

Russia

Russia has been researching hypersonic weapons technology since the 1980s, but its efforts accelerated significantly following two major geopolitical events: the deployment of U.S. missile defence systems in the United States and Europe, and the U.S. withdrawal from the Anti-Ballistic Missile Treaty in 2001. President Putin articulated Russia's concerns about U.S. missile defence capabilities, stating that the expansion and improvement of these systems could neutralize Russia's nuclear deterrence. In response, Russia sought to develop hypersonic weapons that could manoeuvre during flight, ensuring that they could penetrate U.S. missile defences, thereby maintaining strategic balance [329].

Russia has focused on three primary hypersonic weapons programs: the Avangard, 3M22 Tsirkon (Zircon), and Kinzhal. The Avangard is a hypersonic glide vehicle that is launched from an intercontinental ballistic missile (ICBM), which grants it an effectively "unlimited" range. This glide vehicle can reportedly reach speeds up to Mach 20 and is designed to carry nuclear warheads. It has been deployed on the SS-19 Stiletto ICBM, with future plans to deploy it on the Sarmat ICBM. Testing of Avangard has seen both successes and failures, but it reportedly entered combat duty in 2019 [329].

The Tsirkon is a ship-launched hypersonic cruise missile capable of speeds between Mach 6 and Mach 8, with a range of around 625 miles. This missile is designed to strike both naval and ground targets and has been tested on several naval platforms, including the Project 22350 frigates and Project 885 Yasen-class submarines. It was reportedly deployed on the frigate Admiral of the Fleet of the Soviet Union Gorshkov in 2023 and was first used in combat during the conflict in Ukraine in 2024 [329].

Kinzhal, a manoeuvring air-launched ballistic missile, is a modification of the Iskander missile and can be launched from MiG-31 interceptor aircraft, as well as potentially from the Su-34 long-range strike fighter and the Tu-22M3 strategic bomber. Kinzhal is reported to have a top speed of Mach 10 and a range of up to 1,200 miles. While Russian media claim that it can carry both conventional and nuclear warheads and strike ground and naval targets, the missile's performance has not been independently verified, leading some experts to express scepticism about its capabilities [329].

In terms of infrastructure, Russia conducts hypersonic testing at specialized facilities such as the Central Aero-Hydrodynamic Institute in Zhukovsky and the Khristianovich Institute of Theoretical and Applied Mechanics in Novosibirsk. It also carries out flight testing at the Dombarovskiy Air Base, the Baykonur Cosmodrome, and the Kura Range. These efforts underscore Russia's commitment to maintaining a strategic edge in hypersonic weaponry amidst a rapidly evolving global arms race [329].

China

China's development of hypersonic weapons is driven by the necessity to counter U.S. missile defence systems and the perceived threat of a pre-emptive strike on China's nuclear arsenal. According to experts like Tong Zhao of the Carnegie-Tsinghua Center for Global Policy, China sees hypersonic technology as a way to maintain a credible deterrent in the face of the growing sophistication of U.S. military capabilities. Chinese leaders worry that U.S. hypersonic weapons, combined with missile defence systems, could degrade their ability to launch retaliatory strikes, destabilizing the strategic balance. This concern, shared by Russia, has pushed China to aggressively pursue hypersonic technologies as part of its national defence strategy [329].

China's hypersonic programs focus on developing capabilities that can penetrate U.S. defences and enhance their nuclear deterrence. According to General Terrence O'Shaughnessy, China is testing a nuclear-capable hypersonic glide vehicle designed to evade U.S. missile defence systems. In 2021, China reportedly tested a fractional orbital bombardment system (FOBS) that launched a hypersonic glide vehicle into space before deorbiting toward its target. This system provides a space-based global strike capability and shortens warning time before an attack, making it a potent threat [329].

China's interest in hypersonic technology is also influenced by Russian advancements. Analysts note that China has studied and referenced Russian hypersonic programs extensively, incorporating lessons learned into their development. China has already successfully tested the DF-17, a medium-range ballistic missile specifically designed to launch hypersonic glide vehicles. The missile has a range of approximately 1,000 to 1,500 miles and is believed to have been deployed. In addition, China has tested the DF-41 intercontinental ballistic missile (ICBM), which could be adapted to carry nuclear or conventional hypersonic glide vehicles [329].

Among its tested systems, the DF-ZF (formerly WU-14) hypersonic glide vehicle has garnered significant attention. It has been tested multiple times since 2014 and can reportedly perform extreme in-flight manoeuvres, with a range of around 1,200 miles. U.S. officials have confirmed that the DF-ZF was deployed in 2020, demonstrating its operational readiness. China is also developing the Starry Sky-2 (Xing Kong-2), a nuclear-capable hypersonic vehicle prototype that operates as a waverider, using powered flight to achieve speeds of Mach 6 or higher. The Starry Sky-2 could be operational by 2025, although U.S. officials have not publicly commented on its progress [329].

China's hypersonic development is supported by a robust infrastructure for research and testing. In 2018, then-USD(R&E) Michael Griffin noted that China had conducted 20 times as many hypersonic tests as the United States. China has tested different vehicle models, such as the D18 series, to explore various aerodynamic properties and is researching hypersonic cruise missile technologies through the Lingyun Mach 6+ engine. China has also invested heavily in ground testing facilities, with the China Aerodynamics Research and Development Center boasting 18 wind tunnels and the China Academy of Aerospace Aerodynamics operating several more, including those capable of reaching speeds of Mach 8, Mach 10, and Mach 12. The JF-12 and JF-22 wind tunnels, which reach speeds of Mach 9 and Mach 30, respectively, further highlight China's cutting-edge hypersonic testing infrastructure [329].

In summary, China's hypersonic weapons programs are driven by the strategic goal of countering U.S. missile defence systems and maintaining its nuclear deterrent. Through the development of glide vehicles like the DF-ZF and cruise missile prototypes like the Starry Sky-2, combined with significant investment in research and testing infrastructure, China aims to assert its position as a key player in the global hypersonic arms race [329].

Implications for Global Power Dynamics

The potential implications of hypersonic missiles for global stability are a subject of considerable debate among defence analysts. On one side of the argument, some experts suggest that hypersonic weapons could increase the risk of conflict escalation due to the unique characteristics of these weapons, particularly in areas such as target ambiguity, warhead ambiguity, and their precision strike capability. On the other side, there are those who believe that the strategic significance of hypersonic missiles has been exaggerated and that they do not fundamentally alter the nuclear balance between major powers [1].

One of the primary concerns raised by analysts is target ambiguity. Hypersonic missiles, due to their high speed and ability to manoeuvre unpredictably, can make it extremely difficult for a targeted state to accurately determine where the missile is headed. This uncertainty reduces the amount of time that decision-makers have to assess the threat and respond appropriately. A 2017 report by the RAND Corporation highlighted this issue, suggesting that such ambiguity could make states more "trigger-happy" in a crisis. In a high-pressure situation where the consequences of an attack are severe, the inability to predict the target of a hypersonic missile might lead to premature or disproportionate responses, escalating a conflict unnecessarily. The compressed decision-making window that hypersonic missiles create could push states toward hasty decisions, increasing the likelihood of unintended escalation [1].

Another key issue is warhead ambiguity—the inability to discern whether an incoming hypersonic missile is armed with a conventional or nuclear warhead. Hypersonic missiles are often dual-capable, meaning they can be fitted with either type of warhead, and this uncertainty could create significant problems during a conflict. If a state cannot determine whether an incoming missile carries a nuclear payload, it might assume the worst and respond with nuclear retaliation. This risk of miscalculation could lead to a nuclear escalation even in situations where the attacking state only intended to use conventional force. While this concern also applies to other dual-capable missiles, the speed and manoeuvrability of hypersonic missiles could exacerbate the problem by further reducing the time available to make a critical decision. As a result, warhead ambiguity increases the risk of escalating conflicts into nuclear exchanges, even when nuclear weapons were not initially intended to be used [1].

The precision strike capability of hypersonic missiles is another aspect that raises concerns. Hypersonic missiles could be used as part of a prompt global strike system, which would allow a state to quickly and precisely target an adversary's nuclear forces or other strategic assets. This capability might enable pre-emptive strikes aimed at disabling an adversary's nuclear weapons before they can be launched. Some analysts argue that this could undermine the principle of deterrence, as it might give states the confidence to attack first, knowing they could disable their adversary's nuclear forces before a retaliatory strike could be launched. This potential could lead to the expansion of nuclear arsenals, as states might build more weapons or diversify their delivery systems to ensure that they could still launch a second strike even in the face of a hypersonic attack. The possibility of pre-emptive strikes by hypersonic missiles introduces new concerns about strategic stability, particularly in the context of nuclear deterrence [1].

However, not all analysts agree that hypersonic missiles pose a significant new threat. Some argue that the strategic balance between nuclear-armed states remains largely unchanged by the introduction of hypersonic technology. These analysts point out that missile defence systems designed to protect against nuclear-armed ballistic missiles are already limited in effectiveness, particularly against large-scale attacks involving multiple intercontinental ballistic missiles (ICBMs). Even without hypersonic missiles, states with nuclear capabilities can still penetrate missile defences and maintain credible deterrence. For instance, a coordinated launch of multiple ICBMs would likely overwhelm existing missile defence systems, meaning that hypersonic weapons are not uniquely capable of defeating missile defences [1].

Moreover, some analysts suggest that nuclear-armed hypersonic weapons do not offer significant advantages over existing nuclear delivery systems, such as ICBMs or submarine-launched ballistic missiles (SLBMs). These traditional delivery systems are already highly effective at delivering nuclear payloads over long distances, and their reliability and precision are well-established. As a result, for many states, hypersonic missiles may not present a new or revolutionary threat beyond what is already posed by current nuclear arsenals. They argue that the introduction of hypersonic missiles is more about maintaining technological parity and modernizing arsenals than fundamentally changing the dynamics of nuclear deterrence [1].

The debate over the impact of hypersonic missiles on global stability highlights differing perspectives on their significance. Some analysts believe that hypersonic missiles introduce new risks of escalation, particularly due to target and warhead ambiguity and their potential use in precision strikes. These factors could lead to miscalculations and heighten the risk of conflict, including the unintended use of nuclear weapons. On the other hand, critics of this view argue that hypersonic missiles do not fundamentally change the strategic balance between nuclear powers, as missile defence systems are already limited in their ability to counter large-scale nuclear attacks. Furthermore, they suggest that hypersonic weapons do not offer substantial advantages over existing nuclear delivery systems, meaning their overall impact on global stability may be overstated. Ultimately, the role of hypersonic missiles in future conflicts will depend on how states choose to integrate them into their military strategies and whether they can be effectively managed through arms control and international agreements [1].

Arms Race Behaviour

The pursuit of hypersonic weapons by major global powers—namely the United States, China, and Russia—has increasingly taken on the characteristics of an arms race. Each country began its hypersonic programs with unique strategic purposes, but as they observe the advances made by their rivals, they have accelerated their own efforts. This competitive dynamic mirrors the classic arms race behaviour seen in earlier eras of military development. In the U.S., the hypersonic developments in China and Russia are often cited by military officials as a cause for concern, with some claiming the U.S. is falling behind in this new technological frontier. For instance, in 2018, Admiral Harry Harris testified before Congress, stating that China's hypersonic weapons development was outpacing that of the U.S. This message has been reiterated by U.S. military leadership in subsequent years to rally support for increased investment in hypersonic technologies [330].

Despite the claims of being behind, there is still debate over whether the U.S. has indeed been overtaken by China and Russia in the hypersonic arms race. Both China and Russia have made bold claims about their hypersonic capabilities, such as the Chinese hypersonic glide vehicle (HGV) test in 2021. However, there are questions about the accuracy and effectiveness of these systems, especially when compared to U.S. developments. For example, the Chinese HGV reportedly missed its target by 24 miles, whereas a U.S. hypersonic vehicle test had a much smaller margin of error [330]. Moreover, the U.S. likely enjoys significant technological advantages in the specific types of hypersonic weapons it seeks to develop for its arsenal, which calls into question the narrative of a lost hypersonic arms race [330].

Nevertheless, the competition to develop hypersonic capabilities has led to a rushed approach to some programs. In the U.S., there are currently efforts to field at least eight new hypersonic weapons by the mid-2020s, despite the fact that their strategic purpose remains unclear. The arms race dynamic is further exacerbated by the U.S. decision to develop hypersonic missiles specifically designed to defend against an adversary's hypersonic weapons, triggering an action-reaction cycle. This cycle, where offensive advances lead to defensive countermeasures, only to prompt further offensive developments, is a hallmark of arms race behaviour and could lead to endless escalation [330].

The involvement of major U.S. defense contractors—such as Raytheon, Lockheed Martin, and Northrop Grumman—further fuels the race. These companies are heavily invested in hypersonic development, expecting significant business opportunities from increased spending in this area. The hypersonic weapons market is projected to be worth billions of dollars, and with the promise of entirely new classes of weapons, defence contractors are vying for lucrative contracts to meet the growing demand for both offensive and defensive hypersonic technologies [330].

Despite the technological advancements in hypersonic weaponry, the strategic value of these systems remains contentious. Current spending trends indicate that nations may prioritize prestige over pragmatic defence needs, raising questions about the actual effectiveness of hypersonic weapons compared to existing systems [2, 331]. The ongoing arms race in hypersonic technology necessitates a critical evaluation of not just the capabilities but also the costs associated with these systems. Ultimately, the true test for hypersonic weapons will be their ability to deliver strategic advantages without incurring disproportionate costs, a challenge that remains to be fully addressed [331].

Hypersonic Weapons and Deterrence Theory

Role in Deterrence and Escalation Management

Hypersonic Weapons as a Deterrent

Hypersonic weapons have emerged as a pivotal element in modern deterrence strategies, primarily due to their unparalleled speed, precision, and capability to evade conventional defence systems. These weapons, which travel at speeds exceeding Mach 5, possess the ability to manoeuvre unpredictably, complicating detection and interception efforts by traditional missile defence systems. The strategic implications of hypersonic technology are profound, as nations like the United States, Russia, and China are actively developing these systems to enhance their military capabilities and deterrent posture [2, 5].

The speed of hypersonic weapons significantly reduces the time available for adversaries to respond to potential strikes, thereby enhancing their deterrent effect. For instance, the U.S. Conventional Prompt Strike (CPS) program aims to enable rapid global strikes within an hour, targeting high-value assets and time-sensitive threats [5]. This capability exerts pressure on adversaries, compelling them to reconsider aggressive actions due to the imminent threat posed by such rapid response capabilities [2]. The ability to neutralize critical infrastructure or command centres within minutes can alter the strategic calculus of potential aggressors, making them more cautious in their military engagements [7].

Moreover, hypersonic weapons are designed to penetrate existing missile defence systems effectively. Traditional defences are primarily optimized for intercepting ballistic missiles that follow predictable trajectories. In contrast, hypersonic weapons can operate at lower altitudes and execute sharp manoeuvres, making them exceedingly difficult to track and intercept [4]. This capability not only enhances their effectiveness but also contributes to a strategic balance, as adversaries must account for the possibility that their defences may be insufficient against such advanced threats [5]. Countries like Russia and China have explicitly focused on hypersonic development to ensure their ability to circumvent U.S. missile defences, thereby reinforcing their deterrent capabilities [5, 77].

The dual-capable nature of many hypersonic systems, which can carry both nuclear and conventional warheads, introduces an element of ambiguity that complicates deterrence dynamics. Adversaries may struggle to determine whether an incoming hypersonic missile is armed with a conventional or nuclear payload, potentially leading to miscalculations and escalatory responses [7]. This "warhead ambiguity" can create a powerful deterrent effect, as the uncertainty surrounding the nature of the threat may dissuade adversaries from initiating conflict [7]. However, it also raises the risk of unintended escalation, particularly in high-stress scenarios where rapid decision-making is required [7].

In addition to their physical capabilities, hypersonic weapons carry significant psychological implications. The mere existence and deployment of such advanced systems can signal technological superiority and military readiness, further deterring adversaries from engaging in military actions due to the perceived risks involved [2, 5]. The prestige associated with hypersonic capabilities can also serve as a strategic signal of a nation's military prowess, reinforcing its deterrent posture [2, 5].

However, the advent of hypersonic weapons also presents challenges to global security and stability. The rapid escalation potential associated with their deployment raises concerns about miscalculation and unintended conflict escalation, particularly given the compressed timelines for decision-making in response to hypersonic threats [7, 77]. Furthermore, the ongoing development of these systems has contributed to an arms race dynamic among major powers, potentially destabilizing existing security frameworks and increasing the likelihood of conflict [2]. Critics argue that the rush to develop hypersonic capabilities may lead to strategic instability, as nations may feel compelled to continuously enhance their offensive and defensive systems in response to perceived threats [7].

Hypersonic Weapons and Deterrence Theory

Deterrence theory serves as a foundational concept in international relations and military strategy, primarily aimed at preventing adversaries from engaging in aggressive actions through the threat of significant retaliation. The core premise is that potential aggressors will weigh the costs of their actions against the anticipated benefits, leading them to refrain from aggression if the perceived costs outweigh any potential gains. This cost-benefit analysis is crucial in shaping the behaviour of states and non-state actors alike, as they navigate the complexities of international interactions and conflicts [332, 333].

Key components of deterrence theory include capability, credibility, communication, and rationality. The deterring party must possess sufficient military and economic resources to impose the threatened retaliation, which often involves advanced weaponry or strategic alliances capable of executing effective counterattacks [333, 334]. Furthermore, the credibility of the threat is paramount; adversaries must genuinely believe that the deterring party is both willing and able to follow through on its threats. If the threat lacks credibility, it is unlikely to deter aggression [335]. Clear communication of the deterrent threat is also essential, as adversaries must understand the specific actions that would provoke retaliation and the nature of that retaliation [332, 333]. Lastly, the assumption of rationality underpins deterrence theory, positing that all actors will engage in a rational calculation of costs and benefits, which influences their decision-making processes [336].

Deterrence can be categorized into various types, including nuclear and conventional deterrence. Nuclear deterrence, particularly prominent during the Cold War, is based on the doctrine of mutual assured destruction (MAD), where both sides maintain sufficient nuclear arsenals to ensure that any first strike would result in catastrophic retaliation [334]. Conventional deterrence extends beyond nuclear capabilities, relying on strong military forces to impose significant losses on potential aggressors [334]. Additionally, deterrence can be further divided into deterrence by punishment, which involves threatening overwhelming retaliation, and deterrence by denial, which focuses on preventing the aggressor from achieving their objectives through robust defensive measures [334, 337, 338].

The evolution of deterrence theory from the Cold War to the present reflects the changing landscape of global threats, including terrorism and cyber warfare. The traditional models of deterrence face challenges in addressing these new forms of conflict, as non-state actors may not respond to conventional deterrent threats in the same manner as state actors [339]. Critics of deterrence theory argue that its reliance on rationality may overlook the motivations of leaders driven by ideology or desperation, which can complicate the effectiveness of deterrent strategies [339, 340]. Furthermore, the risks of escalation and misinterpretation of intentions can lead to unintended conflicts, undermining the very objectives deterrence seeks to achieve [340].

While deterrence theory remains a critical framework in understanding military strategy and international relations, it is essential to recognize its limitations and the evolving nature of global threats. The interplay of capability, credibility, communication, and rationality continues to shape the effectiveness of deterrent strategies, necessitating ongoing adaptation to address contemporary security challenges [332, 333, 340].

Hypersonic weapons present new challenges and dynamics when examined through the lens of deterrence theory. These weapons, which can travel at speeds exceeding Mach 5 and possess in-flight manoeuvrability, introduce complexities that both strengthen and undermine traditional notions of deterrence. The strategic implications of hypersonic weapons in the context of deterrence theory can be assessed through their effects on capability, credibility, communication, and rationality—the key pillars of deterrence.

In deterrence theory, capability refers to the ability of a nation to effectively retaliate or threaten punishment, while credibility refers to the adversary's belief that the deterrent threat will be carried out. Hypersonic weapons enhance deterrence by increasing a nation's offensive capabilities. With their high speed, manoeuvrability, and ability to bypass conventional missile defence systems, hypersonic missiles increase the capability to strike high-value targets with little warning, including mobile or hardened targets that were previously difficult to hit with conventional weapons. This creates a powerful deterrent effect, as adversaries recognize the advanced capabilities of hypersonic missiles to inflict significant damage.

However, credibility becomes more complex in the hypersonic context. Because these weapons reduce the time available for decision-making, adversaries may question whether a nation could realistically execute a proportionate and timely retaliatory strike, especially in high-stress conflict scenarios. Moreover, if the weapon's use remains confined to conventional warheads, adversaries may perceive a lack of credibility if the deterrent relies primarily on non-nuclear capabilities against nuclear-armed adversaries, as nuclear deterrence has historically been more credible due to the devastating consequences of a nuclear strike.

Clear communication of deterrence threats is essential, but hypersonic weapons can complicate this by introducing ambiguity in both warhead type (nuclear vs. conventional) and target intent. For instance, a country could launch a hypersonic missile with a conventional payload, but the adversary, due to the high speed and short detection time, might assume a nuclear strike is imminent. This could lead to an overreaction or escalation. The compressed decision-making window caused by hypersonic weapons adds to the difficulty of communication, as adversaries have less time to confirm the nature and target of the attack, potentially leading to miscalculation.

In terms of rationality, the deterrence calculus becomes more uncertain with hypersonic weapons. Traditional deterrence assumes that actors are rational and will avoid aggression if they calculate that the cost of retaliation outweighs the benefits. Hypersonic weapons, by introducing unpredictability in flight paths and significantly reducing response time, can undermine rational decision-making processes. The fear of a surprise attack or pre-emptive strike may push decision-makers toward hasty or escalatory actions, increasing the risk of unintended conflict, especially in tense geopolitical environments.

Hypersonic weapons also alter the balance between deterrence by punishment and deterrence by denial. Deterrence by punishment relies on the threat of overwhelming retaliation, as seen in nuclear deterrence strategies. Hypersonic weapons, particularly nuclear-capable ones, strengthen this form of deterrence because they make counter-strikes more effective by evading traditional missile defence systems, thus enhancing the retaliatory threat. This can make adversaries more cautious about launching a first strike, knowing that a hypersonic missile could reach critical targets swiftly and with devastating force.

Conversely, deterrence by denial, which involves preventing an adversary from achieving its objectives by defending against an attack, becomes more difficult with hypersonic technology. The unique speed and manoeuvrability of these weapons mean that existing missile defence systems are largely ineffective against them. This erodes the deterrence by denial framework because adversaries may believe they can strike without being intercepted, increasing the temptation to initiate aggressive actions.

Hypersonic weapons have the potential to destabilize strategic stability, a core objective of deterrence theory. In a strategic arms race, each side seeks to gain or maintain a technological advantage, and hypersonic weapons accelerate this process. The introduction of hypersonic capabilities by one nation can prompt others to develop similar systems or enhance their defences, creating an action-reaction cycle that undermines long-term stability. For example, if the United States develops and deploys hypersonic missiles, adversaries like China and Russia are likely to accelerate their own hypersonic programs or develop countermeasures, leading to a new arms race.

Additionally, the deployment of hypersonic weapons raises concerns about escalation risks, particularly in crises where the distinction between conventional and nuclear warfare may become blurred. The risk of entanglement—where conventional and nuclear command-and-control systems are co-located—exacerbates this issue. If an adversary perceives that their nuclear forces are at risk from a hypersonic strike, they may be more likely to respond with nuclear force, even if the initial strike was conventional.

Strategic Implications

The strategic implications of hypersonic weapons have sparked debate among analysts, with differing views on how these advanced weapons affect global stability. Hypersonic weapons, characterized by their short flight times and unpredictable trajectories, pose unique challenges to strategic stability. Their ability to compress response timelines means that the time available for decision-makers to assess a threat and determine an appropriate response is significantly reduced. This reduced timeframe can lead to hasty decisions and increase the risk of miscalculation, particularly in high-stakes conflict situations. The unpredictable flight paths of hypersonic weapons add another layer of uncertainty, as it becomes difficult to ascertain the intended target, which could lead to further complications in the event of a conflict [329].

One of the most concerning aspects of hypersonic weapons is the potential for unintended escalation due to warhead ambiguity. In situations where it is unclear whether a hypersonic weapon is armed with a conventional or nuclear payload, the targeted state may assume the worst-case scenario and respond with a disproportionate or nuclear response. Even if a state knows that the weapon is conventionally armed, it may still perceive the attack as strategic in nature, potentially prompting a strategic-level response. This ambiguity creates a dangerous environment, especially in countries where nuclear and conventional capabilities are co-located, as an attack on such a site could be perceived as targeting a state's nuclear forces, triggering a nuclear escalation [329].

The United Nations has raised concerns about how states might misinterpret an attack involving hypersonic weapons, noting that different threat perceptions and escalation thresholds can lead to unintended consequences. These risks have prompted some policymakers, including the U.S. Congress, to restrict funding for programs like Conventional Prompt Strike (CPS) out of concern for the potential escalation risks [329].

On the other hand, some analysts argue that the strategic impact of hypersonic weapons may be overstated. According to experts like Pavel Podvig, hypersonic weapons do not significantly alter the strategic balance because states like China and Russia already possess intercontinental ballistic missiles (ICBMs) capable of overwhelming U.S. defences. These analysts contend that traditional principles of nuclear deterrence still apply in the hypersonic era, as no rational state would risk an attack on the U.S. or its allies, given the catastrophic consequences of nuclear retaliation [329].

In response to these concerns, Section 1671 of the FY2021 National Defense Authorization Act (NDAA) requires the U.S. Department of Defense to assess the escalation risks associated with hypersonic weapons. This assessment is expected to include an analysis of both U.S. and foreign hypersonic systems, with a particular focus on dual-use (nuclear and conventional) capabilities, and the potential for escalation in the event of a conflict involving these weapons. These efforts aim to address the complex challenges posed by hypersonic technology and ensure that escalation risks are properly managed in future strategic planning [329].

Hypersonic weapons introduce significant escalation risks due to their speed, precision, and the ambiguity surrounding their intended targets and payloads. These risks are compounded by the dual-use nature of many command and control (C2) systems and missiles, which can serve both nuclear and conventional functions. When hypersonic weapons are designed primarily for offensive use, targeting critical assets early in a conflict, they can accelerate the escalation of hostilities, turning a minor skirmish into a full-blown war. This rapid escalation could result from a perceived need to strike pre-emptively to gain an advantage, which may compel both sides to act swiftly, reducing opportunities for diplomacy or de-escalation [330].

A key concern in hypersonic weapons deployment is warhead ambiguity. Defending nations, given the short response times—sometimes as little as five minutes—might struggle to determine whether an incoming hypersonic missile carries a conventional or nuclear warhead. This uncertainty could push the defending nation toward worst-case assumptions, potentially triggering a nuclear response to what was intended as a conventional strike. This issue is particularly acute in countries like Russia and China, where dual-use hypersonic missiles, such as Russia's Kinzhal or China's DF-17, can carry either conventional or nuclear payloads [330].

Target ambiguity further exacerbates the risk of unintended nuclear escalation. Hypersonic missiles, even when armed with conventional warheads, might target C2 systems or mobile missile launchers that are also integral to a nation's nuclear arsenal. Strikes on these facilities could be misinterpreted as the first step in a nuclear disarming strike, prompting the defending nation to launch its nuclear weapons pre-emptively. The interconnectedness, or "entanglement," of nuclear and conventional C2 systems heightens this risk, as attacks on one could easily be misread as an attack on the other [330].

The implications of these risks are magnified in regions like Europe, where U.S./NATO forces face Russian forces along a broad perimeter, or in Asia, where tensions between the U.S. and China are already high. The mere presence of hypersonic weapons in these theatres could induce leaders to escalate early in a conflict, believing that speed and surprise are key to ensuring success. This would increase the likelihood of large-scale military engagements and potentially nuclear escalation [330].

A final concern is that the complexity and speed of hypersonic conflicts may prompt military leaders to rely more heavily on artificial intelligence (AI) for decision-making. The fast-paced nature of hypersonic warfare, with vast amounts of data pouring in from various sensors and systems, could overwhelm human decision-makers. As a result, there may be a growing temptation to delegate critical decision-making processes, including potentially escalatory actions, to AI systems. This reliance on automated systems increases the risk of miscalculation, as AI could misinterpret data or fail to account for the nuanced judgments that humans might bring to the battlefield, potentially leading to unintended and rapid escalation of conflicts [330].

Possibility of Escalation

Hypersonic weapons present significant risks of conflict escalation due to various ambiguities and the extremely short time available for decision-making. One of the most critical issues is target ambiguity. Hypersonic weapons are capable of striking targets that were previously out of reach for conventional weapons, such as road-mobile missile launchers or deeply buried command and control centres. This capability increases the risk of escalation, as these strikes could be misinterpreted as attempts to degrade a state's nuclear forces, especially if the target houses both nuclear and conventional assets. For example, an attack on a dual-use site, like Russia's Kaliningrad enclave, could lead to misunderstandings about whether the goal was to neutralize conventional or nuclear forces, potentially triggering a nuclear response [341].

Another key issue is warhead ambiguity, which refers to the uncertainty about whether a hypersonic weapon is carrying a conventional or nuclear payload. This ambiguity is particularly dangerous when dealing with dual-capable missiles, as the targeted state might not have enough time to differentiate between a nuclear or conventional strike, prompting a worst-case response. The rapid speed of hypersonic missiles, which can traverse thousands of miles in minutes, exacerbates this problem by reducing decision time for defensive action. China and Russia, who are developing dual-capable hypersonic glide vehicles (HGVs) and missiles, add to this risk, especially in the context of a great-power conflict where misinterpretation could lead to inadvertent nuclear escalation [341].

The high speeds of these weapons introduce additional time-pressure risks, as nations must detect, identify, and decide on countermeasures within a matter of minutes or even seconds. This compressed decision-making timeline increases the likelihood of overreactions or mistaken interpretations, particularly when military command structures feel their retaliatory forces are under imminent threat. According to estimates, hypersonic missiles traveling at speeds of Mach 10 could reduce response times for nearby adversaries to as little as six minutes, leaving minimal time for a thoughtful or measured response [341].

Finally, the development of hypersonic weapons threatens to trigger an arms race, as states may feel pressured to develop more advanced offensive and defensive technologies to maintain strategic balance. This cycle of competition could be destabilizing, much like the Cold War-era arms race. For example, China and Russia's hypersonic advancements are partially motivated by their desire to evade U.S. missile defences, while the U.S. is rapidly increasing its hypersonic development to counter these threats. The financial costs of this arms race are substantial, with the U.S. having allocated billions for hypersonic weapon and defence systems. Other emerging technologies, such as artificial intelligence (AI) and cyber capabilities, further compound the risks, as these technologies could misinterpret signals or inadvertently trigger catastrophic responses [341].

Taken together, the speed, ambiguity, and escalation risks associated with hypersonic weapons create a highly unstable environment in which miscalculations could easily lead to devastating consequences, including nuclear conflict.

Balancing Conventional vs. Nuclear Hypersonic Systems

The balance between conventional and nuclear hypersonic systems is a pivotal concern in contemporary military strategy, significantly influencing global security dynamics. Hypersonic systems possess both conventional and nuclear strike capabilities, each presenting unique advantages and challenges. The strategic implications of these systems are profound, as they intersect with issues of deterrence, escalation risks, and arms control.

The emergence of hypersonic weapons has been characterized as a potential game-changer in military conflicts, particularly due to their speed, manoeuvrability, and ability to evade existing missile defence systems. Raza notes that major powers, including the US, China, and Russia, are actively pursuing hypersonic technologies, which could undermine conventional military capabilities and destabilize existing strategic balances [5]. This rapid advancement in hypersonic technology raises concerns about strategic stability, as the integration of these systems into military arsenals could lead to an arms race, exacerbating tensions among nuclear-armed states [2, 35].

Moreover, the dual-use nature of hypersonic systems complicates deterrence strategies. Mclaughlin [342] emphasizes that the versatility of hypersonic weapons allows them to perform multiple roles—strategic nuclear strikes, nuclear counterforce operations, and conventional tactical strikes—without the need for reconfiguration. This ambiguity can undermine traditional deterrence theories, as adversaries may struggle to ascertain the intent behind a hypersonic launch, increasing the risk of miscalculation and escalation [7]. Coetzee [7] further argues that the speed and agility of hypersonic missiles significantly reduce response times for nuclear states, potentially encouraging pre-emptive strikes and destabilizing the nuclear deterrence framework.

The implications for arms control are equally significant. Borrie et al. [35] highlight that hypersonic weapons present both challenges and opportunities for strategic arms control, necessitating new frameworks to address the unique characteristics of these systems. The need for enhanced verification measures and international cooperation is underscored by Bassey [343], who suggests that technological advancements are reshaping the landscape of nuclear stability and arms control negotiations. As nations grapple with the integration of hypersonic capabilities into their military doctrines, the potential for an arms race looms large, necessitating urgent dialogue and innovative approaches to mitigate risks [344].

Conventional Hypersonic Systems: Conventional hypersonic systems are designed to carry non-nuclear payloads and offer several strategic advantages. These weapons provide the ability to strike time-sensitive, high-value targets quickly and with precision. For example, they can target enemy command-and-control centres, missile batteries, or mobile targets that might otherwise be difficult to destroy with conventional forces. Conventional hypersonic weapons, such as the U.S. Air Force's Hypersonic Attack Cruise Missile (HACM) or the Army's Long-Range Hypersonic Weapon (LRHW), are part of a broader strategy known as "Prompt Global Strike," which seeks to enable the U.S. to deliver precise strikes anywhere in the world within minutes.

One key advantage of conventional hypersonic weapons is that they reduce the risk of nuclear escalation in conflicts. They provide a powerful deterrent and strike capability without resorting to nuclear options, which helps in situations where a full-scale nuclear response would be disproportionate. Moreover, these weapons are seen as critical in countering anti-access/area denial (A2/AD) systems, which are designed to prevent conventional military forces from entering contested areas. Hypersonic weapons' ability to penetrate such defences enhances a nation's ability to project power and respond to threats in contested environments.

However, the development and deployment of conventional hypersonic systems also present challenges. The high costs of research, development, and production make these weapons accessible only to a few nations, leading to a technological and military imbalance globally. Moreover, the sheer speed and manoeuvrability of these weapons compress the decision-making time for the adversary, increasing the risk of miscalculation and unintended escalation, even in a conventional conflict.

Nuclear Hypersonic Systems: Nuclear-capable hypersonic systems pose an entirely different set of strategic and operational challenges. Nations like Russia and China have already developed nuclear-capable hypersonic weapons, such as Russia's Avangard hypersonic glide vehicle and the Kinzhal missile, which can be fitted with nuclear warheads. These weapons enhance a nation's deterrence capability by providing a more credible second-strike option. Their ability to evade current missile defence systems and deliver nuclear payloads at high speeds makes them particularly dangerous.

Nuclear hypersonic systems are also viewed as a way to ensure strategic stability by providing a survivable deterrent in the face of advances in missile defence technology. Russia, for example, accelerated its hypersonic weapons program in response to U.S. missile defence developments, fearing that these systems could undermine its nuclear deterrent. By developing hypersonic nuclear weapons, Russia and China aim to ensure their second-strike capability remains intact, deterring a potential first strike by adversaries.

Despite their deterrent value, nuclear-capable hypersonic systems raise significant escalation risks. The compressed timelines for decision-making in the face of a hypersonic missile launch could lead to a "use-it-or-lose-it" mentality, where a nation under attack might feel compelled to launch a retaliatory nuclear strike before fully understanding the nature of the incoming threat. This issue is further complicated by warhead ambiguity—the difficulty in distinguishing whether an incoming hypersonic missile is carrying a conventional or nuclear warhead. This ambiguity could prompt a nuclear-armed nation to respond with nuclear weapons to a conventional hypersonic strike, fearing that it is under nuclear attack.

The balance between conventional and nuclear hypersonic systems is a critical aspect of contemporary military strategy, particularly in the context of global strategic stability. Conventional hypersonic weapons provide a means for precision strikes and deterrence in conventional conflicts, allowing nations to engage in military operations without escalating to nuclear warfare. Conversely, nuclear hypersonic systems, while enhancing deterrence capabilities, introduce significant risks of unintended nuclear escalation, particularly due to their speed and agility which can complicate crisis decision-making [7, 8].

To effectively balance these two types of systems, nations must prioritize clear communication and doctrine regarding their military strategies. This clarity is essential to mitigate the risks of misinterpretation or overreaction during crises. As highlighted by Chyba [8], the rapid advancement of hypersonic technologies necessitates a reassessment of existing deterrence frameworks, emphasizing the importance of transparent communication between states to avoid misunderstandings that could lead to catastrophic outcomes. Furthermore, the development of robust military doctrines that delineate the use of conventional versus nuclear hypersonic capabilities is crucial in establishing a stable strategic environment [7].

In addition to communication, the development of hypersonic defences is imperative. As hypersonic weapons proliferate, nations must invest in defensive systems capable of countering these threats. This investment can help alleviate the perceived need for rapid escalation in response to a hypersonic strike, thereby providing decision-makers with additional time to assess the situation and respond appropriately [5]. The concept of deterrence by denial, which focuses on preventing adversaries from achieving their objectives through defensive measures, is particularly relevant in this context [345].

Moreover, the establishment of international arms control agreements addressing both conventional and nuclear hypersonic systems is essential. Such agreements could set limits on the development and deployment of these weapons, thereby reducing the risk of an arms race and enhancing global stability. The potential for hypersonic weapons to disrupt existing strategic balances underscores the necessity for cooperative measures among nations to manage these emerging threats effectively [5].

Finally, a focus on deterrence by denial can further balance the reliance on hypersonic offensive capabilities. By enhancing missile defence systems and other countermeasures, nations can reduce their dependence on hypersonic weapons for deterrence, shifting the emphasis toward defensive strategies that maintain stability without escalating tensions [345]. This approach not only fosters a more stable security environment but also encourages a more nuanced understanding of deterrence that encompasses both offensive and defensive dimensions [346].

The Future of Global Stability

The development of hypersonic weapons introduces significant implications for the future of global stability. These advanced weapons, which can travel at speeds greater than Mach 5, provide nations with the capability to strike targets quickly and with unprecedented precision, bypassing traditional missile defences. However, their integration into military arsenals, particularly by major global powers, brings both potential benefits and heightened risks that could reshape international security dynamics.

Strategic Implications for Global Stability

Hypersonic weapons possess several characteristics that challenge existing frameworks of strategic stability, which has traditionally relied on deterrence theories and arms control agreements. Their ability to manoeuvre unpredictably and travel at ultra-high speeds greatly reduces the time available for nations to detect, assess, and respond to a threat. This compressed decision-making window is a crucial factor that could destabilize global security, particularly during times of crisis.

- **Compressed Decision Timelines**: Hypersonic weapons drastically reduce the time between detection and impact, leaving decision-makers with minutes, or even seconds, to evaluate the threat and determine a course of action. This creates a highly pressurized environment where nations may feel compelled to respond pre-emptively, increasing the risk of miscalculation or unintended escalation. In a situation where nations are already on high alert, the presence of hypersonic weapons could exacerbate tensions and lead to rapid escalation, potentially triggering a conflict before diplomacy or de-escalation efforts can take effect.

- **Warhead and Target Ambiguity**: One of the most significant risks associated with hypersonic weapons is warhead ambiguity—the difficulty of determining whether an incoming missile is armed with a conventional or nuclear payload. Since hypersonic missiles can carry either, a defending nation may assume the worst, fearing a nuclear strike even if the incoming weapon is conventional. This could lead to a disproportionate response, including the possible use of nuclear weapons, which would have catastrophic consequences for global stability.

Additionally, target ambiguity arises when hypersonic weapons are used to strike command-and-control (C2) systems or other dual-use infrastructure that serve both conventional and nuclear forces. An attack on such facilities could be misinterpreted as the start of a broader nuclear offensive, prompting the targeted nation to retaliate with nuclear force, even if the initial strike was not intended to escalate to that level.

Impact on Arms Control and Deterrence

Hypersonic weapons also complicate existing arms control agreements and deterrence strategies. Traditional arms control frameworks, such as the New START treaty between the U.S. and Russia, do not fully account for hypersonic weapons, as these systems differ significantly from ballistic missiles in terms of their flight trajectories and capabilities. Current arms control agreements focus on limiting the number of strategic nuclear delivery systems, but hypersonic weapons—especially conventional variants—fall outside these definitions. As a result, these treaties may need to be renegotiated or expanded to cover hypersonic systems to prevent an arms race.

Emerging Arms Race: An arms race involving hypersonic weapons is already underway, with the U.S., Russia, and China at the forefront of developing and deploying these systems. Each nation views hypersonic weapons as a critical component of its future military capabilities, particularly for countering adversaries' advanced air and missile defences. The fear of falling behind in this technological race could prompt even more countries to develop hypersonic systems, leading to a broader proliferation of these weapons. This competition could destabilize global security as nations rush to develop, test, and deploy hypersonic technologies without fully considering their strategic implications.

The deployment of hypersonic weapons also intensifies the existing offence-defence competition. For example, nations may seek to develop counter-hypersonic defences, such as directed energy weapons (e.g., lasers) or advanced missile interceptors, to protect against hypersonic threats. However, this only escalates the arms race, as each side continues to develop more advanced offensive and defensive technologies, perpetuating a cycle of competition that can destabilize regional and global security.

Potential for Escalation and Crisis Instability: Hypersonic weapons, by their very nature, increase the likelihood of escalation in a conflict. Their speed, manoeuvrability, and precision make them ideal for pre-emptive strikes against high-value, heavily defended targets such as radar installations, missile batteries, and critical C2 centres. In a crisis, a nation may be tempted to use hypersonic weapons early, believing that a quick and decisive strike could neutralize an adversary's military capabilities and prevent further escalation. However, such actions could backfire, provoking a massive retaliatory response, potentially involving nuclear weapons, and leading to uncontrollable escalation.

The risks of escalation are further compounded when considering the potential use of artificial intelligence (AI) in decision-making processes. As military systems become increasingly reliant on AI for rapid analysis and response, the risk of machines making critical decisions in the heat of a crisis grows. This introduces the possibility of an automated system misinterpreting data and launching an inappropriate or escalatory response, leading to unintended conflict.

The Need for New Arms Control Measures and International Cooperation: To mitigate the risks posed by hypersonic weapons, there is a growing recognition of the need for new arms control measures. Expanding or renegotiating existing treaties like New START to include hypersonic weapons is one option, but multilateral cooperation is essential. Given that multiple nations are pursuing hypersonic capabilities, any meaningful arms control effort must include not only the U.S. and Russia but also China and other nations developing hypersonic systems.

In addition to formal arms control agreements, confidence-building measures (CBMs), such as increased transparency, data sharing, and joint testing initiatives, could play a vital role in reducing the risks of miscalculation and unintended escalation. For example, nations could agree to share information about the deployment and capabilities of hypersonic systems, as well as establish protocols for communicating during crises to avoid misunderstandings.

Hypersonic weapons represent a transformative shift in military technology, offering nations new capabilities but also introducing significant risks to global stability. The development and deployment of these weapons challenge existing deterrence theories and arms control frameworks, creating the potential for rapid escalation in conflicts and an intensified arms race. To ensure that the introduction of hypersonic systems does not undermine global security, it is essential for nations to pursue new arms control measures, engage in multilateral dialogue, and prioritize crisis management mechanisms that can reduce the risk of unintended conflict.

Part IV - Challenges and Countermeasures

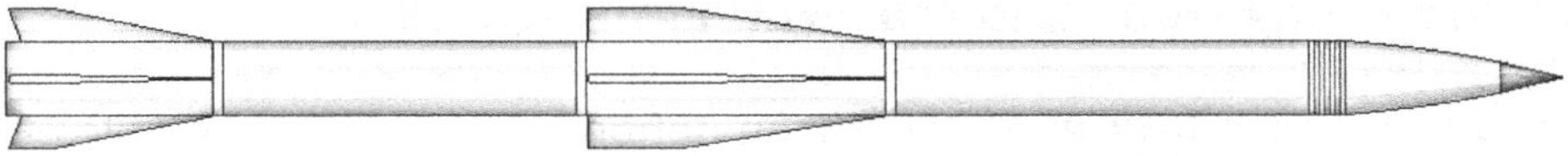

TECHNICAL CHALLENGES IN HYPERSONIC WEAPON DEVELOPMENT

Overcoming Heat and Material Limitations

Overcoming heat and material limitations in hypersonic missiles is one of the most critical challenges in their development, given the extreme environments these weapons must endure. At speeds exceeding Mach 5 (around 3,800 mph or 6,100 km/h), the friction between the missile and the surrounding atmosphere generates intense heat, leading to temperatures that can reach thousands of degrees Celsius. This heat can degrade the materials used in the missile's structure, reducing its performance, accuracy, and survivability. Addressing these challenges involves developing advanced materials, innovative thermal management techniques, and specialized design solutions.

High-Temperature Materials Development

The development of hypersonic missiles necessitates advanced materials capable of withstanding extreme thermal environments. Traditional metals and alloys have proven inadequate for the high temperatures encountered during hypersonic flight, prompting engineers to explore high-temperature-resistant materials such as ceramic matrix composites (CMCs), carbon-carbon composites, and ultra-high-temperature ceramics (UHTCs).

Ceramic Matrix Composites (CMCs)

CMCs are engineered by embedding ceramic fibres within a ceramic matrix, providing exceptional thermal resistance and structural integrity at elevated temperatures. These materials are lightweight and exhibit high melting points, making them ideal for applications in hypersonic missiles, particularly in thermal protection systems (TPS) and control surfaces. The unique properties of CMCs, including their ability to maintain performance under thermal shock conditions, have been highlighted in various studies. For instance, Yang et al. [246] discuss the performance of C/C-SiC composites, emphasizing their suitability for hypersonic and re-entry vehicles due to their high-temperature resistance and structural capabilities. Additionally, Zhao et al. [347] note that CMCs possess excellent oxidation resistance and heat insulation performance, which are critical for hypersonic applications.

The motivation for developing ceramic matrix composites (CMCs) arises from the limitations of conventional technical ceramics, such as alumina, silicon carbide, aluminium nitride, silicon nitride, and zirconia. These materials are known for their brittleness and susceptibility to fracture under mechanical or thermo-mechanical loads, primarily due to the initiation of cracks from small defects or scratches. The crack resistance of these conventional ceramics is notably low, which significantly limits their application in demanding environments [348, 349].

To enhance the crack resistance and fracture toughness of these ceramics, researchers have incorporated particles, such as monocrystalline whiskers or platelets, into the ceramic matrix. However, this approach has yielded only limited improvements, and such products have found application primarily in specific ceramic cutting tools [350]. The introduction of long multi-strand fibres into the matrix has proven to be a more effective strategy, drastically increasing crack resistance, elongation, and thermal shock resistance. This advancement has led to a broader range of applications for CMCs, particularly in high-temperature and high-stress environments, such as aerospace and automotive industries [348, 350].

The reinforcements used in CMCs, particularly continuous-length ceramic fibres, play a crucial role in enhancing the fracture toughness of the composite material system. These fibres, which typically possess an elastic modulus lower than that of the ceramic matrix, contribute to the overall mechanical performance in two significant ways. First, they increase the stress required for the propagation of micro-cracks through the matrix, thereby enhancing the energy expended during crack propagation. Second, when through-thickness cracks begin to form under higher stress conditions, these fibres can bridge the cracks without fracturing, thus providing the CMC with a high ultimate tensile strength (UTS). This mechanism not only improves the initial resistance to crack propagation but also mitigates the risk of abrupt brittle failure, a common characteristic of monolithic ceramics [348, 349].

Continuous fibre-reinforced ceramic matrix composites (CFRCMCs) utilize different fibre architectures to enhance their mechanical properties, especially in terms of strength, toughness, and thermal resistance. These architectures are crucial for determining the performance of the composite under various loads and high-temperature environments. Here's an overview of the main fibre architectures used in CFRCMCs:

1. Unidirectional (1D) Fiber Architecture: In this architecture, continuous fibres are aligned in a single direction throughout the ceramic matrix. This configuration provides high strength and stiffness along the direction of the fibres but lacks reinforcement in other directions. Unidirectional fibre composites are most suitable for applications where the primary load is applied in a single direction, offering significant tensile strength and load-bearing capabilities along the fibre axis. However, they are more susceptible to damage from out-of-plane stresses.

2. Bidirectional (2D) Fiber Architecture: Bidirectional or 2D architecture consists of fibres woven or arranged in two orthogonal directions, typically in a woven or braided pattern. This architecture offers improved strength in two dimensions—both the primary and secondary fibre directions—making it more resistant to multi-directional stresses compared to unidirectional fibre composites. The woven nature of 2D composites also improves their resistance to crack propagation and enhances toughness in applications like aircraft components and high-temperature structural parts.

Types of Bidirectional Architectures:

- **Plain Weave**: A simple weaving pattern where fibres are alternately woven over and under each other, offering balanced strength and stability.

- **Satin Weave**: A more flexible weave with fewer crossovers, providing better surface smoothness and resistance to deformation under stress.

- **Twill Weave**: A diagonal weave pattern that provides increased flexibility and impact resistance.

3. Triaxial (3D) Fibre Architecture: In a triaxial fibre architecture, fibres are aligned in three directions—two in-plane directions (similar to the 2D weave) and one out-of-plane. This creates a 3D reinforcement structure that significantly improves the composite's resistance to delamination, increases interlaminar strength, and enhances damage tolerance. Triaxial composites are highly suitable for applications in aerospace and defence industries, where multi-directional loading and thermal cycling are prevalent. This type of architecture increases the composite's overall toughness and makes it capable of handling complex loads.

4. Braided and Knitted Architectures: Braiding and knitting techniques are used to create continuous fibre networks with complex geometries. These architectures allow fibres to be interwoven in multiple directions, providing uniform strength in various directions and allowing for more intricate and customizable shapes. Braided CFRCMCs are often used in components with tubular or cylindrical shapes, such as rocket nozzles or pressure vessels. Knitted fibre architectures, on the other hand, provide flexibility and are well-suited for curved or intricate surfaces.

5. 3D Woven or 3D Orthogonal Architectures: 3D woven architectures feature fibres interwoven in three orthogonal directions (x, y, and z axes), creating a highly integrated and robust structure. These composites exhibit superior resistance to mechanical stress, both in-plane and out-of-plane, making them ideal for load-bearing components. The z-direction reinforcement improves the interlaminar shear strength and reduces the risk of delamination. These composites are often used in high-performance aerospace and turbine applications due to their outstanding strength and durability.

6. Multilayer (Laminated) Fiber Architecture: In a laminated fibre architecture, multiple layers of continuous fibres are stacked and bonded together, with each layer having fibres aligned in a specific direction. The layers may be oriented at different angles (e.g., 0°, 45°, 90°) to create a structure that can handle multi-directional stresses. Laminated composites provide tailored mechanical properties depending on the specific layer orientations, making them versatile for applications with complex load requirements. However, they may be more susceptible to delamination compared to 3D-woven architectures.

7. Z-Pinning: Z-pinning is a reinforcement technique where additional fibres or pins are inserted perpendicular to the plane of 2D fibre-reinforced layers. This method increases the through-thickness strength and enhances resistance to delamination by providing additional out-of-plane reinforcement. Z-pinning is often used in combination with other fibre architectures to improve the overall durability of the composite structure.

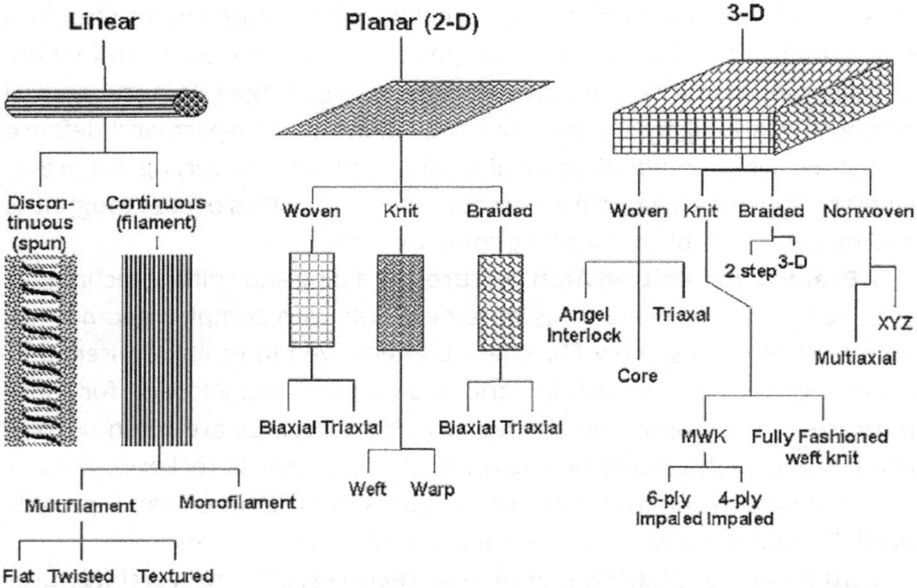

Figure 29: Schematic diagram of fibre architectures used for continuous fibre-reinforced ceramic matrix composites. James K. Wessel, CC BY-SA 4.0, via Wikimedia Commons.

The choice of fibre architecture in continuous fibre-reinforced ceramic matrix composites depends on the specific performance requirements, such as the type of load, temperature conditions, and the desired balance between strength and toughness. Unidirectional fibres offer strength in one direction, while more complex architectures like 3D weaves, braided structures, and z-pinned composites provide superior multi-directional strength and damage tolerance, making CFRCMCs highly versatile for demanding applications like aerospace, defence, and high-temperature industrial settings.

The behaviour of ceramic fibres in CMCs is distinct from their role in polymer matrix composites (PMCs) and metal matrix composites (MMCs). In PMCs and MMCs, fibres typically fracture before the matrix due to the higher failure strain capabilities of these matrices. Conversely, in CMCs, the fibres are designed to withstand higher stress levels, thereby enhancing the overall toughness of the composite [350]. Commonly used reinforcement materials for CMCs include carbon (C), silicon carbide (SiC), alumina (Al2O3), and mullite (Al2O3–SiO2), with the matrix materials often being the same as the reinforcements [348, 349].

They are primarily classified based on the combination of the fibre and matrix materials used, with their names reflecting this relationship. Some of the most common CMC types include:

1. **C/C (Carbon/Carbon)**: This type uses carbon fibres within a carbon matrix. It is widely used in aerospace applications, particularly in components like brake discs and thermal protection systems, due to its excellent strength at high temperatures and thermal shock resistance.

2. **C/SiC (Carbon/Silicon Carbide)**: In this composite, carbon fibres are embedded in a silicon carbide matrix. C/SiC offers the advantages of high stiffness and thermal resistance, making it suitable for high-performance applications like aerospace structures and heat shields.

3. **SiC/SiC (Silicon Carbide/Silicon Carbide)**: This is a fully silicon carbide composite, where both the fibres and the matrix are made from silicon carbide. SiC/SiC composites are highly resistant to high temperatures and harsh chemical environments, often used in gas turbine engines and other high-temperature settings.

4. **Al2O3/Al2O3 (Alumina/Alumina)**: In this composite, alumina fibres are used in an alumina matrix. These materials are commonly found in applications that require insulation and corrosion resistance but do not need the extreme high-temperature capabilities of C/SiC or SiC/SiC composites.

Ceramic matrix composite thrust cells are specialized components used in high-temperature aerospace applications, particularly in propulsion systems such as rocket engines and hypersonic vehicles. They play a crucial role in enhancing the performance of engines and thrust mechanisms by offering significant advantages over traditional materials, especially when operating under extreme conditions like high temperatures and stresses.

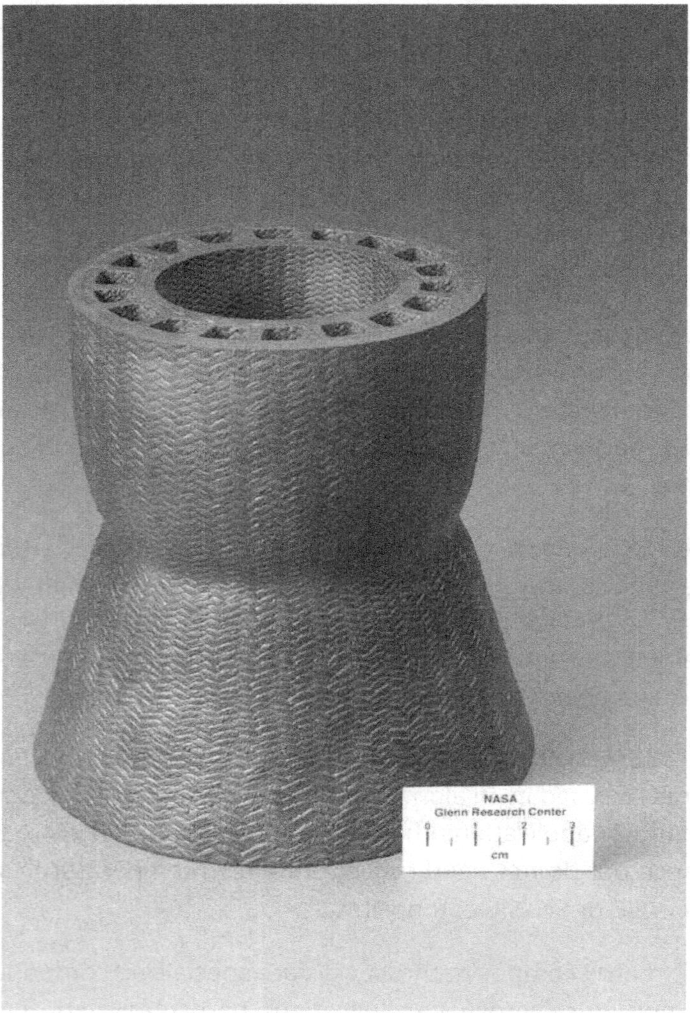

Figure 30: Ceramic matrix composite thrust cell. Defense Visual Information Distribution Service, Public Domain, via Picryl.

Ceramic matrix composites (CMCs) are advanced materials made from ceramic fibres embedded within a ceramic matrix. CMC thrust cells refer to the sections of propulsion systems, such as engine nozzles or combustion chambers, that are constructed using these materials to manage the thrust generated during propulsion.

Thrust cells are the areas within engines or propulsion systems where the thrust is generated and directed. In the case of rocket engines, these include critical areas exposed to combustion or high-speed exhaust, which experience intense heat and pressure. The use of ceramic matrix composites in these areas provides several performance benefits:

1. **High-Temperature Resistance**: CMCs can operate at much higher temperatures (over 1,300°C) than conventional metal alloys, allowing for improved engine efficiency by tolerating greater thermal loads without degrading.

2. **Lightweight Properties**: CMCs are significantly lighter than traditional high-temperature metals like superalloys, reducing the overall weight of the propulsion system. This can lead to improvements in fuel efficiency and flight range.

3. **Durability and Strength**: The ceramic fibres within the CMC structure enhance the material's strength and toughness, making the thrust cells more durable under mechanical and thermal stress.

4. **Oxidation and Corrosion Resistance**: CMCs are highly resistant to oxidation and chemical corrosion, which is essential in the harsh environments of rocket engines, where materials are exposed to highly reactive exhaust gases and other corrosive elements.

Recent advancements have also explored the integration of ultra-high temperature ceramics (UHTCs) into a new class of CMCs known as Ultra-high Temperature Ceramic Matrix Composites (UHTCMCs). These materials are designed to withstand extreme thermal environments, further expanding the potential applications of CMCs in fields such as aerospace and energy generation [348]. The nomenclature for CMCs typically reflects the type of fibre and matrix used, such as C/C for carbon-fibre-reinforced carbon or C/SiC for carbon-fibre-reinforced silicon carbide, which aids in identifying their specific properties and applications [350].

Manufacturing Process

The manufacturing process of continuous fibre-reinforced ceramic matrix composites (CFRCMCs) typically follows three main steps, each critical to the production of high-performance components designed to withstand extreme environments such as high temperatures and mechanical stresses.

1. Lay-up and Fixation of Fibers: In the first step, fibres, commonly referred to as "rovings," are arranged in the desired shape of the final component. The fibres are generally continuous and can be made from various materials like silicon carbide (SiC), alumina, or carbon. Techniques such as lay-up of fabric, filament winding, braiding, or knitting are used to arrange and fix the fibres. These methods are similar to those used in fibre-reinforced plastic materials, ensuring that the fibres are aligned in the optimal orientation for mechanical strength and durability. The resulting structure is called the "fibre preform" or simply "preform." This step is fundamental because the alignment of the fibres significantly impacts the composite's final mechanical properties, including strength, toughness, and load-bearing capabilities.

2. Infiltration of the Matrix Material: Once the fibre preform is shaped, the matrix material is infiltrated into the gaps between the fibres to form a solid composite. There are several methods for infiltrating the ceramic matrix:

- **Chemical Vapor Infiltration (CVI):** A chemical vapor is introduced into the fibre preform, where it decomposes and deposits ceramic material between the fibres. This method is common for producing non-oxide ceramics like carbon-carbon (C/C) or silicon carbide (SiC) composites. One example is the deposition of carbon using a hydrocarbon gas in the presence of a carbon-fibre preform.

- **Polymer Infiltration and Pyrolysis (PIP):** A polymer precursor, which contains carbon or silicon, is infiltrated into the preform and then pyrolyzed to form the ceramic matrix. This method requires multiple cycles of infiltration and pyrolysis to achieve a dense matrix, as the initial polymer infiltration leaves a porous structure.

- **Reaction Bonding:** This process involves a chemical reaction between the elements in the preform and the infiltrating material, such as silicon reacting with carbon to form silicon carbide. It is commonly used in processes like Liquid Silicon Infiltration (LSI), which is used to produce materials like carbon-silicon carbide (C/C-SiC).

- **Sintering:** In this method, the preform is infiltrated with ceramic powders and then heated at temperatures between 1000°C and 1200°C to sinter and bond the materials. This is mainly used for oxide ceramics like alumina or mullite composites.

- **Electrophoretic Deposition:** A developing technique where electrically charged ceramic particles are deposited onto the fibres through an electric field. This method is still in its experimental stages but shows promise for creating dense ceramic matrices.

Each of these methods results in a certain level of porosity, which can affect the performance of the composite. Porosity is generally reduced with additional processing cycles or by using techniques like impregnation or coating.

3. Final Machining and Further Treatments: The final step in the manufacturing process involves machining and any necessary post-processing treatments. CFRCMCs are extremely hard and brittle, requiring specialized tools like diamond-coated drills and saws for operations like grinding, drilling, or milling. Alternatively, techniques like water-jet cutting, laser machining, or ultrasonic machining may be used.

Further treatments, such as applying protective coatings or filling intrinsic porosity with additional materials, can improve the composite's thermal resistance or structural integrity. These steps are critical for optimizing the composite's performance in extreme environments, such as aerospace or turbine engine components.

The performance and characteristics of CMCs are also influenced by the manufacturing process, which is sometimes indicated in their naming. The common methods for producing CMCs include:

- **LPI (Liquid Polymer Infiltration)**: This involves infiltrating the fibre preform with a polymer precursor that is subsequently pyrolyzed to form a ceramic matrix. The process can be repeated multiple times to reduce porosity and improve the mechanical properties of the material. An example is LPI-C/SiC, where the carbon fibres are embedded in a SiC matrix through liquid polymer infiltration.

- **LSI (Liquid Silicon Infiltration)**: In this process, a carbon fibre preform is infiltrated with liquid silicon, which reacts with the carbon matrix to form silicon carbide. LSI-C/SiC is commonly used in applications where high mechanical strength and resistance to oxidation are required.

- **CVI (Chemical Vapor Infiltration)**: In this method, ceramic materials are deposited onto the fibre preform using chemical vapours, resulting in a uniform coating of the fibres. CVI is widely used to produce CVI-SiC/SiC, where silicon carbide is deposited as the matrix material, producing a high-strength, high-temperature resistant composite.

- **CVD (Chemical Vapor Deposition)**: Similar to CVI, CVD involves the deposition of ceramic materials from gaseous precursors, but the deposition occurs without the fibres. This method can be used to form the matrix material in composites.

Each of these processes has its own advantages and limitations, affecting factors like porosity, strength, and temperature resistance of the final composite.

The manufacturing of CMCs generally follows a multi-step process, with Polymer Infiltration and Pyrolysis (PIP) being the most common method. This method involves infiltrating a ceramic matrix precursor into a fibre reinforcement material and then converting it into ceramic through a process called pyrolysis.

Pyrolysis refers to the thermal decomposition of an organic substance, which is achieved by exposing the material to high temperatures in an inert atmosphere, typically nitrogen, argon, or ammonia. In the context of CMC production, pyrolysis converts polymer precursors into ceramic matrices such as silicon carbide (SiC) or carbon. During this decomposition, volatile products like CO, CO2, and H2 are released. The process is advantageous over other methods like sintering and hot pressing, which can degrade the fibres, introduce porosity, and weaken the material at the fibre-matrix interface. Pyrolysis avoids these issues, resulting in a stronger and more durable CMC.

The Polymer Infiltration and Pyrolysis (PIP) process is widely used to create CMCs, particularly with SiC or carbon matrices. In PIP, a low-viscosity polymer infiltrates the fibre preform, which is often woven or in tape form. The fibres are typically made from materials like carbon, SiC, or alumina, depending on the application. These fibres can be arranged in different architectures (continuous, non-continuous, or woven) to achieve specific material properties.

The polymer is then subjected to pyrolysis at temperatures ranging from 1,472°F to 2,372°F in an oxygen-free environment, causing it to decompose into a ceramic. The process might require multiple cycles of infiltration and pyrolysis (usually 4 to 10 cycles) to achieve a low-porosity, high-density composite.

Key stages of the PIP process:

1. **Prepreg Fabrication**: The ceramic fibre preform is impregnated with a resin (preceramic polymer) and partially cured. A material may be applied to the fibres to weaken the fibre-matrix bond,

allowing the fibres to pull out during crack propagation, enhancing toughness.

2. **Lay-up**: Multiple layers of the prepreg material are stacked in specific orientations to control the final material's mechanical properties.

3. **Moulding**: The layered prepreg is shaped into the final component's geometry.

4. **Polymer Infiltration**: The preform is infiltrated with a low-viscosity preceramic polymer to fill the pores in the material.

5. **Pyrolysis**: The infiltrated material is heated in a controlled atmosphere, converting the polymer into a ceramic matrix.

6. **Re-infiltration and Pyrolysis**: These stages are repeated until the desired material properties are achieved.

Alternatives to PIP (Polymer Infiltration and Pyrolysis) for CMC Fabrication

While Polymer Infiltration and Pyrolysis (PIP) is the most common method for manufacturing ceramic matrix composites (CMCs), there are several alternative techniques, each with its own advantages and applications. These alternatives include Electrophoretic Deposition (EPD), Chemical Vapor Infiltration (CVI), Chemical Reactions, and Sintering. Below is a detailed explanation of these methods [351]:

1. Electrophoretic Deposition (EPD): Electrophoretic deposition is a relatively new technique for fabricating CMCs. In this process, electrically charged ceramic particles are suspended in a liquid solution and are deposited onto the fibre preform by applying an electric field. The ceramic particles, such as silicon carbide or alumina, are driven to migrate and adhere to the surface of the fibres, forming the ceramic matrix.

This method offers several advantages, including the ability to achieve uniform deposition, control over the thickness of the ceramic layer, and the potential to coat complex geometries. However, EPD is still in the early stages of development and has not yet been widely adopted in industrial production of CMCs. Its promise lies in its precision, but it requires further refinement to be competitive with more established techniques like PIP and CVI.

2. Chemical Vapor Infiltration (CVI): Chemical Vapor Infiltration (CVI) is a gas-phase process commonly used for non-oxide CMCs, particularly in the fabrication of SiC/SiC composites. During CVI, a fibre preform is placed in a chamber, and a gaseous precursor, such as methyltrichlorosilane (MTS), is introduced. At elevated temperatures, the gas decomposes and deposits a ceramic material (such as silicon carbide) onto the fibres and throughout the preform. The process occurs at relatively low pressures, allowing the ceramic to infiltrate the interior of the fibre structure.

One of the key advantages of CVI is the ability to produce a dense, high-quality ceramic matrix with minimal fibre damage. However, it tends to be a slow and costly process, with significant porosity (10–15%) remaining in the final product. This method is often used in applications requiring high-temperature resistance and high mechanical strength, such as aerospace components.

3. Chemical Reaction: In some fabrication processes, a ceramic matrix is formed via chemical reactions between elements. For example, reaction-bonded silicon carbide (RBSiC) is produced through the reaction of silicon powder with nitrogen gas. Another example is the liquid silicon infiltration (LSI) method, used to create C/C-SiC composites. In this process, a preform made from carbon fibers and a porous carbon matrix is infiltrated with molten silicon. The silicon reacts with the carbon to form silicon carbide.

This approach offers a cost-effective way to produce CMCs with low porosity and high thermal conductivity. However, the matrix formation through chemical reactions can also lead to inclusions of secondary phases, such as free silicon in LSI, which can affect the mechanical and thermal properties of the final material. Nevertheless, this method is widely used in applications like ceramic brake discs due to its high temperature and wear resistance.

4. Sintering: Sintering is primarily used for fabricating oxide CMCs, such as alumina or mullite-based composites. In this process, ceramic powders are mixed with a binder and applied to the fibre preform. The assembly is then subjected to a relatively low-temperature heating process (around 1,000–1,200 °C). The heat causes the ceramic particles to bond together, forming a continuous matrix around the fibres.

While sintering is a well-established technique in conventional ceramics, it is less commonly used in CMC manufacturing because many ceramic fibres cannot withstand the high temperatures required for full densification. Instead, specialized precursor liquids are often used to lower the sintering temperature. The final CMCs produced using this method typically have higher porosity (about 20%), which can limit their mechanical performance in some applications.

Each of these alternative methods for CMC fabrication—Electrophoretic Deposition (EPD), Chemical Vapor Infiltration (CVI), Chemical Reactions, and Sintering—has its own advantages and limitations. While PIP remains the most widely used method, these alternatives offer unique pathways for creating specialized CMCs with varying properties and performance characteristics, making them suitable for different industrial applications. Whether it's the speed of LSI, the precision of EPD, or the high-performance capabilities of CVI, these techniques expand the possibilities for producing advanced materials in aerospace, automotive, and other high-temperature environments.

Ceramic Fibres in CMCs

Ceramic fibres, utilized as reinforcements in ceramic matrix composites (CMCs), are characterized by their polycrystalline, amorphous, or inhomogeneous compositions. These fibres are engineered to withstand high manufacturing and operational temperatures, typically exceeding 1000°C, making them suitable for demanding applications such as aerospace and automotive industries. Among the most common fibres are silicon carbide (SiC), alumina, mullite, and zirconia, each offering unique properties that enhance the performance of CMCs.

Silicon carbide fibres, in particular, have gained significant attention due to their impressive elongation capabilities, often exceeding 2%. This flexibility is crucial in textile manufacturing processes, where fibres must be shaped into complex, three-dimensional forms. SiC fibres are known for their high thermal stability and mechanical strength, which are essential for applications in high-temperature environments, such as gas turbine engines, where they can operate at temperatures around 1250°C [352, 353]. The ability of SiC fibres to maintain structural integrity under extreme conditions is a key factor in their widespread use in advanced composite materials [354].

The mechanical properties of CMCs are significantly influenced by the type of ceramic fibre used. For instance, zirconia fibres exhibit high strength and toughness, which contribute to the overall durability of the composite [355]. Alumina and mullite fibres also play important roles, providing additional thermal and mechanical stability [353]. The combination of these fibres within a ceramic matrix allows for the creation of materials that not only withstand high temperatures but also possess enhanced fracture toughness and resistance to thermal shock [353, 354].

Moreover, the architecture of the fibres, particularly in textile forms, enhances the versatility of CMCs in manufacturing processes. The ability to create complex fibre orientations—from one-dimensional to three-dimensional structures—enables the production of components that meet specific design requirements while maintaining high performance under operational stresses [354]. This adaptability is crucial in industries where lightweight and high-strength materials are essential for efficiency and performance.

Properties of Ceramic Matrix Composites (CMCs)

Mechanical Properties and Fracture Toughness: CMCs are distinct in their ability to resist cracking compared to conventional ceramics, thanks to the fibre-reinforcement mechanism. When subjected to load, the ceramic matrix cracks at an elongation of about 0.05%, as is typical with brittle materials. However, the embedded fibres within CMCs act as crack bridges, preventing catastrophic failure. This process works because the fibres can slide along the matrix, a feature ensured by having a weak bond between the fibres and matrix. If this bond were too strong, the fibres would also break upon matrix cracking, resulting in brittle fracture. To achieve this weak bonding, a layer of pyrolytic carbon or boron nitride is deposited on the fibres, enabling them to pull out of the matrix when cracks form.

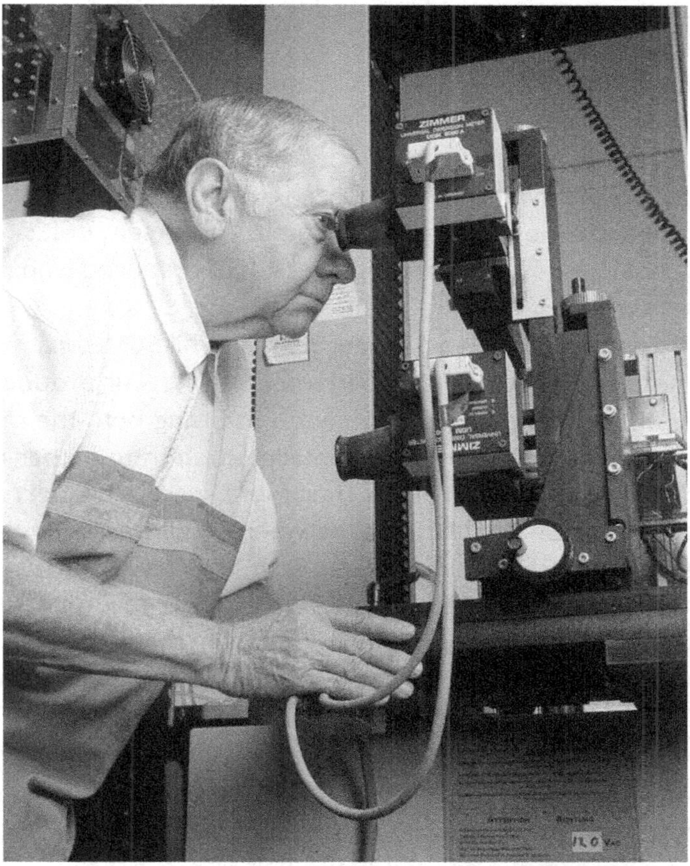

Figure 31: Instron Testing Machine studying the strength of Ceramic Matrix Composite (CMC) Material to develop and improve their mechanical properties. Defense Visual Information Distribution Service, Public Domain, via Picryl.

This fibre-matrix interface is a critical factor in the mechanical performance of CMCs, influencing properties like crack resistance and fracture toughness. When tested under conditions like notched single-edge-notch-bend (SENB), CMC materials show much higher energy absorption (toughness) compared to traditional ceramics like silicon-infiltrated silicon carbide (SiSiC). This means that CMCs can tolerate cracks without immediate failure, a crucial property for high-performance applications where materials are subject to severe mechanical stresses.

Tensile and Bending Behaviour: CMCs exhibit unique stress-strain curves during tensile tests, often showing non-linear behaviour that mimics plastic deformation. However, this is not true plasticity but rather a result of microcrack formation and fibre bridging within the composite. As the load increases, more fibres carry the stress, causing a gradual reduction in the stiffness of the material (Young's modulus).

The performance of CMCs under tensile and bending loads depends on the fibre and matrix composition. For example, CMCs with a high fibre content generally exhibit better tensile strength, close to the fibber's own tensile strength, but lower bending strength. Conversely, CMCs with more matrix material have higher bending strength but reduced elongation and tensile strength.

Creep Resistance at High Temperatures: While CMCs can withstand high temperatures, creep deformation can still occur around 1000°C. Creep in CMCs is a complex phenomenon involving both the matrix and the fibres. The component with the smaller strain rate (either matrix or fibre) bears most of the load and is thus more susceptible to creep. Over time, internal stresses redistribute between the fibres and matrix, and as temperature and applied stress increase, the material may fail either through matrix cracking or fibre fracture. In oxidizing environments, matrix cracking can lead to oxidation, further weakening the material.

Fiber Architectures and Anisotropy: CMCs typically use two-dimensional (2D) fibre weaves, making them anisotropic—stronger in directions aligned with the fibre orientation. However, this creates weaknesses in the interlaminar shear strength, meaning the layers can delaminate under certain mechanical stresses. To combat this, three-dimensional (3D) fibre architectures are sometimes used, which help improve the material's overall strength and reduce the risk of delamination.

Thermal and Electrical Properties: The thermal and electrical properties of CMCs are influenced by the composition of their fibres and matrix, as well as their porosity. Oxide-based CMCs are good electrical insulators, while non-oxide-based materials (like those using carbon fibres) can conduct electricity. The silicon carbide (SiC) matrix, used in many CMCs, is a good thermal conductor but a semiconductor in terms of electrical properties. The amorphous nature of the SiC fibres used in some CMCs reduces their heat and electrical conductivity compared to crystalline SiC.

Thermal shock resistance is a significant advantage of CMCs. Unlike conventional ceramics, which crack under thermal stress due to their high Young's modulus and low elongation capability, CMCs are more resilient because their fibres bridge cracks and prevent catastrophic failure, even under extreme temperature variations.

Corrosion Resistance: CMCs are generally resistant to corrosion, but their behaviour depends on the specific materials used in the matrix and fibres. For example, alumina-based composites resist most chemical corrosion but can creep at high temperatures, limiting their application. Carbon fibres are susceptible to oxidation at temperatures above 500°C, while silicon carbide fibres provide better corrosion resistance in most environments. However, SiC can still react with bases, oxygen, and molten metals at high temperatures.

Carbon-Carbon Composites

Carbon-carbon composites are another critical material for hypersonic applications, characterized by their reinforcement of carbon fibres within a carbon matrix. These composites are particularly effective in areas exposed to extreme heat, such as the leading edges of wings or nosecones, where temperatures can exceed 2,000°C. The exceptional heat resistance of carbon-carbon composites has been well-documented. Paul et al. [247] emphasize their historical significance and recent resurgence in interest for thermal protection systems in hypersonic vehicles, noting their ability to withstand severe thermal and structural stresses. Furthermore, Zou et al. [356] highlight the mechanical and ablation-resistant properties of continuous carbon fibre reinforced UHTC composites, which fundamentally overcome the brittleness of traditional UHTCs, making them promising candidates for hypersonic vehicle thermal structures.

Carbon Fiber Reinforced Composites (CFRC) have a rich historical development, primarily initiated for aerospace applications, particularly in re-entry vehicles such as intercontinental ballistic missiles (ICBMs). These vehicles encounter extreme thermal conditions during atmospheric re-entry, necessitating materials that can withstand significant thermal loads without deforming. CFRCs, known for their high thermal resistance, were crucial in constructing components like the nose cones and wing leading edges of the Space Shuttle orbiter, which utilized Reinforced Carbon-Carbon (RCC) panels capable of enduring temperatures exceeding 1,500°C (2,732°F) [357].

Carbon Fiber Reinforced Carbon (CFRC), also known as Carbon-Carbon (C/C) or Reinforced Carbon-Carbon (RCC), is a specialized composite material that integrates carbon fibres within a graphite matrix. This composite is particularly esteemed for its remarkable thermal and mechanical properties, which render it suitable for extreme environments characterized by high temperatures and significant thermal stresses. Applications of CFRC span various high-performance sectors, including aerospace, motorsports, and defence industries, where the material's unique attributes are essential for operational efficiency and safety.

The exceptional thermal stability and low density of CFRC make it a prime candidate for aerospace applications, particularly in components that must withstand extreme thermal conditions, such as heat shields and leading edges of spacecraft. Studies have demonstrated that CFRC composites exhibit superior ablation resistance, which is critical for materials exposed to high-velocity airflow and thermal loads during re-entry into the Earth's atmosphere [358]. The lightweight nature and high specific strength of CFRC also contribute to its utility in motorsports, where performance and weight reduction are paramount [358].

Moreover, the mechanical properties of CFRC are enhanced by the incorporation of carbon fibres, which provide significant improvements in tensile strength and modulus. These enhancements are crucial for applications that require materials to maintain structural integrity under high-stress conditions [359, 360]. The unique combination of properties such as high thermal conductivity, chemical resistance, and low thermal expansion further solidifies CFRC's position as a material of choice in demanding applications [361].

In addition to its mechanical and thermal advantages, CFRC is also being explored for its recyclability and sustainability. Recent advancements in the development of recyclable CFRCs, utilizing bio-based matrices and innovative processing techniques, have opened new avenues for reducing environmental impact while maintaining performance standards [360, 362, 363]. These developments are particularly relevant in the context of increasing regulatory pressures and societal demands for sustainable materials in high-tech industries.

Composite materials, particularly carbon fibre-reinforced composites (CFRC), are engineered by combining two or more distinct components to achieve superior mechanical properties compared to the individual constituents. In CFRC, carbon fibres are utilized for their exceptional strength and stiffness, while the matrix, often a polymer, contributes thermal stability and resilience. This synergistic combination results in a material that can endure extreme environmental conditions while maintaining structural integrity, making it suitable for various applications, including aerospace and automotive fields [364].

The unique characteristics of CFRC, such as thermal shock resistance, low thermal expansion, and high-temperature stability, enable it to withstand conditions that would typically degrade or destroy other materials. These properties are particularly advantageous in applications requiring dimensional stability across a wide range of temperatures, such as in aerospace components and high-performance automotive parts [365, 366]. Moreover, CFRC exhibits high chemical inertness and biocompatibility, which expands its applicability to medical devices and implants. For instance, the incorporation of carbon fibres into biocompatible matrices has been shown to enhance mechanical properties while ensuring compatibility with biological tissues [366].

The interfacial properties between the carbon fibres and the matrix play a crucial role in determining the overall performance of CFRC. Enhanced interfacial adhesion can significantly improve mechanical properties, such as tensile strength and impact resistance, thereby optimizing the composite's performance in demanding applications [367]. Recent studies have demonstrated that modifications to the fibre surface, such as grafting dendritic structures or applying plasma treatments, can improve wettability and bonding with the matrix, leading to superior mechanical performance [368, 369].

Figure 32: Carbon/carbon fibre panel. Defense Visual Information Distribution Service, Public Domain, via Picryl.

Carbon Fiber Reinforced Composites (CFRC) exhibit several key properties that make them suitable for high-performance applications, particularly in aerospace and automotive sectors. These include:

- **High Thermal Stability:** CFRC maintains its mechanical integrity at elevated temperatures, which is crucial for applications such as re-entry vehicles and rocket nozzles. The thermal stability of CFRC is attributed to its composition and the inherent properties of carbon fibres, which can withstand extreme thermal conditions without significant degradation [370]. Additionally, studies have shown that the incorporation of ceramic materials can further enhance the thermal stability of CFRC, making it suitable for high-performance braking systems [370].

- **Thermal Shock Resistance:** The ability of CFRC to endure rapid temperature fluctuations without cracking is a critical property for applications involving high thermal stress, such as spacecraft re-entry. Research indicates that the thermal shock resistance of CFRC can be significantly improved by optimizing its microstructure and reducing the coefficient of thermal expansion [371, 372]. This resistance is essential for maintaining structural integrity under conditions of rapid thermal cycling [373].

- **Low Coefficient of Thermal Expansion:** CFRC exhibits a minimal coefficient of thermal expansion, which ensures that the material remains dimensionally stable under varying thermal conditions. This property is vital for aerospace applications where structural stability is paramount during extreme temperature changes [371]. The low thermal expansion characteristic helps in minimizing thermal stresses that could lead to failure in critical components [372].

- **Lightweight:** One of the most significant advantages of CFRC is its lightweight nature compared to traditional metal-based composites. This property is particularly beneficial in aerospace applications, where reducing weight can lead to improved fuel efficiency and performance [374]. The lightweight characteristics of CFRC are attributed to the high strength-to-weight ratio of carbon fibres, which allows for the design of structures that do not compromise on strength while minimizing mass [374].

- **Moderate Brittleness:** Despite its many advantages, CFRC does exhibit moderate brittleness, which can limit its impact resistance. This brittleness was notably highlighted during the Space Shuttle Columbia disaster, where damage to the Reinforced Carbon-Carbon (RCC) panels led to catastrophic failure [375]. While CFRC is less brittle than some ceramics, its susceptibility to impact damage necessitates careful consideration in design and application [376].

Material Characterization of C/C Composites

The material characterization of carbon fibre-reinforced carbon (C/C) composites presents unique challenges due to their inherently heterogeneous microstructure. This microstructure primarily consists of carbon fibre reinforcements, carbon matrix, and various defects such as voids and cracks. Accurately assessing and deciphering the mechanical properties of C/C composites requires comprehensive testing and evaluation because the microstructural complexity influences both short-term and long-term material behaviour [377].

The properties of C/C composites are significantly influenced by their manufacturing processes, as various parameters—such as resin choice, impregnation methods, carbonization temperatures, and densification techniques—can be adjusted based on specific application requirements. By tailoring these parameters, the microstructure of the composite can be optimized to achieve the desired performance. For instance, controlling the resin type and carbonization temperature can affect the porosity and mechanical strength of the final composite, impacting its performance in high-temperature or high-stress environments [377].

Characterization of C/C composites involves a variety of testing methods aimed at evaluating their mechanical, thermal, and microscopic properties. Common methods include [377]:

- **Mechanical Testing:** Includes tensile, compressive, flexural, and shear testing to assess properties such as ultimate tensile strength, compressive strength, flexural strength, and interlaminar shear strength.

- **Fatigue Testing:** Examines the composite's durability under cyclic loads, helping to understand how it will perform under repeated stresses.

- **Thermal Properties Testing:** Techniques such as thermogravimetric analysis, differential scanning calorimetry, and thermal conductivity tests assess how the material behaves under thermal stress, including heat flow, decomposition temperatures, and conductivity.

- **Microscopic Analysis:** Tools like scanning electron microscopy (SEM) and transmission electron microscopy (TEM) provide insights into the composite's surface morphology and fine structural characteristics. Other methods like X-ray diffraction (XRD) and Raman spectroscopy are used to examine crystallographic structure and bonding configurations.

The architecture of the carbon fibres within the composite, often referred to as yarn architecture, is a key factor in determining the mechanical and thermal properties of the C/C composite. Common architectures include 3-D orthogonal, 4-D in-plane, and 4-D diagonal structures, each offering distinct properties based on fibre arrangement. Adjusting fibre volume fraction, spacing, and yarn size allows engineers to tailor the composite's thermal and mechanical performance to meet specific application requirements.

For example, in aerospace applications, optimizing the fibre architecture enhances properties like thermal conductivity, tensile strength, and flexural strength, while minimizing porosity and improving resistance to thermal and mechanical stresses [377].

C/C composites exhibit exceptional high-temperature performance, with properties such as Young's modulus and thermal conductivity being significantly greater along the fibre direction than in the transverse direction. The material's anisotropic nature allows for the optimization of properties like thermal stability and electrical conductivity, which can be crucial in applications like aerospace, motorsports, and energy sectors [377].

Key mechanical properties such as tensile strength, compressive strength, and flexural strength can be further enhanced through processes like re-baking and impregnation. For example, repeated cycles of impregnation and pyrolysis allow the matrix to become denser, reducing porosity and improving the composite's structural integrity [377].

One major limitation of C/C composites is their vulnerability to oxidation at high temperatures, which restricts their use in certain environments. To mitigate this, protective coatings, such as silicon carbide (SiC) or silicon nitride (Si3N4), are often applied to the surface. These coatings form protective films like SiO2, which act as a barrier to oxidation. In some cases, inhibitors like halogens and borate-based substances are incorporated into the composite during manufacturing to further reduce oxidation rates [377].

In high-temperature applications, achieving high-density C/C composites is critical to ensuring performance. However, the manufacturing process is labour-intensive, often taking weeks to complete. Techniques like slurry impregnation and pre-densification are employed to reduce processing time and costs while maintaining high material quality.

Protective Coatings for Carbon-Carbon Composites

Carbon-carbon (C/C) composites are known for their exceptional mechanical properties and high-temperature performance. However, their key limitation is susceptibility to oxidation at elevated temperatures, which makes protective coatings essential for ensuring their longevity in extreme environments. The primary goal of protective coatings on C/C composites is to prevent oxidation, which can cause the composite to degrade rapidly when exposed to oxygen-rich atmospheres at high temperatures [377].

When C/C composites are exposed to oxygen at high temperatures, the surface oxidation process follows several stages. These stages include [377]:

1. **Boundary Layer Diffusion of Gas Phase**: Oxygen must diffuse across the boundary layer at the composite's surface before reaching the underlying material.

2. **Diffusion Through Cracks**: If the oxide layer cracks or fractures, oxygen can diffuse through these gaps, reaching the substrate underneath the coating. The size of these cracks determines the rate of oxygen diffusion, which can be slowed by Knudsen diffusion (when gas molecules collide with the walls of tiny cracks).

3. **Condensed Phase Diffusion**: Once oxygen permeates the cracks, it diffuses through the solid oxide coating. The transport process is influenced by the oxide's fault structure.

4. **Interface Reaction**: At the substrate-oxide interface, oxygen reacts with the carbon to form oxides, continuing the oxidation process.

The effectiveness of a protective coating depends on its ability to limit oxygen permeation, particularly in environments above 500°C, where the oxidation of C/C composites is most problematic.

Silicon-Based Coatings: Silicon carbide (SiC) is one of the most common protective coatings used for C/C composites due to its ability to form a dense SiO2 glass layer during oxidation. This SiO2 layer acts as a barrier to further oxygen penetration, significantly slowing down the oxidation process. Silicon-based coatings offer protection at temperatures above 1,000°C by forming a protective glassy surface layer, but they have limitations in environments that experience rapid thermal cycling. Thermal expansion mismatch between the SiC coating and the C/C composite can lead to the formation of microcracks, reducing the effectiveness of the coating [377].

To improve the performance of SiC coatings, materials like chromium silicide ($CrSi_2$) or molybdenum disilicide ($MoSi_2$) are sometimes introduced. These materials exhibit high melting points and provide enhanced oxidation resistance. For instance, $CrSi_2$ can be injected into SiC coatings to fill cracks and improve oxidation resistance at high temperatures.

Mullite Coatings: Another effective coating material is mullite ($3Al_2O_3 \cdot 2SiO_2$), a ceramic known for its thermal stability and low thermal expansion. Mullite coatings on C/C composites provide protection by forming a dense, protective barrier that limits oxygen diffusion. Studies have shown that mullite-coated C/C composites can withstand high temperatures for extended periods, with minimal weight loss due to oxidation.

Mullite-coated C/C composites have demonstrated successful insulation against oxidation for up to 45 hours at 1,873 K (around 1,600°C), reducing the material's weight loss to only 1.86% during that time. This makes mullite coatings ideal for high-temperature aerospace applications where long-term oxidation resistance is required [377].

Zirconium Carbide (ZrC) Doping: ZrC-doped C/C composites are another development in improving oxidation resistance. The addition of ZrC increases the material's ablation resistance, which is crucial in high-temperature environments such as aerospace and defence applications. ZrC coatings provide a robust barrier to oxygen diffusion, and their ablation resistance increases as the ZrC content is increased, though excess ZrC can weaken the bond between the carbon fibres and the matrix [377].

Advanced Coating Architectures: In recent years, advanced multi-layered coatings have been developed to address the limitations of traditional coatings. These systems typically consist of a porous base layer and a dense ceramic top layer. The porous base layer helps accommodate the different thermal expansions between the C/C composite and the coating, while the dense top layer provides chemical and oxidation resistance. For example, a system that combines SiC with additional top layers of materials like ZrC or $MoSi_2$ can offer enhanced performance in high-temperature environments, providing both oxidation protection and improved mechanical properties [377].

Despite the significant advances in protective coatings for C/C composites, challenges remain. For example, thermal expansion mismatches and the brittleness of certain coatings can lead to cracks, reducing their effectiveness. Future research focuses on improving the durability of coatings under thermal cycling conditions, enhancing the mechanical properties of the coatings themselves, and developing new materials that offer both high-temperature performance and oxidation resistance. Additionally, research is being conducted on reducing the cost and complexity of coating processes, making these advanced materials more accessible for a broader range of applications.

Corrosion and Wear in Carbon-Carbon (C/C) Composites

Carbon-Carbon (C/C) composites, known for their remarkable high-temperature properties, exhibit excellent corrosion and wear resistance due to their inert nature. Unlike metals that can suffer significant deterioration under extreme thermal conditions, C/C composites maintain dimensional stability even at temperatures exceeding 3,000°C. This property makes them highly suitable for environments requiring resistance to corrosion, oxidation, and mechanical stress, as is often needed in aerospace and automotive industries. The durability of these composites under frictional wear is also superior compared to advanced ceramics, making them a preferred material in applications like brakes and high-heat environments [377].

The wear resistance of C/C composites is particularly notable in conditions where high friction and thermal expansion are a concern. For example, the low coefficient of thermal expansion (CTE) of carbon-carbon composites, ranging from 0.3–0.5 mm in the fibre direction and 0.5–0.8 mm perpendicular to the fibre direction, allows them to absorb and withstand heat without significant deformation. The load applied to the composite is absorbed by the matrix phase and transferred to the reinforcing fibres, enhancing overall strength. These properties, coupled with their low wear rates (0.05–0.3 mm depending on direction), make C/C composites ideal for high-performance applications like motorsports and aerospace [377].

Tribological Properties

C/C composites tend to exhibit less desirable tribological behaviour in wet or corrosive environments. However, in dry, high-temperature conditions, their performance is more reliable. For automotive and aerospace braking systems, where high friction and rapid braking are needed, C/C-SiC composites (a variation with silicon carbide) perform exceptionally well. The wear and friction properties of C/C composites are influenced by factors such as the adsorption of water vapor, wear due to friction, and the partial pressure of vapor, with behaviour changing drastically at temperatures around 400°C [377].

Studies have shown that the tribological properties of C/C composites can be improved by incorporating additives like hydrocarbons or inorganic compounds, which alter the material's behaviour under high-temperature, dry friction conditions. Additives have been shown to raise the peak transition temperature (where friction and wear peak) beyond the standard 200°C, helping the composite maintain its structural integrity in extreme environments. The use of these additives helps counteract wear and maintains the material's friction stability even in the vacuum conditions [377].

Creep and Fatigue Failure

C/C composites are also known for their exceptional creep resistance, which is the material's ability to resist gradual deformation under constant high stress and temperature. At temperatures above 1,000°C, uniaxial C/C composites exhibit creep behaviour that follows a steady-state response after an initial transient stage. Studies have demonstrated that their creep activation energy is high, reflecting their ability to resist deformation over time in extreme thermal environments. However, under repeated loading and high temperatures, C/C composites can experience fatigue failure through various mechanisms depending on stress levels [377].

At lower stress levels, damage to the interface between fibres and the matrix can improve the composite's residual strength. However, at higher stress levels, fatigue failure manifests similarly to that observed in static tensile testing, with fibre-matrix debonding being a common failure mode. These insights into the fatigue properties are essential for optimizing the design and application of C/C composites in high-stress environments, particularly in aerospace components subjected to repeated thermal cycling and mechanical loading [377].

Corrosion Resistance and Applications

Although C/C composites offer significant advantages, their performance in corrosive and wet environments is limited. They are rarely used in such conditions unless enhanced with specific coatings or impregnated with additional materials like silicon carbide (C/C-SiC composites), which improve their resistance to both oxidation and moisture. This has allowed them to find applications in automotive and aerospace braking systems, where components are exposed to high-speed, high-temperature friction forces [377].

Fabrication Methods of Carbon-Carbon (C/C) Composites

The production of Carbon-Carbon (C/C) composites involves two primary methods: Chemical Vapor Deposition (CVD) and Polymer Infiltration and Pyrolysis (PIP). Both techniques allow for the creation of highly specialized materials designed for extreme temperature applications, but they differ in their approach to forming the carbon matrix around the reinforcing fibres.

Chemical Vapor Deposition (CVD): In the CVD method, carbon is deposited from a gas-phase onto a preform of carbon fibres. The process starts by heating a gas (typically a hydrocarbon like methane or acetylene) in the presence of the fibre preform. As the gas decomposes at high temperatures, it releases carbon atoms, which then deposit onto the fibres, forming a matrix that surrounds and reinforces the fibres. This process is carried out in a controlled environment, often requiring vacuum chambers or inert atmospheres to prevent contamination and ensure high-purity carbon deposition.

A carbon preform stringer is a component made from carbon fibres that are shaped and arranged in a specific pattern to reinforce composite structures. It acts as a structural element in composite materials, providing additional strength and rigidity. These stringers are typically used in aerospace, automotive, and high-performance applications, where lightweight but strong materials are required.

The "preform" refers to the arrangement of carbon fibres into a shape or configuration before the matrix (typically a resin or carbon matrix) is added during the composite fabrication process. Stringers are elongated, beam-like structures that add stiffness and distribute loads across the composite, reinforcing areas where higher structural support is needed, such as aircraft wings, fuselage, and other high-stress regions.

Key aspects of carbon preform stringers include:

- **Strength and Stiffness**: Carbon fibres are well known for their high tensile strength and stiffness, which improve the load-bearing capacity of composite materials.

- **Lightweight**: Despite their strength, carbon preform stringers are lightweight, making them ideal for applications where reducing weight is essential, such as in aerospace engineering.

- **Customizability**: Preforms can be designed in various shapes and sizes depending on the specific requirements of the structure. This customization helps optimize the performance of the final composite material.

- **Application in Composite Structures**: Carbon preform stringers are integrated into larger composite components and are typically embedded during the fabrication process using techniques like resin infusion, polymer infiltration, or chemical vapor deposition.

Figure 33: Dry preform stringer from carbon fabric. AlexSmith1905, CC BY-SA 4.0, via Wikimedia Commons.

These stringers are crucial in advanced engineering fields where strength, weight, and durability are critical, helping structures maintain integrity under extreme conditions.

CVD is known for its ability to produce composites with highly uniform and dense matrices. However, the process is slow, often taking several days to achieve the necessary deposition thickness, which drives up manufacturing costs. Despite this, CVD remains a popular choice for applications that demand superior material properties, such as aerospace components exposed to extreme temperatures and stresses.

Polymer Infiltration and Pyrolysis (PIP): PIP is a more widely used and cost-effective method for producing C/C composites. In this technique, a polymer that contains carbon, such as a thermosetting resin, is infiltrated into the fibre preform. Once the polymer saturates the fibre structure, the composite undergoes pyrolysis—a high-temperature heat treatment in an oxygen-free atmosphere (usually nitrogen or argon). During pyrolysis, the polymer thermally decomposes into carbon, which forms the matrix material around the fibres.

Since the polymer shrinks during pyrolysis, multiple cycles of infiltration and pyrolysis are often necessary to achieve the desired density and mechanical properties. Typically, 5 to 10 cycles are performed, each further densifying the matrix. The repeated process of infiltration and pyrolysis allows for the fine-tuning of the composite's properties, such as thermal conductivity, mechanical strength, and resistance to thermal shock. PIP is frequently used for producing large, complex shapes or components where precise control of material properties is required.

In summary, both CVD and PIP are crucial techniques for fabricating C/C composites, each offering distinct advantages depending on the application. CVD is favoured for its precision and the high-quality carbon matrix it produces, while PIP offers a more cost-effective solution for producing composites with excellent high-temperature performance and customizable properties.

Challenges and Limitations

Carbon Fiber Reinforced Composites (CFRC) are widely recognized for their advantageous properties, including high strength-to-weight ratios and excellent thermal stability. However, they also present significant challenges and limitations, particularly concerning impact resistance and production costs. The catastrophic failure of the Space Shuttle Columbia, attributed to damage from foam insulation, underscores the vulnerabilities associated with CFRCs in aerospace applications. This incident highlighted that while CFRCs are designed to withstand extreme conditions, they can still be compromised by relatively minor impacts, such as foam shedding during launch, which led to severe structural damage and ultimately the loss of the shuttle and its crew [378-380].

The production of CFRCs involves complex manufacturing processes that can be costly and require precise control over various parameters. The three primary stages of production include the lay-up of materials, pyrolysis, and densification. In the lay-up stage, carbon fibres are arranged and combined with an organic binder, which is then subjected to pyrolysis—a process that involves heating to decompose the binder into carbon, creating voids that must be filled in subsequent steps [381, 382]. Densification follows, where carbon-forming gases are introduced at high temperatures to fill these voids, ultimately resulting in a dense composite material. Each of these stages demands meticulous control over temperature, pressure, and chemical reactions, which can significantly increase production costs and complexity [381, 382].

Moreover, the impact resistance of CFRCs is a critical concern. Research indicates that while CFRCs exhibit improved mechanical properties, including enhanced impact strength compared to other materials, they still face limitations under certain conditions. For instance, the impact resistance of CFRCs can be significantly influenced by factors such as fibre length and orientation, which can alter the material's response to dynamic loads [376, 383]. The Columbia disaster serves as a poignant reminder of these limitations, as the foam insulation that detached during launch caused catastrophic damage to the reinforced carbon-carbon wing leading edges, demonstrating that even advanced materials like CFRCs are not immune to failure under specific impact scenarios [378-380].

Ongoing research in the field of carbon-carbon (C/C) composites is focused on reducing production costs while enhancing the material's performance through improved manufacturing methods and tailored microstructures. A significant aspect of this research involves optimizing the interactions between fibres and matrices, as well as controlling porosity, which are critical factors influencing the mechanical and thermal properties of C/C composites. For instance, advancements in carbon fibre technology, including the development of high-performance carbon fibres and cost-effective production techniques, are pivotal in making these materials more accessible for various applications beyond traditional sectors like aerospace and motorsports [384, 385].

The optimization of fibre-matrix interactions is particularly important, as these interactions directly affect the composite's overall performance. Research has indicated that manufacturing methods, such as vacuum infusion, can significantly improve the fibre volume fraction in composites, thereby enhancing their mechanical properties [386]. Furthermore, the control of porosity is essential for achieving the desired thermal stability and lightweight characteristics that make C/C composites attractive for high-temperature applications [387].

As the demand for high-performance materials grows, the potential applications for C/C composites are expanding into industries such as automotive, energy, and civil infrastructure. This shift is facilitated by ongoing innovations in composite manufacturing processes that aim to lower costs while maintaining or improving performance characteristics [388]. The unique combination of high-temperature resilience, thermal stability, and lightweight properties of C/C composites positions them as indispensable materials in various high-performance applications.

Ultra-High-Temperature Ceramics (UHTCs)

UHTCs, such as zirconium diboride (ZrB_2) and hafnium diboride (HfB_2), are being extensively researched for their potential in extreme thermal environments, with melting points exceeding 3,000°C. These materials are particularly suited for the leading edges of hypersonic missiles, where they can maintain structural integrity under dynamic flight conditions. Research indicates that UHTCs can be integrated into composite structures to enhance their thermal shock resistance and mechanical properties. For example, Li et al. [389] discuss the incorporation of UHTCs into CMC matrices to improve high-temperature performance, making them viable candidates for TPS in hypersonic vehicles. Additionally, Nisar et al. [390]explore the enhanced oxidation resistance and mechanical integrity of multicomponent UHTCs, further supporting their application in thermal protection systems.

Ultra-High-Temperature Ceramics (UHTCs) are a class of materials that can endure extreme thermal environments, often above 2,000°C. These ceramics are primarily composed of borides, carbides, nitrides, and oxides of early transition metals like zirconium (Zr), hafnium (Hf), and tantalum (Ta). Due to their outstanding thermal properties, UHTCs are highly resistant to thermal shock and are ideal for applications exposed to rapid temperature changes, such as re-entry vehicles, heat shields, and hypersonic aircraft.

UHTCs are known for their high thermal conductivity, which allows them to dissipate heat effectively, and low thermal expansion, which prevents cracking under thermal stress. These properties, combined with their chemical inertness, make them excellent candidates for use in environments with extreme heat fluctuations, like space re-entry vehicles and nuclear reactors. UHTCs are used in the aerospace and defence industries, where components like heat shields, nose cones, and furnace linings need to maintain structural integrity at high temperatures.

The most widely studied UHTCs are zirconium diboride (ZrB_2) and hafnium diboride (HfB_2), often combined with silicon carbide (SiC) to improve oxidation resistance. Silicon carbide helps form a protective SiO_2 layer on the surface when exposed to oxygen, which enhances the material's resistance to high-temperature oxidation.

Ultra-high-temperature ceramics (UHTCs) such as zirconium diboride (ZrB_2) and hafnium diboride (HfB_2) have garnered significant attention due to their exceptional thermal and chemical stability, particularly in high-temperature environments. ZrB_2 is noted for its high melting point (approximately 3245 °C) and excellent oxidation resistance, making it suitable for applications in aerospace and hypersonic vehicles [391-393]. The incorporation of silicon carbide (SiC) into ZrB_2 composites has been shown to enhance oxidation resistance further. SiC contributes to the formation of a protective SiO_2 layer on the surface when exposed to oxygen, which significantly mitigates the oxidation process at elevated temperatures [394, 395].

Research indicates that ZrB_2 begins to oxidize at temperatures above 600 °C, leading to material degradation unless suitable additives are introduced [396, 397]. The addition of SiC not only improves the mechanical properties of the composite but also plays a crucial role in enhancing its oxidation resistance. Specifically, SiC reacts with oxygen to form a stable SiO_2 layer, which acts as a barrier against further oxidation of the underlying ZrB_2 [394, 395]. This protective mechanism is vital for maintaining the integrity of UHTCs in environments where they are subjected to high temperatures and reactive oxidizing conditions [394, 397].

Similarly, HfB_2 exhibits comparable properties to ZrB_2, including a high melting point and excellent oxidation resistance [398]. The combination of HfB_2 with SiC also results in improved thermal stability and mechanical performance, making these composites suitable for extreme applications [393]. The synergistic effect of SiC in both ZrB_2 and HfB_2 composites underscores the importance of this material in enhancing the performance of UHTCs under oxidative conditions.

Historical Development

The historical development of Ultra-High Temperature Ceramics (UHTCs) can be traced back to the 1960s, primarily driven by the aerospace industry's demand for materials capable of withstanding extreme thermal environments, particularly during re-entry of spacecraft. The U.S. Air Force Materials Laboratory played a pivotal role by funding research into refractory ceramics, which led to the identification of early transition metal borides, carbides, and nitrides as promising candidates due to their exceptional thermal resistance, mechanical strength, and oxidation resistance [399-401]. Among these materials, zirconium diboride (ZrB_2) and hafnium diboride (HfB_2) were particularly noted for their performance under high-temperature conditions, which facilitated their selection for critical aerospace applications, including the Space Shuttle's heat shield [399, 402].

Despite the initial enthusiasm, interest in UHTCs diminished following the conclusion of the Space Shuttle program. However, this interest was rekindled in the 1990s when NASA initiated the development of new hypersonic spaceplanes, recognizing the need for advanced materials that could endure the harsh conditions associated with hypersonic flight [403-405]. The resurgence of research into UHTCs was marked by significant advancements in material composition and fabrication techniques, as well as a deeper understanding of their mechanical properties and thermal behaviour under extreme conditions [406, 407].

NASA's extensive testing of UHTCs has been instrumental in advancing the field. Notable milestones include the SHARP-B1 and SHARP-B2 flight experiments conducted in 1997 and 2000, respectively. These experiments involved the deployment of UHTC materials on re-entry vehicles launched via Minuteman III ICBMs, providing critical data on the performance of these materials under extreme heat fluxes [404, 408, 409]. The SHARP-B2 test, in particular, allowed for the recovery and analysis of multiple UHTC strakes made from various materials, leading to valuable insights regarding the effects of grain size and composite structure on mechanical strength and oxidation resistance [410, 411]. Such empirical data have been crucial for refining the design and application of UHTCs in future aerospace vehicles, ensuring their reliability in the demanding environments of hypersonic flight [390, 412].

Material Properties

Research over the past two decades has primarily focused on enhancing the properties of zirconium diboride (ZrB_2) and hafnium diboride (HfB_2), two of the most promising compounds in the UHTC family. These materials are part of the group four and five transition metal borides, nitrides, and carbides, which are known for their ability to withstand extreme temperatures, typically above 2,000°C. Despite their high melting points, UHTCs generally display strong thermal shock resistance because they combine high thermal conductivity with other favourable properties. However, they are often prone to oxidation at elevated temperatures, limiting their effectiveness in environments with high oxygen exposure.

Thermal Properties

One of the key properties of UHTCs is their thermal conductivity, which is notably higher in borides like ZrB_2 and HfB_2 when compared to nitrides and carbides. Although diborides have a lower melting point than their carbide and nitride counterparts, they exhibit better performance in applications where thermal shock resistance is vital. This makes diborides particularly suitable for applications like heat shields for spacecraft and furnace linings, where sudden temperature changes can occur.

For instance, HfB_2-SiC composites (containing silicon carbide) have shown excellent thermal conductivity, measured at 62 W/mK at 1,000°C, making them well-suited for use in hypersonic vehicles and atmospheric re-entry applications. In contrast, HfC, a common carbide-based UHTC, has a significantly lower thermal conductivity of around 30 W/mK at 800°C. This trade-off between higher conductivity in borides and higher melting points in carbides allows UHTCs to be tailored to specific applications.

Structure and Bonding

UHTCs benefit from their covalent bonding, which provides high structural stability at extreme temperatures. The strong bonds between carbon, boron, and nitrogen atoms contribute to the materials' mechanical strength and resistance to degradation under thermal stress. For example, HfC and ZrC have very high melting points due to their covalent carbon networks. These networks provide excellent thermal stability, but they also make the materials brittle, which can limit their applications under mechanical stress.

Similarly, the structure of metal borides like HfB_2 and ZrB_2, which consist of alternating hexagonal layers of metal and boron atoms, contributes to their anisotropic strength. The strong bonding between boron atoms and metal-boron bonds gives these materials exceptional thermal and mechanical properties, but it also makes them difficult to process and machine.

Mechanical Properties

The mechanical properties of UHTCs are a critical factor in determining their usefulness in high-temperature environments. UHTCs generally have high hardness, often exceeding 20 GPa, due to the covalent nature of the bonds. However, the exact hardness and flexural strength can vary significantly depending on how the material is processed. For example, adding silicon carbide (SiC) to HfB_2 or ZrB_2 composites has been shown to enhance the fracture toughness by up to 20%, increasing its flexural strength and reducing its brittleness. This is primarily due to the material densification and smaller grain sizes produced during processing.

UHTC composites, especially those incorporating SiC, can achieve flexural strengths above 200 MPa even at temperatures exceeding 1,800°C. This makes them ideal for aerospace and hypersonic applications, where materials must retain structural integrity under high thermal and mechanical stress.

Oxidation Resistance

Despite their desirable thermal and mechanical properties, UHTCs are susceptible to oxidation at high temperatures. For example, boron-based UHTCs like ZrB_2 and HfB_2 form boron oxide (B_2O_3) when exposed to oxygen, which melts at around 490°C and vaporizes at temperatures above 1,100°C. This vaporization leads to material degradation and limits the operational lifespan of UHTCs in oxidative environments.

To mitigate this issue, UHTC composites are often reinforced with SiC, which forms a protective silicon oxide (SiO_2) layer on the surface during high-temperature exposure. This glassy layer acts as a barrier, reducing the material's oxidation rate and improving its overall durability in oxygen-rich environments. Research has shown that the inclusion of 30% SiC in HfB_2 composites significantly enhances oxidation resistance, allowing these materials to be used in extreme conditions like re-entry vehicles and nuclear reactors.

Manufacturing Processes

Ultra-high-temperature ceramics (UHTCs) can be fabricated using several techniques, with each method providing unique advantages in terms of density, grain size control, and material properties. The most commonly employed methods are hot pressing, spark plasma sintering (SPS), and chemical vapor deposition (CVD).

Hot Pressing: Hot pressing is one of the most widely used methods for producing UHTCs. It involves applying high pressure and temperature simultaneously to compact ceramic powders into dense, solid forms. This technique is favoured because it promotes grain boundary diffusion, leading to the formation of a highly densified material with low porosity. Hot pressing is typically performed at temperatures above 1,800°C under a pressure of around 30 MPa. The combination of high pressure and temperature encourages the particles to rearrange and bond more effectively, reducing voids and achieving near-theoretical densities. However, hot pressing requires specialized equipment capable of withstanding extreme conditions and is often limited to producing relatively simple shapes because of the high-pressure environment needed.

Spark Plasma Sintering (SPS): Spark plasma sintering (SPS) is another advanced method used for the fabrication of UHTCs. It differs from traditional sintering processes by utilizing pulsed direct electrical currents to generate heat within the ceramic powder itself. The current is passed through graphite dies, which apply uniaxial pressure while simultaneously heating the material. This method allows for rapid densification at lower temperatures (around 1,500°C to 2,000°C), typically in a matter of minutes compared to hours for conventional sintering. The short processing time helps suppress grain growth, resulting in finer microstructures, which can enhance the mechanical properties of UHTCs, such as fracture toughness and thermal shock resistance. The application of electrical current also helps remove surface impurities and enhances grain boundary diffusion, further improving the material's density and mechanical integrity.

Chemical Vapor Deposition (CVD): CVD is a method in which a material is deposited onto a substrate from a gaseous precursor. In the case of UHTCs, metal halides (such as $TiCl_4$ for titanium-based UHTCs) and boron halides (such as BCl_3) are often used as gaseous precursors. These react with hydrogen gas or other reducing agents at high temperatures, leading to the deposition of thin layers of UHTCs on a substrate. CVD is primarily used to produce coatings or thin films rather than bulk materials. It offers precise control over the composition and structure of the deposited material, making it useful for applications like thermal barrier coatings on aerospace components. However, CVD is typically more time-consuming than hot pressing and SPS and is mainly used for specific applications requiring high-purity films.

Among these methods, hot pressing and spark plasma sintering are the most commonly used for fabricating UHTCs because they can achieve high density and low porosity, which are critical for the optimal performance of UHTCs in harsh environments. Hot pressing is known for its ability to produce bulk materials with minimal porosity, making it ideal for structural components. SPS, on the other hand, is praised for its efficiency in producing finer microstructures due to its rapid processing times. Both techniques allow UHTCs to reach their full potential in applications that demand high thermal stability, oxidation resistance, and mechanical strength, such as aerospace, nuclear reactors, and hypersonic vehicles. CVD, while important for coating applications, is not as commonly used for bulk UHTC fabrication but plays a critical role in enhancing surface properties.

Synthesis of Diboride (Zr, Hf, Ti) Ultra-High-Temperature Ceramics (UHTCs)

UHTCs like zirconium diboride (ZrB_2), hafnium diboride (HfB_2), and titanium diboride (TiB_2) are notable for their ability to withstand extreme temperatures, often exceeding 2,000°C. These materials, composed of transition metal borides, are synthesized using various methods, each offering specific advantages and challenges depending on the scale, cost, and desired material properties.

<u>Stoichiometric Reaction</u>

One of the most straightforward methods for synthesizing UHTCs is by direct stoichiometric reaction between the constituent elements. For example, ZrB_2 can be synthesized by reacting zirconium (Zr) with boron (B) in a stoichiometric ratio. This process allows for precise control over the material's composition. The reaction is highly exothermic at temperatures around 2,000 K, making it favourable for self-propagating high-temperature synthesis (SHS). SHS leverages the heat generated by the exothermic reaction to drive rapid combustion, producing high-purity ceramic materials. However, the rapid heating rates can result in porosity and incomplete reactions, leading to the formation of stable oxides and limiting the process for large-scale industrial applications.

Metallothermic Reduction

Another approach involves the reduction of metal oxides, such as ZrO_2 or HfO_2, using magnesium (Mg) in combination with B_2O_3. This reaction produces the desired diborides along with magnesium oxide (MgO) as a byproduct. After the reaction, the unwanted MgO is removed through acid leaching, leaving behind high-purity ZrB_2. The metallothermic reduction process can also be conducted via SHS, but incomplete conversion of reactants often occurs. A variation known as double SHS (DSHS) can improve conversion rates by repeating the reaction under slightly modified conditions, typically using H_3BO_3 as a secondary reactant.

Boron Carbide Reduction

Boron carbide reduction is one of the most popular and cost-effective methods for synthesizing UHTCs. In this process, boron carbide (B_4C) is reacted with metal oxides such as ZrO_2 or TiO_2 at temperatures greater than 1,600°C. The reaction produces UHTCs, such as ZrB_2, along with carbon monoxide (CO) as a byproduct. Excess boron is often used to prevent the formation of zirconium carbide (ZrC), which can occur under certain conditions. This method is widely used due to the availability of less expensive raw materials, and it can produce UHTCs with finer grain sizes and better sinterability when performed at lower temperatures.

Borothermic Reduction

Borothermic reduction is another method for producing UHTCs, which involves reacting metal oxides like ZrO_2, HfO_2, or TiO_2 with boron at temperatures above 1,600°C. Similar to boron carbide reduction, excess boron is needed to account for boron losses during the process. Mechanical milling can be employed to reduce the particle sizes of the reactants, which lowers the reaction temperature and enhances the mixing of particles. This method is often used for producing nanoscale UHTC particles but is less practical for large-scale applications due to the high cost of boron and the formation of boron oxide byproducts.

Chemical Vapor Deposition (CVD) and Plasma Enhanced CVD (PECVD)

For coating applications, UHTCs can be synthesized using chemical vapor deposition (CVD). In this method, a metal halide (such as $TiCl_4$ or $ZrCl_4$) and a boron halide (like BCl_3) are used as precursors, with hydrogen (H_2) acting as a reducing agent. The resulting diborides, such as ZrB_2, are deposited as thin films on surfaces such as copper or zircalloy at temperatures ranging from 700°C to 900°C. Plasma Enhanced CVD (PECVD) is a variation of CVD that uses plasma to enhance the reaction process at lower temperatures (as low as 600°C). This method has been used to prepare high-quality coatings with finer grain structures and improved uniformity.

Processing and Densification Techniques

Diboride-based UHTCs often require high temperatures and pressures for densification. Hot pressing and spark plasma sintering (SPS) are two common techniques used to achieve densified UHTCs. Hot pressing involves heating the material under high pressure, typically between 1,800°C and 2,000°C, to promote grain boundary diffusion and achieve high densities. However, this method can result in large grain sizes and poor mechanical properties, such as reduced fracture toughness.

To overcome these issues, additives such as silicon carbide (SiC) can be introduced to lower the sintering temperature and improve both densification and oxidation resistance. SiC reacts with surface oxides to form a protective layer and enhances the mechanical properties of the diborides. For example, ZrB_2 composites with 20 vol% SiC exhibit increased fracture toughness and flexural strength. In addition, spark plasma sintering can achieve near-ideal densities in UHTCs at lower temperatures and shorter processing times compared to hot pressing. SPS utilizes pulsed electric currents to enhance grain boundary diffusion, reducing grain growth and producing fine-grained UHTCs with superior mechanical properties.

The synthesis of diboride UHTCs involves a range of techniques, each tailored to specific applications and desired properties. While stoichiometric reactions offer precise control over composition, methods like boron carbide reduction and metallothermic reduction provide more cost-effective solutions for large-scale production. Additive-enhanced sintering techniques, such as the incorporation of SiC, further improve the mechanical properties and oxidation resistance of these materials, making them ideal for extreme temperature applications in aerospace, defence, and nuclear industries. Ongoing research aims to refine these methods, reduce costs, and expand the industrial viability of UHTCs.

Applications

Ultra-high-temperature ceramics (UHTCs), particularly hafnium diboride (HfB_2) and zirconium diboride (ZrB_2), are being developed to handle the extreme conditions experienced by vehicles during atmospheric re-entry and sustained hypersonic flight. These materials must withstand temperatures above 2,500°C while exposed to high-temperature, high-flow-rate oxidizing plasma. This makes them essential for applications such as the sharp leading edges of hypersonic vehicles, which experience not only intense heat but also immense mechanical forces.

The development of UHTCs addresses a key challenge: the relationship between curvature radius and temperature on a vehicle's leading edge during re-entry. The smaller the radius of curvature, the higher the temperature. While blunt bodies in re-entry vehicles benefit from a bow shock that creates a dense, cooler plasma layer protecting the surface, sharp leading edges face much harsher conditions. For hypersonic vehicles like DARPA's HTV-3 and experimental spaceplanes like Reaction Engines Skylon or Boeing X-33, sharp edges are critical for achieving high lift-to-drag ratios, which improve fuel efficiency and operational flexibility. UHTCs, with their high melting points and resistance to oxidation, are uniquely suited for these demanding environments.

Beyond aerospace, zirconium diboride (ZrB_2) is utilized in boiling water reactors due to its excellent refractory properties and its high-neutron absorption cross-section (759 barns). Boron, a key component of ZrB_2, serves as a "burnable" neutron absorber, protecting other reactor components from becoming highly radioactive by absorbing neutrons and transforming into stable nuclear reaction products (such as helium and lithium). ZrB_2 also helps manage fuel temperatures by preventing helium evolution from causing structural gaps between the fuel coating and uranium dioxide (UO_2) pellets, a method employed in nuclear reactors like the Westinghouse AP-1000.

However, there are trade-offs associated with boron in nuclear applications. While it serves as an effective neutron absorber, the presence of helium from boron transmutation can create issues, including fuel pellet swelling. For this reason, hafnium diboride (HfB_2), which also has a high neutron capture cross-section (113 barns), is explored as an alternative material for control rods. Its higher melting point and lower reactivity with refractory metals such as tungsten make it a potential material for nuclear control rods, especially in space nuclear power systems.

Titanium diboride (TiB_2) is widely used in the aluminium industry, particularly in molten aluminium electrolysis, due to its electrical conductivity and refractory properties. TiB_2 electrodes offer a superior electrical interface with molten aluminium while resisting contamination from boron or titanium. However, implementing TiB_2 in industrial processes faces challenges such as bonding the material to graphite substrates and ensuring long-term material stability during operation.

UHTC composites, particularly ZrB_2/SiC composites, have been explored for use in conducting ceramic heaters due to their high oxidation resistance and metal-like conductivity. These materials are beneficial because they do not exhibit the negative temperature coefficient resistance of pure silicon carbide (SiC). The incorporation of ZrB_2 into SiC composites enhances thermal stability, with conductivity remaining stable at high temperatures, making these materials ideal for aerospace and high-temperature industrial applications.

The development of UHTCs like HfB_2, ZrB_2, and TiB_2 plays a critical role in high-temperature applications ranging from hypersonic flight and spacecraft re-entry to nuclear power systems and industrial processes. Their unique combination of high melting points, oxidation resistance, and thermal conductivity make them indispensable for handling the extreme conditions of modern engineering challenges. However, issues such as material degradation in nuclear environments and the difficulty of achieving durable material bonds in industrial applications continue to drive ongoing research and development.

Limitations and Challenges

Ultra-high temperature ceramics (UHTCs) are recognized for their remarkable thermal stability and mechanical properties, making them suitable for extreme environments such as aerospace applications. However, despite these advantages, UHTCs face significant limitations, particularly their inherent brittleness and challenges associated with machining. The brittleness of UHTCs renders them susceptible to catastrophic failure under mechanical stress, which poses a considerable barrier to their widespread application in various industries [413, 414]. This brittleness is primarily attributed to the covalent bonding within the ceramic structure, which limits the plastic deformation mechanisms that are often seen in metals, thus resulting in a lack of ductility [414].

Moreover, the hardness of UHTCs complicates machining processes, making them not only difficult but also costly. Traditional machining methods, such as grinding and diamond machining, are often employed to shape these materials; however, these methods are limited by the ceramics' brittleness and the precision required to achieve desired geometries [415, 416]. As a result, the manufacturing of complex shapes or large components from UHTCs is constrained, which further limits their application in high-performance environments [413].

Ongoing research aims to mitigate these limitations by enhancing the mechanical properties of UHTCs through various processing techniques. For instance, studies have shown that reducing the grain size in UHTC composites can significantly improve their mechanical strength and resistance to cracking [417, 418]. Additionally, the development of UHTC composites, such as those incorporating carbon or silicon carbide fibres, has shown promise in overcoming the brittleness associated with monolithic UHTCs [408, 417]. These composite materials leverage the toughness benefits of fibres while maintaining the high-temperature stability characteristic of UHTCs, thus broadening their potential applications [413, 419].

Research into ultra-high temperature ceramics (UHTCs) is advancing rapidly, particularly in enhancing their mechanical properties and refining manufacturing processes. A significant focus is on reducing the grain size in UHTC composites, which has been shown to markedly improve their strength and resistance to cracking. Studies have demonstrated that smaller grain sizes can lead to enhanced mechanical performance due to the Hall-Petch effect, where finer microstructures yield greater strength [420, 421]. Additionally, innovative manufacturing techniques such as spark plasma sintering (SPS) and robocasting are being explored to fabricate UHTC components with complex geometries more efficiently. SPS, in particular, allows for better control over microstructural features, which is crucial for optimizing the mechanical properties of UHTCs [422, 423].

The role of UHTCs in aerospace applications is becoming increasingly critical, especially for next-generation vehicles such as hypersonic aircraft and reusable spaceplanes. Their ability to withstand extreme temperatures and thermal shock makes them ideal candidates for thermal protection systems (TPS) in these applications. Research highlights that UHTCs can endure temperatures exceeding 3000 °C, making them suitable for environments characterized by intense heat and mechanical stress during various flight phases, including take-off, flight, and re-entry [404, 424, 425]. The development of UHTCs is driven by the need for materials that not only resist oxidation and thermal degradation but also maintain structural integrity under extreme conditions [426, 427].

Moreover, advancements in UHTC research are not limited to mechanical property enhancements; they also encompass the exploration of high-entropy ceramics and composite materials that combine multiple phases to achieve superior performance [428, 429]. These multi-component systems have shown promise in providing better thermal and mechanical properties compared to traditional UHTCs. The ongoing research into the processing and characterization of these materials is essential for their successful application in aerospace technologies, where reliability and performance are paramount [413, 430].

Thermal Protection Systems (TPS)

Even with advanced materials, a comprehensive thermal protection system (TPS) is essential for managing the heat generated by hypersonic flight. TPS involves a combination of passive and active methods to protect the missile's internal components from overheating. Thermal Protection Systems (TPS) are critical for managing the extreme heat generated during hypersonic flight, where temperatures can exceed 1000 °C due to aerodynamic heating. A comprehensive TPS employs both passive and active methods to safeguard the internal components of hypersonic vehicles, including missiles, from overheating. This requires various advanced materials and techniques utilized in TPS, focusing on ablative heat shields, insulative layers, and active cooling techniques.

Ablative heat shields are among the most widely utilized techniques in hypersonic missile technology. These materials are designed to erode or "ablate" during flight, effectively carrying away heat and preventing it from penetrating into the missile's critical components. The sacrificial nature of ablative materials has been successfully demonstrated in re-entry vehicles, such as space capsules, and can be adapted for hypersonic applications to protect vulnerable areas like nosecones and fuselages [232, 431]. The effectiveness of ablative materials is underscored by their ability to withstand extreme thermal environments, making them indispensable for TPS in hypersonic flight [432].

In addition to ablative shields, insulative layers play a vital role in minimizing heat transfer to internal components. Materials such as silica-based composites and aerogels are commonly employed to provide thermal insulation, ensuring that the missile's electronics and warhead remain operational under high external temperatures [127]. Research indicates that the optimization of multilayer thermal insulations is crucial for meeting the stringent requirements of new hypersonic vehicles, particularly in high-temperature air conditions [127, 433]. The integration of these insulative materials contributes significantly to the overall effectiveness of the TPS by maintaining the integrity of internal systems during flight.

Active cooling techniques are also essential for certain components of hypersonic vehicles, particularly engines. These systems often involve circulating a coolant, typically fuel or liquid, through channels located near the hottest surfaces of the missile. For instance, scramjet engines may utilize regenerative cooling, where the fuel absorbs heat before entering the combustion chamber, thus ensuring the engine's functionality while mitigating the risk of material failure due to overheating [233, 434]. The combination of active cooling with passive methods enhances the resilience of hypersonic vehicles against thermal stresses, thereby improving their operational reliability [233, 435].

Ablative Heat Shields

Ablative heat shields are critical components in the design of hypersonic missiles, functioning primarily to protect the missile's structural integrity from extreme thermal environments encountered during flight. These shields utilize ablative materials that erode or "ablate" in response to heat, effectively carrying away thermal energy and preventing it from penetrating into the missile's critical components. This sacrificial mechanism is essential for maintaining the operational capability of hypersonic vehicles, which experience severe thermal loads due to high-speed atmospheric entry.

Ablative thermal protection systems (TPS) are becoming increasingly relevant for hypersonic weapons, paralleling their use in space exploration. Hypersonic weapons, which travel at speeds exceeding Mach 5, encounter extreme thermal stresses due to the friction generated as they move through the atmosphere at high velocities. Ablative TPS provides an essential solution to managing these stresses by shielding the weapon's structure from the intense heat.

Similar to spacecraft re-entering Earth's atmosphere, hypersonic weapons experience severe aerodynamic heating as they reach extreme speeds. Blunt body architecture, which was developed to manage high re-entry temperatures, is also applicable to hypersonic weapons. This architecture generates a bow shock that diverts much of the heat away from the surface, reducing the thermal load on the vehicle [436]. Ablative materials further enhance this protection by decomposing and dissipating heat as they erode, providing a layer of defence that is crucial for single-use missions like those undertaken by hypersonic weapons [436]. These materials sacrifice themselves during flight, making them suitable for non-reusable systems like intercontinental ballistic missiles (ICBMs) and certain types of cruise missiles.

Modern advancements in computational models have significantly improved our ability to predict how ablative TPS will perform under the extreme conditions experienced by hypersonic weapons. By leveraging simulations based on porous media theory, researchers can better understand how materials degrade under high heat loads, such as during a hypersonic glide. These models simulate the complex interactions between gas flows and ablative materials, providing critical insights into how different configurations and materials will behave in real-world scenarios [436]. This predictive capability is crucial for optimizing hypersonic weapon designs and ensuring that they can withstand the extreme heat of sustained hypersonic flight.

Materials like phenolic impregnated carbon ablator (PICA), which has been used in space exploration, also show potential in hypersonic weapons. PICA's low density and high thermal resistance make it suitable for withstanding the extreme aerodynamic heating that occurs during hypersonic flight. The material's porous structure reduces heat transfer, while the decomposition process allows cool gases to flow through the structure, further mitigating heat buildup [436]. Such materials are not only effective in space missions but also offer a promising solution for protecting hypersonic weapons during high-speed atmospheric manoeuvres.

For hypersonic weapons, as with space missions, flight data is essential for refining thermal protection models. As more hypersonic vehicles are tested, the data gathered on ablative material performance under real-world conditions will help improve the accuracy of existing models. This continuous refinement is necessary to predict the behaviour of these materials accurately and ensure the reliability and effectiveness of thermal protection systems in future hypersonic weapon systems.

Ablative TPS technologies developed for space exploration have significant applications in hypersonic weapons. The extreme heat generated during hypersonic flight mirrors the conditions experienced by spacecraft re-entering the atmosphere, making ablative materials an essential solution for protecting the structural integrity of these advanced weapons. With ongoing advancements in predictive modelling and material science, the future of hypersonic weaponry will likely continue to leverage ablative TPS to ensure their successful operation under extreme thermal conditions.

The effectiveness of ablative materials, such as Phenolic Impregnated Carbon Ablator (PICA), is well-documented. PICA has been extensively used in various aerospace applications, including re-entry vehicles, due to its ability to withstand high heat fluxes and its favourable thermal properties [437, 438]. The thermal protection system (TPS) provided by such materials is designed to mitigate heat through a combination of phase changes, chemical reactions, and material removal, which is crucial for maintaining the integrity of the vehicle during hypersonic flight [438, 439]. The unique properties of these ablators allow them to dissipate significant amounts of heat, thereby protecting the underlying structures from thermal damage [440, 441].

Research has shown that the thermal performance of ablative materials is influenced by their composition and structure. For instance, the thermal conductivity of phenolic resin composites can vary significantly based on the filler materials used and their distribution within the matrix [442, 443]. Additionally, advancements in material science have led to the development of hybrid ablative systems that combine traditional ablators with ceramic matrices, enhancing their thermal stability and mechanical performance under extreme conditions [444, 445]. Such innovations are essential for the next generation of hypersonic vehicles, which face increasingly challenging aerothermodynamic environments [442, 446].

Moreover, the design of ablative heat shields must consider the specific thermal loads and environmental conditions expected during flight. For example, the integration of lightweight porous ceramic materials has been explored to improve thermal insulation while maintaining structural integrity [447]. These materials exhibit low thermal conductivity and high thermal erosion resistance, making them suitable for use in hypersonic applications where weight and performance are critical [295, 448].

In the aerospace field, the development of composites with high thermomechanical and thermochemical stability is critical, especially for use in high-temperature applications like hypersonic missiles and aircraft. Carbon fibre-reinforced silicon-based (C/SiM) composites have emerged as a leading candidate for these conditions due to their excellent oxidation resistance, ablative properties, and mechanical strength. These composites primarily rely on carbon fibre for reinforcement, offering good mechanical resistance and thermal stability, making them ideal for ultrahigh-temperature environments such as the leading edges and nose cones of hypersonic vehicles [449].

A key feature of C/SiM composites is their oxidation resistance. Silicon carbide (SiC), commonly used in these composites, forms a protective glassy layer of silicon dioxide (SiO_2) when exposed to high temperatures. This SiO_2 layer acts as a barrier, preventing further oxidation of the carbon fibres beneath. However, as temperatures rise above 1600°C, other ultrahigh-temperature ceramics (UHTCs), such as zirconium diboride (ZrB_2) and hafnium diboride (HfB_2), are incorporated into the matrix to improve thermal protection at these extreme conditions. These materials enhance the ablation resistance of the composite by creating additional protective oxide layers [449].

Ablation mechanisms in these composites operate through a combination of thermochemical, thermomechanical, and thermophysical processes. As hypersonic missiles travel at speeds exceeding Mach 5, they encounter intense aerodynamic heating. The surface of the missile is subject to extreme temperatures, causing the outer layers of the composite to undergo ablation. The ablative process involves the gradual erosion of the material as it absorbs and dissipates heat through pyrolysis and thermochemical reactions. In Si-based composites, the protective SiO_2 layer forms a detached shock in hypersonic flight, reducing the heat load transferred to the vehicle [449].

For improved performance, refractory metal oxides such as zirconia (ZrO_2) and hafnia (HfO_2) are used in conjunction with SiO_2 to further insulate the carbon fibres from oxidation. These oxides form complex, multi-layered protective barriers that seal cracks and prevent oxygen penetration. The addition of materials like zirconium diboride helps to create borosilicate glass layers during ablation, which further enhance the ablative resistance by sealing surface defects and reducing the impact of thermal and mechanical stress [449].

Manufacturing of C/SiM composites involves processes such as chemical vapor infiltration (CVI), pressureless infiltration (PI), and reactive melt infiltration (RMI). These methods enable the production of dense composites with minimal porosity, essential for maintaining structural integrity at high temperatures. Moreover, textile technology plays a crucial role in reinforcing the composite's structure, with 3D-braided or needled carbon fibre preforms offering superior mechanical and thermal stability compared to traditional 2D woven fabrics [449].

The materials used in these heat shields are specifically engineered to withstand high temperatures and provide thermal protection through ablation. Among the most prominent materials utilized in this context are

Phenolic Impregnated Carbon Ablator (PICA), Carbon-Carbon Composites, and Ultra-High Temperature Ceramics (UHTCs). Phenolic Impregnated Carbon Ablator (PICA) is a lightweight material composed of carbon fibres impregnated with phenolic resin. This combination results in a structure that can withstand temperatures exceeding 1600°C during atmospheric re-entry, making it suitable for hypersonic applications [450, 451]. PICA has been extensively used in various space missions, including the Mars Science Laboratory, where it demonstrated effective thermal protection under extreme conditions [452, 453]. The material's ability to undergo controlled ablation allows it to dissipate heat while maintaining structural integrity, which is crucial for the performance of hypersonic missiles [454, 455].

Carbon-Carbon Composites are another category of materials that exhibit exceptional thermal resistance and mechanical strength. These composites consist of carbon fibres embedded in a carbon matrix, providing a robust structure that can withstand high temperatures and mechanical stresses [456]. Their lightweight nature and high thermal conductivity make them ideal for hypersonic applications, where both weight and thermal management are critical [456]. However, these materials are susceptible to oxidation, which poses challenges in high-oxygen environments typically encountered during hypersonic flight [456].

Ultra-High Temperature Ceramics (UHTCs), such as zirconium diboride (ZrB_2) and hafnium diboride (HfB_2), are also employed in hypersonic missile heat shields due to their remarkable melting points and oxidation resistance [456]. These materials can endure temperatures above 3000°C, making them suitable for the most extreme thermal environments [456]. UHTCs are particularly valuable in applications where conventional materials would fail, providing an additional layer of thermal protection for hypersonic vehicles [456].

The fabrication process of ablative heatshields, particularly those utilizing materials like Phenolic Impregnated Carbon Ablator (PICA), involves several critical steps that ensure the structural integrity and thermal performance necessary for aerospace applications.

The initial step in the fabrication of ablative heatshields is the construction of a carbon fibre preform. This preform acts as the structural backbone of the heatshield, providing necessary rigidity and support. The use of carbon fiber is crucial due to its high strength-to-weight ratio and thermal stability. Research indicates that the design and arrangement of carbon fibres significantly influence the mechanical properties of the resulting composite materials [436, 457]. The preform must be meticulously crafted to ensure that it can withstand the extreme conditions encountered during re-entry into the Earth's atmosphere or other planetary bodies.

Following the construction of the carbon fibre preform, the next step is resin impregnation. This process involves saturating the carbon fibres with a thermosetting resin, typically phenolic resin, to enhance the material's thermal and mechanical properties. Various methods such as vacuum impregnation and pressure infiltration are employed to ensure thorough resin penetration into the porous structure of the carbon fibres [458, 459]. The impregnation process is critical, as incomplete saturation can lead to weak points in the heatshield. Studies have shown that the viscosity of the resin and the method of application significantly affect the impregnation efficiency, which is essential for achieving optimal performance in high-temperature environments [460, 461].

Once the resin has been adequately impregnated, the composite undergoes a curing process. This step involves heating the material under controlled conditions to facilitate the chemical reaction that solidifies the resin, thereby bonding it to the carbon fibres. Curing is a vital process that determines the final mechanical properties of the heatshield, including its thermal resistance and structural integrity [462, 463]. The curing conditions, such as temperature and duration, must be carefully controlled to prevent defects such as voids or incomplete curing, which could compromise the heatshield's performance [464].

The final stage in the fabrication process is rigorous testing and simulation of the heatshield's performance under hypersonic conditions. This testing is essential to evaluate the material's ability to withstand extreme temperatures and erosion rates during flight. Advanced simulation techniques and experimental setups are utilized to replicate the conditions experienced during re-entry, allowing researchers to assess the thermal dissipation and erosion characteristics of the material [465, 466]. The results from these tests are critical for validating the design and ensuring the reliability of the heatshield in actual flight scenarios.

Advanced manufacturing techniques are pivotal in the development of ablative heatshields for hypersonic vehicles, which must endure extreme thermal and mechanical stresses during flight. These techniques facilitate the precise customization of materials to enhance thermal protection and performance, with Spark Plasma Sintering (SPS), Chemical Vapor Deposition (CVD), and Additive Manufacturing (AM) being three key methods.

Spark Plasma Sintering (SPS) is a sophisticated technique that consolidates powdered materials, including Ultra-High Temperature Ceramics (UHTCs), into dense components through the application of electric current and pressure. The SPS process allows for rapid heating, often exceeding 1600°C, while simultaneously applying mechanical pressure, which significantly reduces sintering time and enhances material densification [467, 468]. This rapid heating leads to fine microstructures, which are critical for improving heat tolerance and mechanical strength in materials used for hypersonic applications [402, 469].

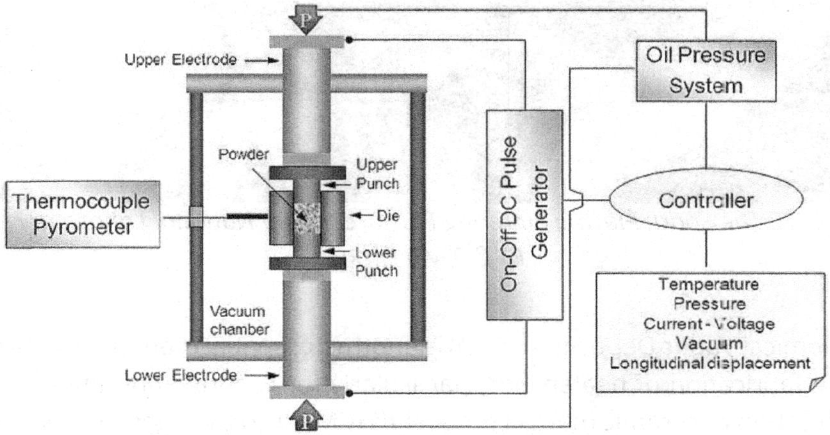

Figure 34: Spark Plasma Sintering process. , CC BY-SA 4.0, via Wikimedia Commons.

Moreover, SPS minimizes grain growth, a phenomenon that can compromise material integrity at elevated temperatures, thus ensuring that the heatshield can withstand the harsh conditions of hypersonic flight [470, 471]. The ability to create complex geometries, such as those required for leading edges and nose cones, further underscores the advantages of SPS in the manufacturing of heatshields [472, 473].

Figure 35: The Spark Plasma Sintering Furnace. Idaho National Laboratory, CC BY 2.0, via Flickr.

Chemical Vapor Deposition (CVD) is another essential method employed in the fabrication of heatshields, particularly for coating applications and the creation of ceramic matrix composites (CMCs). In CVD, gaseous precursors react in a high-temperature environment to form solid materials that deposit uniformly onto substrates, thereby providing excellent thermal protection [474]. The process allows for precise control over material properties at the microstructural level, enabling engineers to tailor the density, composition, and thickness of coatings [475]. This capability is particularly beneficial for developing multi-layer heatshields that can endure both thermal and chemical erosion, which is critical in hypersonic environments where oxidation resistance is paramount [476]. CVD techniques can also facilitate the creation of graded structures, optimizing the heatshield's performance under varying stress conditions [477].

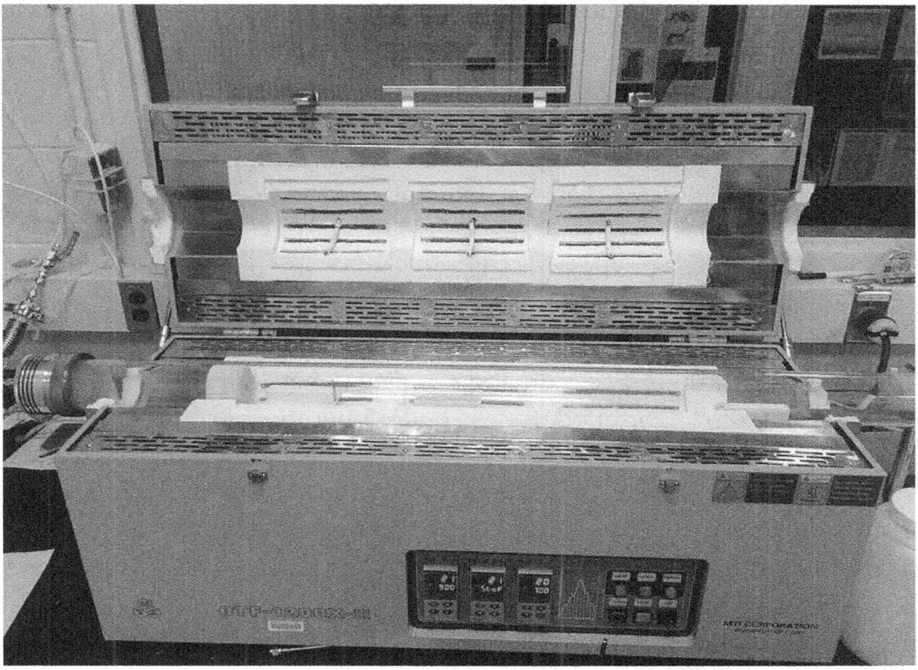

Figure 36: CVD is used for synthesizing materials in a controlled manner. Here is a 3-zone furnace with the requisite gas lines to transport the gas. Aksy88, CC BY-SA 4.0, via Wikimedia Commons.

Additive Manufacturing (AM), commonly known as 3D printing, is an innovative technique gaining traction in the production of ablative materials for heatshields. AM builds objects layer by layer from digital models, allowing for the creation of intricate internal structures and complex geometries that traditional manufacturing methods cannot achieve [478]. This method offers significant advantages, including the customization of internal structures for improved heat dissipation and the reduction of weight without compromising performance—both critical factors for hypersonic vehicles [478]. Furthermore, AM facilitates rapid prototyping, enabling quicker iterations of heatshield designs and reducing overall development costs [478]. Techniques such as Selective Laser Sintering (SLS) and Direct Metal Laser Sintering (DMLS) are particularly useful for experimenting with new material combinations, enhancing thermal management capabilities [478].

Layered and multi-material structures are critical in the design of ablative heat shields, particularly for aerospace applications. These structures typically consist of three main components: an outer ablative layer, insulating layers, and a structural core. Each layer plays a significant role in managing thermal loads and maintaining the integrity of the heat shield during high-speed flight.

The outer ablative layer is engineered to withstand extreme heat loads, absorbing thermal energy and gradually eroding away. This process not only protects the underlying materials but also helps dissipate heat through the removal of material, which is a fundamental characteristic of ablative materials. Studies have shown that materials such as phenolic matrix composites exhibit excellent ablation resistance and high char yielding, making them suitable for use as outer layers in heat shields [479]. Furthermore, the performance of these ablative materials is often characterized by their recession and charring behaviour, which are crucial for understanding their effectiveness under operational conditions [480].

Beneath the ablative layer, insulating materials are employed to prevent excessive heat transfer to the missile's structural components. These insulating layers can include lightweight materials such as silica tiles or advanced composites that provide thermal insulation. The effectiveness of these insulating layers is enhanced by their multi-layered architecture, which can significantly improve thermal performance by reducing heat flux through reflective and absorptive mechanisms [481]. For instance, multi-layer materials have been shown to enhance infrared shielding performance, which is vital for maintaining the structural integrity of the underlying components during high-temperature exposure [481].

The structural core of the heat shield is typically made from robust materials such as carbon-carbon composites or ceramic matrix composites. These materials provide the necessary mechanical strength and stability to withstand the stresses encountered during flight. The layered architecture of these composites contributes to their fatigue performance and crack resistance, which are essential for maintaining the overall integrity of the heat shield under dynamic loading conditions [482, 483]. Additionally, the combination of hard and soft materials in layered structures allows for a balance between strength and flexibility, which is crucial for high-performance applications [482].

Insulative Layers

Besides ablative coatings, which actively degrade to dissipate heat, insulative layers are another essential defence mechanism. These materials, such as silica-based composites, are designed to minimize heat transfer to critical internal systems like the missile's electronics, sensors, and warhead. Without effective insulation, these internal components could overheat, malfunction, or be destroyed, jeopardizing the missile's performance and mission success.

Silica-based composites, such as silica aerogels and fibrous silica materials, are commonly used for their excellent thermal insulation properties. These materials have an exceptionally low density and thermal conductivity, making them ideal for environments where reducing heat transfer is critical. Silica's structure includes microscopic air pockets, which act as barriers to heat conduction, making it a highly effective thermal insulator even under extreme conditions. These materials can withstand temperatures as high as 1600°C, ensuring the protection of sensitive missile components during high-speed flight through the atmosphere.

The layered structure of thermal insulation in missiles typically consists of multiple layers, each serving a specific function. The outermost layer, often composed of ablative materials, handles the direct heat from re-entry or atmospheric friction by degrading and absorbing heat. The inner layers, which include silica-based composites, provide passive insulation by blocking further heat transfer to the core components. These layers are particularly critical for protecting the missile's electronics and warhead, as overheating could cause system failures or premature detonation.

One of the primary challenges in designing thermal insulation systems for hypersonic missiles is achieving a balance between thermal protection and weight minimization. The missile's speed and manoeuvrability are heavily influenced by its overall mass, so the insulation must provide adequate thermal protection without adding unnecessary weight. Silica-based composites, due to their low density, are well-suited to this task, offering high-performance insulation without significantly increasing the missile's mass.

Active Cooling Techniques

Active cooling techniques are vital in hypersonic vehicles, particularly for components such as engines that endure extreme heat due to high-speed flight and combustion processes. In hypersonic missiles and vehicles, traditional passive thermal protection methods, such as ablative coatings and insulative layers, may not be sufficient to handle the intense heat generated during sustained flight at Mach 5 or above. For critical areas, such as the scramjet engine or other high-heat sections, active cooling systems become necessary to prevent material failure and ensure optimal performance.

Active cooling operates by circulating a coolant, often a liquid or the vehicle's own fuel, through channels embedded near the missile's hottest surfaces. This coolant absorbs the intense heat, preventing it from reaching and damaging vital components. One particularly important application of active cooling is in scramjet engines (supersonic combustion ramjets), where the fuel itself is often used as the coolant. This process, known as regenerative cooling, involves passing the fuel through cooling channels in the engine walls before it is injected into the combustion chamber. As the fuel absorbs the heat, it cools the engine and increases the temperature of the fuel, which aids in combustion efficiency once the fuel enters the engine [286, 484].

The principle behind regenerative cooling is to leverage the vehicle's fuel as a dual-purpose material, both as a coolant and as a propellant. By absorbing the heat from the engine walls, the fuel prevents overheating, reducing the thermal stress on the engine's components [286, 484]. This is crucial because, at hypersonic speeds, temperatures around the engine can easily reach several thousand degrees Celsius. Without active cooling, these temperatures could cause the engine materials to fail, leading to catastrophic loss of the vehicle.

In addition to scramjet engines, other hot spots on hypersonic vehicles, such as the nose cone or leading edges, may also employ active cooling methods. For these regions, coolants can circulate through narrow channels designed to manage the intense aerodynamic heating. The coolant absorbs and transports the heat away from these surfaces, where it is either radiated away or further dissipated within the system [286, 484].

One of the key challenges in designing active cooling systems for hypersonic vehicles is managing the heat load without adding too much weight to the system. Engineers must carefully balance the amount of coolant and the complexity of the cooling channels with the need to maintain the vehicle's aerodynamics and structural integrity. Additionally, designing effective cooling systems for hypersonic conditions requires materials that can withstand both high heat and pressure while remaining lightweight.

China's National University of Defence Technology has reportedly developed a new cooling device aimed at improving the performance of hypersonic missiles. This development addresses the critical challenge of managing the intense heat generated during hypersonic flight, a major limiting factor for missile endurance and speed. The new cooling device claims to extend the flight time of hypersonic missiles by almost two and a half hours—significantly longer than previous solutions—potentially enabling travel from one side of the Earth to the other [485].

The new cooling device developed by Chinese scientists is designed with simplicity and cost-effectiveness in mind, eliminating the need for complex systems such as heat pipes or pumps. Instead, it operates autonomously, using the thermal energy generated by aerodynamic heating as the driving force for the cooling cycle. This system allows the device to pressurize itself and circulate cooling water through its structure without external energy sources [485].

During flight, heat from the missile's surface is transferred into the device, causing the cooling water to flow in a clockwise direction. This circulation is maintained by pressure generated from temperature-sensitive components, which helps manage the heat without the need for external booster pumps. A key feature of the device is the use of a water storage container filled with an aerogel, which absorbs and stores water for gradual release [485].

The device's design allows it to maintain the temperature inside the missile's cabin below 100°C (212°F) for the first 50 minutes of flight. After this point, the water coolant begins to boil, which significantly enhances the heat transfer efficiency. By the 66th minute of flight, the system reaches its peak cooling performance, at which point high-temperature steam is gradually released through a pressure relief valve. The porous structure of the aerogel helps control this process, ensuring the system can maintain lower cabin temperatures even after the water has fully evaporated [485].

This innovative approach contrasts with traditional high-temperature heat pipes, which often rely on liquid metals as a cooling medium. The scientists claim that their device is not only simpler but also easier to implement than more complex active cooling systems. It utilizes phase changes in the cooling medium to absorb and release heat efficiently without adding complexity to the system [485].

Despite its impressive performance, this new cooling device is not a standalone solution. Hypersonic missiles require a comprehensive thermal management system that includes various protective technologies. For optimal performance, this cooling device will need to be integrated with other high-temperature-resistant coatings, insulation materials, and advanced thermal protection systems to ensure the missile's structural integrity and operational functionality throughout long-duration flights [485].

Optimization of Active and Passive Thermal Protection Systems for a Hypersonic Vehicle

The optimization of active and passive thermal protection systems (TPS) for hypersonic vehicles is a critical area of research due to the extreme thermal environments these vehicles encounter during flight. Hypersonic vehicles, which operate at speeds exceeding Mach 5, face severe aerodynamic heating that can result in temperatures exceeding 2000 °C, particularly at the leading edges and other critical surfaces [172]. This necessitates the development of advanced thermal protection systems that can effectively manage and mitigate these thermal loads.

Active thermal protection systems (ATPS) utilize mechanisms such as cooling fluids or heat exchangers to dissipate heat from the vehicle's surface. For instance, Liu and Jiang propose a non-ablative thermal protection system that integrates a spike-blunt body structure with lateral jets to reconstruct shock waves, thereby reducing drag and thermal loads [486, 487]. This innovative approach not only addresses thermal management but also enhances aerodynamic performance, which is crucial for maintaining vehicle integrity during hypersonic flight.

In contrast, passive thermal protection systems (PTPS) rely on materials and structural designs that inherently resist heat transfer. Materials such as metallic honeycomb structures and ceramic matrix composites are commonly employed due to their high thermal resistance and lightweight properties [488]. The use of multilayer thermal insulations has also been explored, as these can provide significant thermal resistance while minimizing weight, which is essential for hypersonic applications [127]. Furthermore, the optimization of sandwich structures with low coefficients of thermal expansion has been shown to enhance the structural integrity of TPS under extreme thermal conditions [125].

The integration of both active and passive systems can lead to a more robust thermal protection strategy. For example, the combination of heat-balance thermal protection with high thermal conductivity materials allows for efficient heat transfer from high to low-temperature areas, thereby protecting critical components from thermal damage [489]. Additionally, the modelling of fluid-thermal-structural interactions is vital for predicting the performance of TPS under hypersonic conditions, as it allows for the assessment of how different materials and designs will behave under extreme thermal and aerodynamic loads [291].

Moreover, the development of advanced materials, such as charring composites, has been a focus of research to improve the thermal protection capabilities of TPS. These materials can undergo pyrolysis, forming a protective char layer that insulates the underlying structure from high temperatures [490]. The effectiveness of these materials can be enhanced through numerical simulations that analyse their thermal behaviour under varying conditions [490].

Aerodynamic Design Solutions

Aerothermodynamic Problem Areas

In the past decade or so, the focus on developing hypersonic systems for military applications has shifted from creating long-range aerospace planes for rapid global force delivery to addressing more immediate, time-critical targets. These include mobile missile launchers, anti-ballistic missile (ABM) defences, and theatre missile defence (TMD). Hypersonic missiles offer significant advantages in such scenarios due to their extreme speed, which can limit the enemy's response time. This advantage is particularly relevant as adversaries continue to improve ground- and air-launched defence systems. By increasing the speed of attack weapons, military forces can exploit the reduced reaction times of enemy defences, making hypersonic technology a critical component of modern warfare [491].

However, while hypersonic technologies for exo-atmospheric engagements (such as space-bound systems) remain unchanged, hypersonic vehicles operating within the upper or lower atmosphere present new sets of challenges. Atmospheric manoeuvres introduce complications such as aerodynamic heating, material limitations, and control authority under extreme conditions. Additionally, research into gun-launched hypersonic kinetic energy projectiles has gained renewed attention for defeating new armour concepts and hardened ground targets. These kinetic projectiles can reach velocities of 2.4 to 2.8 km/sec using conventional or electromagnetic gun technologies, which are being explored in both the United Kingdom and the United States [491].

Looking ahead, solid fuel rockets are expected to remain the primary propulsion system for tactical hypersonic missiles in the near future. However, air-breathing propulsion systems, particularly scramjets (supersonic combustion ramjets) using hydrocarbon fuels, hold promise for higher-speed, long-range missiles. Scramjets could potentially achieve speeds up to Mach 8 and benefit from integrated engine-airframe designs. These propulsion systems offer advantages in fuel efficiency and extended flight times, but the technology still faces significant challenges [491].

One of the most notable limitations of hypersonic missiles is related to manoeuvrability. At hypersonic speeds, vehicles must generate greater lateral acceleration to maintain constant-radius turns, which places extreme demands on aerodynamic control systems. The faster a vehicle moves, the more difficult it becomes to maintain control, especially when pulling sharp turns. This need for improved control authority has led to the exploration of non-axisymmetric missile bodies, which can generate more effective normal or side forces during manoeuvres [491].

In addressing hypersonic design challenges, one major focus is on mitigating kinetic heating and ablation, especially at critical points like missile nose cones and control surfaces. For instance, missile seekers (used for targeting) must be protected from excessive heat. Engineers also face the challenge of managing interference heating, which occurs when aerodynamic control surfaces like fins or reaction jets disturb airflow, causing flow separation and reattachment that can significantly increase localized heating. Complex flow interactions make predicting heat transfer difficult, particularly when turbulent or transitional boundary layers are involved [491].

Experimental studies, such as those conducted at the University of Southampton, have provided insight into interference heating effects in hypersonic flow conditions. These studies highlight the intense heating experienced in regions where airflow reattaches after separation, especially on control surfaces like deflected flaps, fins, and flares. For example, deflected flaps can cause shock-induced flow separation, leading to intense heat transfer upon reattachment downstream. Moreover, reaction control jets—used as alternatives to aerodynamic surfaces—introduce similar flow disturbances, with jets causing heat transfer peaks in regions downstream of the interaction. These challenges highlight the complexity of designing hypersonic systems and the need for continued research into flow behaviour, heat transfer, and material resilience under extreme conditions [491].

Design

Aerodynamic design is important in managing heat and reducing drag in hypersonic missiles. At these extreme velocities, the interaction between the missile and the atmosphere generates significant friction, leading to extreme thermal loads that can compromise the integrity of critical components. To address these challenges, engineers implement various aerodynamic design strategies, notably the use of blunted nose cones and advanced shockwave management techniques.

Blunted Nose Cones: The design of the nose cone is pivotal in hypersonic missile aerodynamics. Traditional sharp, pointed nose cones create intense localized heating due to the formation of strong shockwaves at the tip, resulting in high thermal stresses that can lead to material fatigue and structural failure [29]. In contrast, blunted nose cones are employed to mitigate these issues. By spreading the shockwave over a larger surface area, blunted designs significantly reduce the intensity of heat generation at any single point, thereby lowering the heat flux experienced by the missile [492]. This design strategy enhances the thermal resistance of the vehicle, allowing for longer flight durations without degradation of materials [150].

The aerodynamic efficiency of blunted nose cones is a trade-off; while they may create more drag compared to sharp designs, their ability to dissipate heat effectively is critical in hypersonic flight [493]. The larger shockwave area associated with blunted nose cones helps ensure that critical components, such as electronics and warheads, remain within operational temperature ranges during flight [494]. This balance between drag and heat management is essential for the successful operation of hypersonic missiles.

The design of nose cones in hypersonic missile technology is crucial for optimizing both aerodynamic performance and thermal management. Traditional sharp nose cones, while aerodynamically efficient, generate significant localized heating due to intense shockwaves at hypersonic speeds. This localized heating can lead to severe thermal stresses, risking material integrity and structural failure under prolonged exposure [495, 496]. In contrast, blunted nose cones distribute the shockwave over a larger surface area, significantly reducing heat concentration and thermal stress, which is vital for maintaining the integrity of missile components during high-speed flight [497, 498].

Blunted nose cones offer a substantial advantage in thermal protection. By spreading the heat flux over a broader area, these designs enable hypersonic missiles to endure extreme temperatures without compromising the functionality of critical components such as guidance systems and warheads [499, 500]. This is particularly important given that hypersonic flight can expose vehicles to temperatures exceeding 2000 °C [499, 500]. The ability to manage thermal loads effectively is a key factor in extending the operational lifetime of these missiles [496, 498].

However, the adoption of blunted nose cones is not without trade-offs. One significant drawback is the increased aerodynamic drag associated with blunted shapes compared to sharper designs. The bluntness creates a larger area of resistance against airflow, which can reduce overall speed and manoeuvrability [501, 502]. Despite this, the benefits of enhanced thermal protection often outweigh the disadvantages of increased drag in the context of hypersonic flight, where thermal management is critical [502, 503]. Research has shown that while blunted designs may incur a drag penalty, they are essential for ensuring that missiles can operate effectively under the extreme conditions encountered during hypersonic flight [400, 496].

Shockwave Management: In addition to the shape of the nose cone, shockwave management is a vital aspect of hypersonic missile design. At hypersonic speeds, the air cannot move out of the way quickly enough, leading to the formation of shockwaves that contribute significantly to the heat load on the missile [504]. Engineers utilize advanced techniques such as waverider designs, which allow the missile to "ride" its own shockwave. This innovative approach reduces friction with the surrounding air and minimizes heating, as the missile effectively harnesses the shockwaves it generates to create lift [505]. These designs enable the missile to effectively "ride" its own shockwave, which not only enhances its aerodynamic profile but also minimizes drag and friction. By optimizing the vehicle's shape to maintain the shockwave attached to its body, wave-rider designs reduce thermal stresses and improve fuel efficiency, thereby extending the missile's operational range [506]. The aerodynamic characteristics of these configurations allow for stable flight at hypersonic speeds, which is crucial for maintaining control and precision during missions [29].

Waverider designs not only enhance aerodynamic efficiency but also improve heat distribution across the missile's surface. By directing heat away from sensitive components, such as propulsion systems and internal electronics, these designs ensure that critical systems remain operational even under extreme conditions [138]. Furthermore, optimizing the geometry of waverider missiles can enhance fuel efficiency, enabling longer ranges and higher speeds [190]. The integration of blunted nose cones and shockwave management techniques is therefore essential for maintaining the structural integrity and performance of hypersonic missiles, allowing them to withstand the intense thermal and aerodynamic challenges posed by hypersonic flight.

The aerodynamic and thermal benefits of wave-rider designs are substantial. By leveraging shockwave management techniques, engineers can significantly reduce the heat load on thermal protection systems, which are vital for safeguarding critical components such as avionics and guidance systems. Furthermore, the enhanced aerodynamic stability provided by these designs is essential for manoeuvrability and accuracy in both offensive and defensive military applications [189]. The ability to maintain a steady altitude and trajectory while minimizing aerodynamic heating is a critical advancement in hypersonic missile technology, ensuring reliable performance during extended missions [507].

Overview of Waverider Design and Hypersonic Aircraft Configuration

Hypersonic aircraft, by design, must address several key performance and structural challenges due to the extreme conditions they encounter. One of the primary challenges is achieving an optimal balance between lift-to-drag (L/D) ratios, structural integrity, and thermal protection. Understanding the characteristics and design issues of general hypersonic aircraft is crucial before diving into the specifics of waverider design, a configuration developed to enhance overall performance in hypersonic conditions.

Configuration Considerations: Hypersonic aircraft must have very streamlined and minimalistic designs to withstand the severe heat and aerodynamic stresses associated with traveling at speeds of Mach 5 or higher. The flat plate is theoretically the most efficient hypersonic lifting surface, with the highest L/D ratio, but it is impractical due to its lack of space for internal components like engines or payloads. As such, more realistic designs, like the generic hypersonic transport configuration, attempt to maximize performance by using:

- **Small frontal areas** and streamlined shapes to minimize total surface drag.

- **Reduced wing area**, where the fuselage is designed to contribute additional lift.

- **Integrated propulsion systems** that merge with the overall airframe to ensure optimal aerodynamic efficiency.

Aerodynamic Performance: Wind tunnel tests on hypersonic configurations reveal significant differences in aerodynamic performance compared to subsonic and low-supersonic designs. Specifically, the lift-to-drag (L/D) ratio of hypersonic vehicles is generally much lower, primarily due to the effects of high wave drag and skin friction associated with strong shock waves at hypersonic speeds. For instance, while traditional aircraft such as the P-51 Mustang and F-111 bomber achieve maximum L/D ratios between 12 and 15, hypersonic vehicles tested by NASA, such as the X-43A, exhibit L/D ratios around 5.5 at Mach 7 [508]. This reduction in aerodynamic efficiency is a critical challenge in the design and operation of hypersonic vehicles.

The aerodynamic characteristics of hypersonic vehicles are influenced by various factors, including the vehicle's shape and the flow conditions it encounters. Research indicates that the high wave drag experienced at hypersonic speeds results from the formation of shock waves, which significantly impacts the overall aerodynamic performance [120]. Additionally, the skin friction drag, which is sensitive to boundary layer behaviour, further exacerbates the decrease in L/D ratio [509]. Computational Fluid Dynamics (CFD) and wind tunnel testing have been instrumental in developing an aerodynamic database that captures these effects, allowing for better predictions and optimizations of hypersonic vehicle designs [510].

Moreover, studies have shown that the aerodynamic shape optimization of hypersonic vehicles is crucial for improving their performance. Techniques such as the Class-Shape-Transformation (CST) method and multi-objective evolutionary algorithms have been employed to enhance the L/D ratio while considering payload constraints [78, 511]. The findings from these optimization efforts highlight the necessity of addressing the unique aerodynamic challenges posed by hypersonic flight, particularly the need to mitigate wave drag and improve overall aerodynamic efficiency [512].

The relationship between Mach number and L/D ratio was empirically described by Kuchemann's equation, which predicts a decline in L/D as Mach number increases. This phenomenon, sometimes referred to as the L/D barrier, highlights the difficulty of optimizing aerodynamic efficiency in hypersonic flight.

Kuchemann's equation, which describes the relationship between the Mach number and the lift-to-drag (L/D) ratio, is a key empirical tool in hypersonic flight analysis. The equation reveals that as Mach number increases, the L/D ratio decreases. This decline in aerodynamic efficiency is often referred to as the L/D barrier, emphasizing the challenges of achieving high performance at hypersonic speeds.

The Impact of Mach Number on L/D Ratio: At lower speeds, aircraft designs can achieve relatively high L/D ratios because of favourable aerodynamic conditions, where lift is efficiently generated with minimal drag.

Hypersonic vehicles experience extreme aerodynamic heating and intense shock waves, contributing to both wave drag and skin friction drag. These two types of drag are the main factors behind the declining L/D ratio. At high speeds, the energy required to overcome these forces becomes disproportionately large, limiting the vehicle's overall aerodynamic efficiency. Kuchemann's empirical analysis supports this observation by showing a sharp drop in L/D ratios as the Mach number increases. In hypersonic flight, typical L/D values might only reach 5.5, a stark contrast to the values seen in slower speeds.

The "L/D Barrier": This declining trend, sometimes referred to as the **L/D barrier**, underscores the difficulty of maintaining an optimal balance between lift and drag at hypersonic speeds. The "barrier" reflects the inherent limitations imposed by the extreme conditions encountered at these speeds, such as high-temperature flows and strong shockwaves. It indicates that there is a performance ceiling in the hypersonic regime where the combination of factors like viscous effects and shockwave formation make further gains in L/D ratio difficult to achieve.

Although hypersonic flight promises advantages like faster travel and military applications, the efficiency in terms of L/D ratio becomes constrained. The empirical nature of Kuchemann's equation, derived from experimental data and wind tunnel tests, has held up in both practical applications and theoretical models, making it a widely accepted framework for analysing hypersonic aerodynamics.

Research and Challenges: Research into hypersonic vehicle designs, including concepts like waveriders and compression lift vehicles, aims to break through this barrier by innovating on vehicle shape, material, and propulsion integration. For instance, waveriders are designed to "ride" their own shockwaves, reducing drag by aligning the vehicle's leading edges with the shockwave, which can potentially improve L/D ratios even at hypersonic speeds.

Despite these innovations, the fundamental trade-off between aerodynamic efficiency and the harsh thermal and dynamic environment of hypersonic flight continues to present a significant engineering challenge. As a result, even advanced designs must contend with the fact that increasing speed tends to degrade aerodynamic performance, reinforcing the constraints described by Kuchemann's L/D barrier.

Compression Lift and Shape Optimization: In hypersonic conditions, achieving higher L/D ratios requires innovative vehicle shapes. One such optimization is based on the momentum principle, where the ideal hypersonic shape is designed to maximize downward momentum while minimizing the forward momentum imparted on the surrounding fluid. This led to the development of a body of revolution, or a cone shape, where the negative lift on the upper surface is cancelled by positive lift on the lower surface. By further refining this design and attaching wing surfaces that capture momentum changes between the body and the shockwave, hypersonic vehicles can harness compression lift.

Compression lift refers to the increased surface pressure realized on the underside of a hypersonic vehicle due to the proximity of the shock layer. This concept is central to the waverider design, where the leading edge of the wing is carefully matched to the shockwave formed by the vehicle's forebody. This design allows the vehicle to "ride" the shockwave, maximizing lift while minimizing drag.

Stability and Control: Stability and control in hypersonic vehicles are significantly more complex than in subsonic designs. Hypersonic designs typically incorporate rudders, elevators, and tip flaps for enhanced lift and control. These features are critical for maintaining balance and manoeuvrability in high-speed flight. Wind tunnel tests of hypersonic designs with both straight and cranked wing panels revealed trade-offs between roll control and lateral-directional stability, with longitudinal instability remaining a challenge. Advanced fly-by-wire systems or precise fuel management may be required to maintain stability.

Propulsion Integration: Unlike subsonic aircraft where engines are mounted as distinct units, hypersonic vehicles must integrate their propulsion systems into the airframe due to the thin shock layers and high-speed conditions. Most hypersonic designs will use ramjet or scramjet engines, which rely on the vehicle's motion to compress incoming airflow, eliminating the need for complex compressor and turbine stages found in traditional engines. The scramjet engine, for instance, is integrated into the vehicle's body to take advantage of the shockwave generated by the forebody, compressing the air before it enters the combustion chamber. The vehicle's aft section is also designed to act as part of the engine's expansion nozzle, further optimizing thrust production.

Heat Transfer Challenges: Heat management is a critical issue in hypersonic flight, as aerodynamic heating can cause significant structural stress. While sharp leading edges minimize drag, they also concentrate heat, requiring blunted surfaces to distribute heat over a larger area. Additionally, cryogenic cooling systems, such as those used in the B-70 Mach 3 bomber and proposed for the X-30 research vehicle, pump fuel through the structure to absorb and dissipate heat. Hypersonic vehicles are likely to require heat-resistant coatings and thermal protection systems to withstand the intense heat generated during flight.

Waverider designs represent a convergence of advanced aerodynamic principles, shockwave management, and thermal protection strategies. They maximize performance by riding shockwaves, integrating propulsion systems, and addressing the extreme heat generated during hypersonic flight. While waveriders offer significant advantages in lift-to-drag ratios, stability, and heat management, they still face challenges that require ongoing research and technological innovation.

Scramjet Engine Materials and Cooling

Hypersonic vehicles that use scramjet engines—a type of air-breathing propulsion system that operates at supersonic combustion speeds—face significant challenges related to managing the extreme heat generated during flight. As scramjets compress incoming air at hypersonic speeds, the combustion chamber temperatures can exceed 3,000°C, requiring specialized materials and cooling techniques to ensure the engine remains functional under such intense conditions.

High-Temperature Combustion Chambers

In the context of high-temperature combustion chambers, particularly in scramjet engines, the materials utilized must exhibit exceptional thermal stability, mechanical strength, and oxidation resistance. Two primary categories of materials that fulfill these stringent requirements are nickel-based superalloys and ultra-high temperature ceramics (UHTCs).

Nickel-Based Superalloys: Nickel-based superalloys are extensively employed in aerospace applications due to their remarkable ability to maintain mechanical integrity under extreme thermal conditions. These alloys, such as Inconel 718, are designed to withstand high temperatures while exhibiting excellent creep resistance and fatigue life [513-515]. The microstructural characteristics of these superalloys, including the precipitation of the γ' phase, significantly contribute to their strength at elevated temperatures [514, 515]. Research indicates that the mechanical properties of nickel-based superalloys can be further enhanced through specific heat treatments, which optimize the microstructure for high-temperature applications [514, 516].

Moreover, the oxidation resistance of nickel-based superalloys is critical for their performance in combustion chambers, where they are subjected to aggressive oxidizing environments. The addition of elements such as chromium and aluminium enhances their oxidation resistance by forming protective oxide layers [516, 517]. Studies have shown that these alloys can endure cyclic loading and thermal stresses, making them suitable for the demanding conditions found in scramjet combustion chambers [518, 519].

Ultra-High Temperature Ceramics (UHTCs): In addition to nickel-based superalloys, ultra-high temperature ceramics (UHTCs) such as zirconium diboride (ZrB_2) and hafnium carbide (HfC) are critical materials for applications requiring resistance to temperatures exceeding 2,000°C.

UHTCs are characterized by their exceptional thermal stability and low thermal expansion coefficients, which make them ideal for use in environments where thermal shock and oxidation are prevalent [520]. The unique properties of UHTCs allow them to maintain structural integrity under extreme conditions, thereby enhancing the overall performance of combustion chambers in scramjet engines [520].

The integration of UHTCs into the design of combustion chambers is increasingly common, as these materials can effectively manage the thermal loads and stresses encountered during hypersonic flight. Their ability to withstand thermal creep and fatigue is essential for ensuring the longevity and reliability of combustion chamber components [520].

Regenerative Cooling

Regenerative cooling is a pivotal technique in managing heat within scramjet engines, primarily due to the extreme thermal environments these engines operate in. This process involves routing the fuel through cooling channels that are strategically placed around the combustion chamber before it is injected for combustion. As the fuel traverses these channels, it absorbs heat from the surrounding structures, effectively cooling the engine and mitigating the risk of thermal failure [521-523]. The significance of regenerative cooling cannot be overstated, as it serves a dual purpose: it not only cools the engine components but also pre-heats the fuel, enhancing combustion efficiency.

The cooling of engine components is critical in preventing overheating and material degradation. The high temperatures within the combustion chamber can lead to severe thermal stresses that could compromise the integrity of the engine materials [522, 523]. By utilizing regenerative cooling, the fuel absorbs excess heat from the combustion chamber walls, thereby maintaining the structural integrity of the engine components. This method is particularly vital in scramjet engines, where even advanced composite materials struggle to withstand the intense heat loads encountered during hypersonic flight [523, 524].

Moreover, the pre-heating of fuel through regenerative cooling significantly improves combustion efficiency. The absorbed heat raises the fuel's temperature before it enters the combustion chamber, facilitating better fuel-air mixing and ignition [521, 524, 525]. In scramjets, where the high-speed airflow limits the time available for complete combustion, this pre-heating process can enhance overall engine performance. The regenerative cooling system thus plays a crucial role in extending the operational lifespan of the materials used in the combustion chamber by alleviating excessive thermal stresses [524, 526].

The concept of regenerative cooling, which utilizes fuel as a coolant to manage extreme heat in engines, has been a significant development in aerospace engineering, particularly relevant in the context of scramjet engines, where managing intense thermal loads during hypersonic flight is crucial. Regenerative cooling works by circulating the fuel through cooling channels adjacent to the combustion chamber, allowing it to absorb heat before being injected into the combustion process, thus preventing overheating of engine components [527-529].

In scramjet engines, the extreme temperatures generated during operation can exceed the thermal limits of conventional materials, necessitating advanced cooling techniques. Regenerative cooling is recognized as one of the most effective methods for thermal protection in these high-speed applications. The fuel, often in a supercritical state, serves both as a propellant and a heat sink, absorbing significant amounts of heat through endothermic reactions such as thermal cracking [71, 528, 530]. This dual function enhances the overall efficiency of the engine while protecting critical components from thermal damage [531, 532].

Recent studies have demonstrated the effectiveness of regenerative cooling systems in scramjets, highlighting the importance of optimizing cooling channel designs to enhance performance. For instance, the use of computational fluid dynamics (CFD) has allowed researchers to analyse flow and heat transfer characteristics in cooling channels, leading to improved designs that maximize cooling efficiency [533, 534]. Additionally, the incorporation of features such as ribs or dimples in the cooling channels has been shown to further enhance heat transfer capabilities [531, 534]. These advancements are vital for ensuring the structural integrity and operational reliability of scramjet engines under the extreme conditions of hypersonic flight.

Overall, regenerative cooling represents a critical technology in the development of advanced aerospace propulsion systems, particularly for hypersonic vehicles. The integration of hydrocarbon fuels as both a coolant and a fuel source addresses the thermal management challenges posed by high-speed flight and contributes to the overall performance and efficiency of scramjet engines [535, 536]. As research continues to refine these systems, regenerative cooling is poised to play an increasingly important role in the future of aviation and space exploration.

The Cooling Process

Regenerative cooling in scramjets is a critical engineering solution to manage the extreme temperatures generated inside the combustion chamber during hypersonic flight. Scramjet engines, which operate by compressing incoming air at supersonic speeds and mixing it with fuel for combustion, face significant thermal challenges. The combustion process can produce temperatures of up to 3,000°C, which could compromise the structural integrity of the combustion chamber if not properly managed.

Figure 37: Regenerative cooling channel for scramjet.

To address this, regenerative cooling works by routing the engine's fuel through channels embedded in the walls of the combustion chamber before it is injected for combustion. This process achieves two key objectives: cooling the combustion chamber walls and preheating the fuel for combustion, which improves overall engine efficiency.

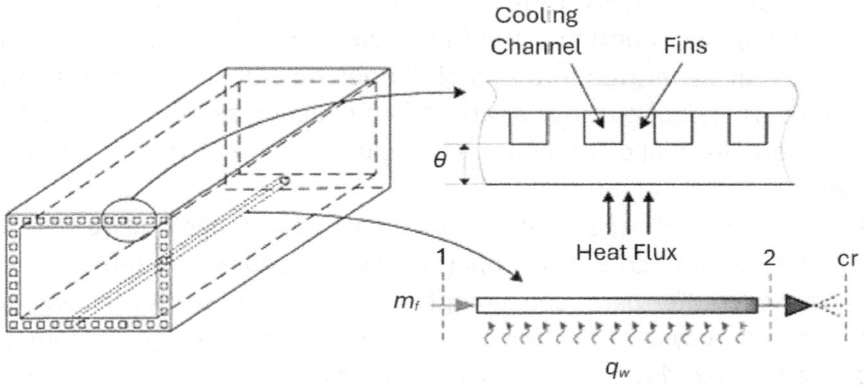

Figure 38: Quasi-One-Dimensional Model of Scramjet Combustor Coupled with Regenerative Cooling [537].

Stages of heat transfer in regenerative cooling:

1. **Convective and Radiative Heat Transfer**: Inside the combustion chamber, hot gases generated by the combustion process transfer heat to the chamber walls through convection (movement of hot gases) and radiation (emission of heat energy). These gases are at extremely high temperatures and exert significant thermal stress on the materials lining the chamber.

2. **Heat Conduction through the Chamber Walls**: Once the heat reaches the chamber walls, it is conducted outward toward the cooling channels. This conduction ensures that the heat generated inside the combustion chamber doesn't remain concentrated on the inner walls, which would otherwise lead to structural degradation or failure.

3. **Convective Heat Transfer to Coolant (Fuel)**: The fuel circulating through the cooling channels absorbs the heat from the combustion chamber walls. This heat transfer is essential not only for cooling the engine but also for preheating the fuel before combustion, making the fuel more reactive and efficient when injected into the chamber. The preheated fuel, now ready for combustion, is then injected into the combustion chamber, completing the cycle.

The continuous cycle of regenerative cooling allows scramjet engines to operate at hypersonic speeds while maintaining the integrity of the combustion chamber. By using fuel as both a coolant and a combustible medium, engineers optimize the fuel's role within the engine, ensuring that the heat generated in the engine does not cause material fatigue or failure. This dual-purpose approach—fuel acting as both coolant and energy source—makes regenerative cooling highly efficient in scramjet engines.

The efficiency of regenerative cooling also enhances engine performance by allowing scramjets to operate over long durations and at higher speeds, where thermal management becomes increasingly challenging. Without such a system, scramjets would struggle to manage the intense heat, leading to reduced flight time and increased risk of engine failure due to overheating.

Regenerative cooling in scramjet engines presents several key concerns and challenges that must be addressed to enhance performance and reliability. One of the primary issues is the significant difference in heat flux experienced in scramjet combustion chambers compared to traditional liquid rocket engines. While scramjets operate with heat fluxes ranging from 2.0 to 20.0 MW/m^2, liquid rocket engines like the Space Shuttle Main Engine (SSME) can experience heat fluxes as high as 160 MW/m^2. Despite the lower heat flux in scramjets, the extreme conditions of hypersonic flight lead to substantial heating, necessitating effective cooling strategies to protect engine components from thermal damage [529, 538, 539].

Another critical concern is the mass flow ratio of fuel in scramjet engines. In liquid rocket engines, the mass flow rate is typically sufficient to manage heat transfer effectively, with fuel temperatures remaining below 250°C. Conversely, scramjets optimize fuel mass flow for combustion efficiency, which complicates cooling efforts. The challenge lies in ensuring that the fuel can absorb enough heat to maintain engine integrity while still being effective for combustion [540-542].

Inherent challenges in regenerative cooling systems for scramjets include managing supercritical heat transfer. At the high temperatures and pressures typical of scramjet operation, fuels can reach supercritical conditions, altering their thermodynamic properties and complicating heat transfer management [521, 543]. This necessitates a sophisticated understanding of thermodynamics and material behaviour to ensure effective cooling under these conditions.

Coking of hydrocarbon fuels is another significant issue. The high temperatures in scramjet engines can lead to the formation of carbon deposits within cooling channels, obstructing fuel flow and potentially leading to engine failure if not adequately managed [544-546]. This coking phenomenon is exacerbated by the supercritical conditions and requires ongoing research into materials and fuel properties to mitigate its effects [547-549].

The optimization of cooling channels is also crucial for effective regenerative cooling. The design must maximize the surface area in contact with the fuel while minimizing pressure losses and maintaining structural integrity. This involves careful engineering to ensure that the cooling channels can handle the thermal loads without adding excessive weight to the engine [529, 550, 551].

The materials and cooling systems used in scramjet engines must be specifically designed to withstand the high temperatures and stresses of hypersonic flight. Nickel-based superalloys and UHTCs play key roles in providing the necessary heat resistance, while regenerative cooling is critical for ensuring that the engine operates safely and efficiently under extreme thermal conditions. Together, these innovations allow scramjet engines to achieve sustained operation at hypersonic speeds, pushing the boundaries of aerospace propulsion technology.

Precision Targeting and Navigation

Precision in hypersonic weapons presents a complex balance between maintaining extreme speed and ensuring accuracy, especially when compared to subsonic cruise missiles or ballistic missiles. Hypersonic missiles, which travel at speeds exceeding Mach 5, face distinct challenges in maintaining accurate targeting during flight due to the high velocities and the physical stresses that these speeds impose on the missile [552].

At hypersonic speeds, manoeuvrability becomes a significant issue. The faster a missile travels, the harder it becomes to make course corrections, particularly in the final phases of flight where precision is most critical. Subsonic cruise missiles, by contrast, can adjust their trajectory with relative ease, allowing them to maintain higher accuracy. In comparison, traditional ballistic missiles are less precise as they cannot easily adjust their path during the terminal phase unless equipped with manoeuvrable re-entry vehicles (MaRVs). Hypersonic weapons generally fall between these two categories: they are more manoeuvrable than ballistic missiles, yet face more difficulties than cruise missiles due to the extreme forces and heat generated during hypersonic flight [552].

One of the major challenges hypersonic missiles face is the "plasma effect," which occurs when the missile travels at such high speeds that the air surrounding it is ionized, creating a cloud of plasma. This plasma can interfere with radio signals and sensors, making communication and guidance systems less effective. As a result, for hypersonic missiles to achieve precision in targeting, they often need to decelerate as they approach their target. This reduction in speed allows the missile to regain full control over its guidance systems, enabling the necessary course corrections for a precise strike on moving or shifting targets [552].

However, even if the missile must slow down during the terminal phase, the overall advantage of reaching the target area quickly—much faster than subsonic or even some ballistic missiles—provides a significant strategic benefit. The high speed allows the missile to bypass defence systems and catch adversaries off-guard before they can respond effectively. This means that although the final approach may sacrifice some of the missile's hypersonic velocity, the speed during most of the flight still provides critical advantages in time-sensitive or heavily defended scenarios [552].

The relevance of the speed-precision trade-off in hypersonic weapons depends heavily on the combat scenario. In cases where the primary objective is to strike a static or hardened target, the need for precise manoeuvring may be less critical, and the missile can maintain its extreme speed throughout most of the flight. On the other hand, when targeting mobile assets, such as enemy vehicles or ships, precision becomes much more important, and the missile must slow down to make the necessary adjustments for a direct hit [552].

In these dynamic combat situations, the missile's ability to quickly enter the operational area before defences can activate remains a significant advantage, even if it must slow down near the end. The fact that a hypersonic missile arrives in the vicinity of its target much faster than conventional weapons forces defenders to react in an extremely compressed timeframe, reducing the chances of an effective defence. Therefore, despite potential sacrifices in terminal-phase speed, the overall lethality of hypersonic missiles remains high due to their speed advantage.

Precision targeting and navigation in hypersonic missiles is a critical component that enables these advanced weapons to hit specific targets accurately, despite traveling at speeds exceeding Mach 5. Due to the extreme velocities involved in hypersonic flight, maintaining precise control over the missile's trajectory and ensuring it reaches its intended target requires sophisticated technologies and methods. Several key elements contribute to the success of precision targeting and navigation in hypersonic missiles:

1. Inertial Navigation Systems (INS): Inertial navigation is a core component of hypersonic missile targeting systems. INS relies on gyroscopes and accelerometers to track the missile's position, velocity, and orientation by measuring its acceleration and rotational movements. As hypersonic missiles travel at extremely high speeds, traditional GPS systems can become unreliable due to signal delays or interference from atmospheric conditions. INS offers an advantage because it doesn't rely on external signals and can continue to provide navigation information autonomously, even in GPS-denied environments.

However, due to the long flight times of some hypersonic missiles, errors in INS can accumulate. To counter this, INS is often used in conjunction with other navigation systems to ensure precision.

2. GPS Augmentation: Although INS is vital, integrating GPS (Global Positioning System) with inertial navigation helps to improve targeting accuracy. GPS updates the missile's INS data by correcting any accumulated drift. This hybrid navigation system ensures that hypersonic missiles can achieve and maintain precise targeting over long distances, even at the extreme velocities and altitudes associated with hypersonic flight. Additionally, GPS-based corrections help the missile adjust for any deviations caused by atmospheric drag, heat, or manoeuvring.

3. Terrain Contour Matching (TERCOM): Terrain contour matching (TERCOM) is a technique used in some hypersonic missiles to enhance precision targeting. By comparing real-time measurements of the terrain beneath the missile with preloaded maps, the missile can adjust its course to follow the terrain more accurately. This method allows for high precision, especially when targeting fixed ground targets. TERCOM is highly resistant to jamming because it doesn't rely on external signals, making it an effective navigation aid in contested or electronically degraded environments.

4. Advanced Radar and Sensors: Hypersonic missiles also employ advanced radar systems and sensors to aid in real-time targeting. Active radar systems can lock onto a target and track its movement, providing guidance as the missile closes in on the target. Electro-optical and infrared sensors may also be used to identify and track targets, especially in the terminal phase of flight when precise targeting is critical. These sensors ensure that the missile can adjust its trajectory in response to target movement or changes in environmental conditions, such as cloud cover or terrain obstructions.

5. Artificial Intelligence and Machine Learning: Recent advancements in artificial intelligence (AI) and machine learning (ML) are being applied to hypersonic missile guidance and targeting systems. AI algorithms can process real-time data from multiple sources (radar, sensors, GPS) to make rapid adjustments to the missile's trajectory. By predicting the movement of mobile targets, AI-enhanced systems can improve the missile's accuracy, especially in dynamic combat environments where targets may change position frequently. This capability is particularly useful in scenarios involving mobile launchers, ships, or other fast-moving assets.

6. Mid-Course and Terminal Phase Guidance: Hypersonic missiles typically employ different guidance strategies during various phases of flight. In the mid-course phase, where the missile is flying at extremely high altitudes and velocities, navigation systems focus on maintaining trajectory with minimal course corrections. As the missile descends into the terminal phase and approaches its target, precision becomes even more critical. Advanced seekers, radar, and sensor systems come into play here, guiding the missile as it makes rapid manoeuvres to evade defences and hone in on the target with pinpoint accuracy.

The accuracy of hypersonic missiles hinges on the balance between their guidance and control systems, as well as the unique challenges posed by their high-speed atmospheric flight. This precision depends on several factors, including how the missile identifies and tracks its target and how it adjusts its flight path during its mission [553].

The guidance system is responsible for determining the missile's position and motion, which can be done through multiple techniques. Inertial guidance systems, for example, track the missile's motion internally using accelerometers and gyroscopes to monitor every movement. These systems are highly reliable because they do not require external input, but their accuracy can degrade over time as errors accumulate during long flight durations [553].

To improve accuracy, external references are often integrated into guidance systems. GPS satellites, celestial navigation (using stars), and terrain mapping are examples of technologies that can be used to ensure that a missile is aware of its exact position. Combining these methods with internal systems allows for greater precision by correcting for any drift that may occur during the flight [553].

Once the missile's guidance system determines its location, it must correct its course to ensure it reaches the target. During the missile's initial ascent, this is usually done by adjusting the thrust of the rockets.

As the missile continues its journey, particularly within the atmosphere, it uses aerodynamic forces—such as fins or wings—to make necessary adjustments [553].

However, several external factors can reduce accuracy. Gravity anomalies (variations in the Earth's gravitational pull) and atmospheric conditions like wind and air density can cause a missile to veer off course. Additionally, at hypersonic speeds, the intense heat from air friction can erode the missile's surface, potentially changing its aerodynamic properties and making it more difficult to control. This is a significant challenge for hypersonic missiles, which endure far more atmospheric exposure than ballistic missiles, as the latter spend most of their flight in outer space where these atmospheric forces are absent [553].

Hypersonic missiles generally have an advantage over ballistic missiles when it comes to manoeuvrability. They spend most of their flight within the atmosphere, allowing them to make aerodynamic adjustments to their trajectory over long periods. Ballistic missiles, on the other hand, can only make limited course corrections during their initial ascent and their final descent, unless equipped with a MaRV (Manoeuvrable Re-entry Vehicle), which allows for limited manoeuvres in the terminal phase [553].

However, this advantage is counterbalanced by the extreme challenges of hypersonic flight. Unlike ballistic missiles, which minimize their time in the atmosphere to reduce exposure to unpredictable forces, hypersonic missiles are subjected to high-speed atmospheric forces for the majority of their flight. This makes them more vulnerable to the effects of wind, variations in air density, and severe thermal stress. The intense heat generated by atmospheric friction can cause material erosion, degrading the missile's ability to manoeuvre [553].

Moreover, hypersonic speeds generate a plasma cloud around the missile due to the ionization of air, which can interfere with its communication systems and external navigation aids, such as GPS. This presents another challenge for guidance accuracy, as the missile may lose connection with essential data sources during critical moments of flight [553]

When considering all these factors, hypersonic missiles face significant obstacles that limit their precision. Although their manoeuvrability within the atmosphere surpasses that of ballistic missiles, the unpredictable atmospheric conditions and extreme heating they experience can degrade their control. Even with the most advanced guidance systems, these atmospheric effects make hypersonic weapons more prone to inaccuracies compared to ballistic missiles equipped with MaRVs, which have demonstrated the ability to strike targets with a precision of a few meters [553].

While hypersonic weapons offer unparalleled speed and some manoeuvring capabilities, their precision is hindered by the very forces that allow them to operate at hypersonic velocities. This makes them less reliable in achieving pinpoint accuracy, especially compared to advanced ballistic missile systems [553].

Inertial Navigation Systems

Inertial Navigation Systems (INS) are essential for the guidance of hypersonic missiles, particularly due to the extreme conditions and high speeds these weapons encounter. INS operates by utilizing gyroscopes and accelerometers to continuously calculate the missile's position, velocity, and orientation from the moment of launch. The accelerometers measure the missile's acceleration, while gyroscopes track its rotational movements, enabling the missile to navigate autonomously without relying on external signals such as GPS. This capability is crucial in environments where GPS signals may be unreliable due to interference or signal delays caused by phenomena like plasma formation around the missile at hypersonic speeds [24].

The operational mechanics of INS involve the continuous integration of acceleration data obtained from the accelerometers. Gyroscopes provide information on changes in orientation, allowing for necessary trajectory adjustments. The extreme speeds of hypersonic missiles, which exceed Mach 5, present unique challenges for traditional navigation systems. For instance, the X-43A hypersonic vehicle employs a mature INS/GPS integrated navigation system, demonstrating the reliance on inertial navigation for precision in high-speed scenarios [24, 554]. Moreover, the self-contained nature of INS makes it particularly effective in GPS-denied environments, where hypersonic missiles often operate, thus ensuring continued navigation based solely on inertial data [555].

Despite the advantages of INS, it is not without limitations. One significant challenge is the accumulation of errors over time, known as "drift." As the missile travels at hypersonic speeds, even minor inaccuracies in initial measurements can compound, leading to substantial navigation errors [556]. To address this issue, INS is frequently integrated with other navigation systems, such as GPS or terrain-matching systems. This hybrid approach allows for periodic corrections from GPS, which helps recalibrate the INS and mitigate accumulated errors, thus enhancing overall navigation accuracy [557]. The combination of INS's autonomy and GPS's precision ensures that hypersonic missiles remain on course during complex missions, highlighting the importance of integrated navigation systems in modern military applications [24, 554].

GPS Augmentation in Hypersonic Missiles

Inertial Navigation Systems (INS) are crucial for the navigation of hypersonic missiles, primarily due to their ability to operate autonomously without external references. However, one of the significant challenges associated with INS is the accumulation of errors over time, commonly referred to as drift. This drift occurs because INS relies on inertial sensors to measure acceleration and orientation, which can lead to inaccuracies in position estimation as the missile travels long distances at high speeds. To mitigate this issue, the integration of Global Positioning System (GPS) technology provides a valuable augmentation to INS. GPS offers continuous updates on the missile's position, allowing for real-time corrections that enhance overall targeting accuracy [133].

The hybridization of INS and GPS is particularly beneficial for hypersonic missiles, which travel at speeds exceeding Mach 5. At such velocities, the precision of targeting is paramount, as even minor deviations can result in significant targeting errors. The atmospheric conditions encountered during hypersonic flight, including intense drag and thermal effects, further complicate navigation. GPS corrections enable the missile to adjust its trajectory dynamically, compensating for any deviations caused by these factors [253]. Moreover, as the missile approaches its target, GPS signals facilitate fine-tuning of its trajectory, which is critical for engaging mobile or dynamic targets that may shift positions before impact [253].

Despite the advantages of GPS, hypersonic missiles face unique challenges, such as the generation of plasma around their surfaces due to extreme speeds. This plasma can interfere with GPS signals, leading to potential data loss during critical phases of flight. However, the hybrid navigation system effectively addresses this vulnerability. During periods of GPS signal loss, the INS can autonomously maintain the missile's course, ensuring that the missile remains on track until GPS signals are restored [133]. This capability underscores the importance of integrating both INS and GPS technologies to create a robust navigation solution capable of overcoming the challenges presented by hypersonic flight conditions.

Terrain Contour Matching (TERCOM)

Terrain Contour Matching (TERCOM) is a navigation technique that enhances the precision of missile targeting, particularly for cruise missiles. This system operates by comparing real-time terrain data collected by the missile with preloaded maps of the Earth's surface. As the missile traverses its flight path, it utilizes altimeters and other sensors to measure elevation data, which is then matched against stored contour maps to adjust its trajectory accordingly. This capability is particularly advantageous in environments where traditional guidance systems may falter due to electronic countermeasures or signal jamming [558-560].

One of the primary strengths of TERCOM lies in its terrain comparison capabilities. The missile's onboard sensors detect variations in the landscape below—such as mountains, valleys, and plains—and relay this information to the missile's computer system. By analysing these data points, the missile can self-correct its trajectory, allowing it to maintain a low altitude and effectively evade radar detection [561-563]. This continuous feedback loop enhances navigational accuracy and ensures that the missile remains on course toward its target, even in complex terrain [564, 565].

TERCOM is particularly effective for targeting fixed ground installations, where the terrain is relatively stable. The system's reliance on high-resolution terrain maps allows for precise course adjustments, which is critical when engaging high-value, stationary targets. This capability is especially important in military operations where the success of a mission often hinges on the ability to strike with pinpoint accuracy [566-568]. Unlike navigation systems that depend on external signals such as GPS, which can be disrupted, TERCOM's self-contained nature ensures operational reliability in contested environments [569, 570].

Moreover, TERCOM exhibits resistance to jamming and electronic warfare tactics. Since it does not rely on external signals, it remains operational even in GPS-denied environments, where satellite signals may be obstructed or interfered with. This self-sufficiency allows the missile to maintain a high degree of precision, even when adversaries deploy electronic countermeasures [571-573]. The ability to navigate based solely on terrain data makes TERCOM a robust solution for modern military applications, particularly in scenarios where traditional navigation aids are compromised [574, 575].

In the context of hypersonic flight, TERCOM presents unique advantages. Hypersonic missiles, which travel at extremely high speeds, face distinct challenges, including the generation of plasma during flight that can interfere with traditional navigation systems. However, TERCOM's capability to operate independently of satellite signals allows for effective mid-course corrections, ensuring that the missile remains aligned with its intended trajectory as it approaches its target [576-578]. This is particularly crucial in the final moments of flight, where precision is paramount for mission success [579].

Advanced Radar and Sensors in Hypersonic Missiles

In the realm of hypersonic missile systems, the integration of advanced radar and sensor technologies is paramount for achieving real-time targeting accuracy, especially at extreme speeds. Hypersonic missiles, which travel at speeds exceeding Mach 5, face significant challenges in maintaining precision due to their rapid movement and the dynamic nature of their operational environments. The combination of active radar systems and electro-optical/infrared (EO/IR) sensors plays a critical role in overcoming these challenges, facilitating dynamic tracking and precise targeting capabilities.

Active radar systems are essential for the guidance of hypersonic missiles. These systems emit radar waves that reflect off targets, enabling the missile to detect and lock onto them. As the missile approaches its target, it continuously receives radar feedback, which allows for real-time adjustments to its flight path. This capability is particularly crucial when engaging mobile targets that may change positions during the missile's flight. The ability of active radar to operate effectively in various weather conditions further enhances the missile's targeting accuracy, ensuring reliable performance despite atmospheric interferences such as rain or fog [19, 580].

Moreover, the integration of EO/IR sensors significantly enhances the missile's targeting capabilities during the terminal phase of flight. These sensors provide high-resolution imaging and heat detection, allowing the missile to identify specific thermal signatures or optical characteristics of targets. EO sensors utilize optical imaging technology to capture visual data, which is invaluable in complex terrains or when radar signals are obstructed by environmental factors such as cloud cover. In contrast, IR sensors detect the heat emitted by targets, making them particularly effective in low-visibility conditions, such as nighttime operations or adverse weather scenarios [581, 582]. This dual-sensor approach ensures that hypersonic missiles can maintain targeting accuracy even in challenging conditions.

Electro-Optical (EO) and Infrared (IR) sensors in hypersonic missiles are critical technologies that enhance targeting accuracy and situational awareness, particularly during the terminal phase of flight. These sensors are designed to provide real-time data on the environment and the target, allowing the missile to adjust its trajectory as it approaches the target at extreme speeds. Below is a detailed explanation of the specific EO/IR sensors used in hypersonic missiles:

1. Electro-Optical (EO) Sensors: Electro-optical sensors rely on capturing visual data from the surrounding environment and target. These sensors operate similarly to digital cameras, capturing light in the visible and sometimes ultraviolet spectrum. The primary advantage of EO sensors is their ability to provide high-resolution imaging, allowing the missile to visually recognize and track targets based on optical characteristics, such as shape, size, and contrast against the background.

Types of EO Sensors:

- **Charge-Coupled Device (CCD) sensors:** These sensors capture detailed images in various lighting conditions, often used in both daytime and nighttime operations.

- **Complementary Metal-Oxide Semiconductor (CMOS) sensors:** CMOS sensors are faster and consume less power compared to CCD sensors, making them suitable for high-speed operations like hypersonic missile flight.

Applications in Hypersonic Missiles: EO sensors are particularly useful in identifying fixed ground targets and in situations where optical clarity is critical, such as areas with terrain obstructions. They can be used in the final moments of flight when precise targeting is needed, especially in visually distinguishing between a target and non-target objects.

2. Infrared (IR) Sensors: Infrared sensors detect heat signatures emitted by objects, rather than relying on visible light. They operate across multiple bands of the infrared spectrum, such as:

- **Near-Infrared (NIR):** Useful in detecting heat from objects that are not extremely hot.

- **Mid-Wave Infrared (MWIR):** Excellent for tracking objects like vehicles and aircraft due to their pronounced heat emissions.

- **Long-Wave Infrared (LWIR):** Particularly effective in detecting temperature differences, making them useful in poor visibility environments, such as night-time or adverse weather.

Types of IR Sensors:

- **Uncooled Infrared Sensors:** These sensors are less sensitive to small temperature changes but offer a more compact and power-efficient solution.

- **Cooled Infrared Sensors:** These are highly sensitive and offer superior image resolution but require more power and are more complex. They are often used in critical hypersonic missile missions where precision is paramount.

Applications in Hypersonic Missiles: IR sensors are particularly effective in detecting and tracking moving targets or objects that emit heat, such as military vehicles, aircraft, or even missile launchers. As these missiles approach their targets at hypersonic speeds, the thermal contrast between the target and the surrounding environment helps the missile home in on its intended target, even in conditions of low visibility.

3. Multi-Spectral and Dual-Band EO/IR Sensors: Hypersonic missiles often employ multi-spectral sensors that can detect both visible and infrared wavelengths. These systems provide redundancy and robustness, ensuring that the missile can track targets even if one type of sensor is compromised due to weather conditions or environmental interferences.

Advantages of Multi-Spectral Sensors:

- They provide enhanced detection capabilities by combining data from both EO and IR spectra, improving tracking accuracy.

- These sensors are resilient to countermeasures such as smoke screens, flares, or camouflage because they rely on different bands of the electromagnetic spectrum.

Both EO and IR sensors are typically employed in the terminal phase of hypersonic missile flight, where pinpoint accuracy is required. During this phase, the missile may experience high levels of turbulence, environmental interference, and jamming attempts. EO and IR sensors provide real-time data, allowing the missile's guidance system to make rapid corrections and home in on the target. For instance:

- EO sensors may visually recognize a building or a vehicle, allowing the missile to focus on a specific fixed target.

- IR sensors may track the heat signature of a moving target like an aircraft or ground vehicle.

Hypersonic missiles equipped with EO/IR sensors are also highly resistant to electronic warfare and jamming techniques, particularly because these systems do not rely on external signals like GPS, which are more vulnerable to interference. By using terrain-based navigation (TERCOM) or onboard sensors that detect real-time environmental data, hypersonic missiles can maintain precision targeting even in contested environments.

The real-time adjustment capabilities afforded by the combination of radar and EO/IR sensors are crucial for effective target tracking. As the missile approaches its target, these sensors provide continuous updates on the target's movements and any environmental changes. For instance, if a target manoeuvres or if atmospheric disruptions occur, the integrated systems work collaboratively to recalibrate the missile's trajectory. This adaptability is vital for maintaining precision during high-speed flight, where even minor deviations can lead to significant targeting errors [22, 507]. The transition from mid-course to terminal guidance phases is particularly critical, as the missile must make last-minute adjustments to ensure a successful strike.

Furthermore, the adaptability of hypersonic missile systems to various environmental conditions is a significant advantage. Advanced sensors mitigate the effects of challenging conditions, such as cloud cover or dust storms. For example, while radar may struggle in certain atmospheric conditions, EO systems can provide visual feedback, and IR systems can continue tracking heat signatures. This multifaceted approach ensures that hypersonic missiles remain effective in a wide range of operational settings, thereby increasing their reliability in contested environments [71, 583].

Artificial Intelligence and Machine Learning

Artificial Intelligence (AI) and Machine Learning (ML) are being applied to enhance hypersonic missile guidance and targeting systems, significantly improving their accuracy and adaptability in complex environments. Here's how these technologies work in hypersonic missile systems:

AI algorithms in hypersonic missiles gather and analyse data from multiple sources, such as radar systems, infrared and electro-optical sensors, and GPS inputs. This multi-source data fusion enables the missile to obtain a comprehensive understanding of its surrounding environment, allowing it to react in real time. For example:

- Radar systems detect the position and speed of nearby targets.
- EO/IR sensors provide visual and heat-based imagery of targets.
- GPS updates ensure location accuracy, especially when combined with other guidance methods.

ML models use this data to predict target movement and adjust the missile's trajectory accordingly, enhancing precision during the flight. This is particularly critical in dynamic combat scenarios where targets may frequently change positions.

One of AI's strengths is its ability to make rapid, autonomous decisions based on patterns in the data it processes. AI can calculate the best course corrections faster than traditional systems by recognizing patterns of movement and predicting future positions of mobile targets such as ships, mobile missile launchers, or moving ground vehicles.

In this context, AI assists by:

- Predicting future target positions based on current data.
- Recognizing potential obstructions or evasive manoeuvres that a target may undertake.
- Adjusting the missile's trajectory or approach to optimize impact accuracy despite target evasions.

ML allows the missile's system to learn from past engagements and ongoing data during a mission. For example, if environmental conditions such as wind speed or heat fluctuations affect the missile's performance, the AI system can adjust on the fly by learning from these conditions in real time. This means:

- Continuous learning algorithms adjust flight parameters (such as angle, speed, or altitude) to maintain accuracy.
- Weather patterns or atmospheric density that could impact flight are accounted for and used to adjust the missile's trajectory.

In scenarios where targets are constantly in motion—like ships or mobile missile launchers—traditional missile systems might struggle to track and adapt. AI and ML algorithms, however, can better handle these complexities by continuously analysing and predicting the most likely path that a target will take. The system dynamically corrects the missile's path in real time, ensuring that the missile can successfully home in on fast-moving targets despite unpredictable changes in trajectory.

In a combat environment, this adaptive approach offers significant advantages, as mobile targets can attempt evasive manoeuvres, use decoys, or try to blend into terrain. AI helps in distinguishing the real target from false signals and maintaining a lock on it.

AI-enhanced systems can also counter electronic warfare (EW) tactics, such as jamming or signal interference. If GPS signals are blocked or radar systems are disrupted, the AI system can:

- Switch between sensors (such as EO or IR) to maintain a visual or heat-based lock on the target.

- Predict target behaviour based on past movement patterns, allowing it to stay on course even without direct updates from external systems.

In certain operational scenarios, AI can help prioritize targets based on strategic importance, fuel constraints, or overall mission objectives. The system can continuously evaluate target options, rerouting if higher-priority or more accessible targets become available. This is particularly useful in multi-target combat scenarios.

By integrating AI and ML, hypersonic missiles gain significant improvements in precision, adaptability, and reliability. The ability to process data in real time, learn from environmental conditions, and predict target movements ensures that these missiles remain accurate and effective, even in the most dynamic and hostile environments. These advancements are crucial for maintaining an edge in modern warfare, where target mobility and electronic interference are key challenges.

Mid-Course and Terminal Phase Guidance in Hypersonic Missiles

Hypersonic missiles utilize distinct guidance strategies during different phases of flight—each phase designed to handle the varying aerodynamic, navigational, and strategic challenges posed by hypersonic speeds. Two key phases in missile guidance are the mid-course phase and the terminal phase.

In the mid-course phase, the missile travels at extremely high altitudes, often near the edge of the atmosphere, at hypersonic speeds exceeding Mach 5. During this phase, the focus is primarily on maintaining a stable trajectory with minimal course corrections to ensure the missile remains on track toward its general target area. Several factors characterize this phase:

1. **Inertial Navigation System (INS):** The missile's guidance system relies heavily on **inertial navigation** at this stage. INS calculates the missile's position and velocity based on internal gyroscopes and accelerometers, providing autonomous navigation without the need for external signals. This is essential, as GPS signals can become unreliable due to interference at high altitudes and speeds.

2. **Minimal Corrections:** In the mid-course phase, the missile is usually operating in a relatively predictable environment with less atmospheric drag, reducing the need for significant adjustments. The goal is to conserve fuel and minimize thermal and mechanical stress by limiting manoeuvres.

3. **Atmospheric Re-entry Considerations:** As the missile travels through the upper atmosphere, thermal management becomes critical. The high velocities generate extreme heat due to air friction. In this phase, the missile may rely on advanced aerodynamic shaping and thermal protection systems to manage this heat, maintaining structural integrity before entering the denser layers of the atmosphere.

4. **External Navigation Aids:** In some cases, GPS augmentation or satellite guidance is used to provide updates on the missile's position and correct any accumulated drift from the INS. This ensures greater accuracy when transitioning to the next phase, though hypersonic missiles can also function in GPS-denied environments.

As the missile descends from high altitudes and enters the terminal phase, its guidance strategy shifts significantly. This phase is critical because the missile is nearing its target, and precision becomes paramount, especially against moving targets or heavily defended areas.

1. **Active Sensors and Seekers:** In the terminal phase, hypersonic missiles often engage advanced radar systems and electro-optical (EO) or infrared (IR) seekers to actively track and lock onto the target. These systems provide real-time data on the target's position and movement, allowing the missile to make final adjustments for precise targeting.

2. **High Manoeuvrability:** Hypersonic missiles must perform rapid and agile manoeuvres in this phase to counter enemy defences such as interceptors or anti-missile systems. Plasma effects—where an electrically charged gas sheath forms around the missile at high speeds—can interfere with communications and sensors, but advanced radar and seeker systems help the missile overcome these challenges, especially when engaging evasive or fast-moving targets.

3. **Precision Targeting:** As the missile closes in on the target, it em-

ploys high-precision targeting algorithms, often enhanced by Artificial Intelligence (AI) and Machine Learning (ML) systems. These technologies allow the missile to predict the movement of mobile targets, further improving accuracy. Real-time data from onboard sensors enable adjustments to the missile's flight path, ensuring it hits its mark with pinpoint precision.

4. **Terminal Manoeuvring:** In this phase, the missile may deploy aerodynamic control surfaces or reaction control systems to execute sharp course corrections and evasive manoeuvres. This is especially important when navigating through dense air at lower altitudes, where atmospheric forces are more unpredictable and can affect the missile's trajectory.

5. **Evasion of Defences:** The missile's ability to make rapid manoeuvres at extreme speeds also serves an additional function—evading interception attempts from enemy missile defence systems. Advanced radar-guided defences may track the missile, but its high speed and erratic manoeuvrability in the terminal phase make it difficult to intercept.

The combination of mid-course inertial guidance and terminal phase active sensors allows hypersonic missiles to adapt to different stages of their flight. While the mid-course phase is relatively stable and requires less active control, the terminal phase relies on sophisticated onboard sensors and real-time adjustments to achieve maximum accuracy. The integration of AI and ML in targeting systems further enhances the missile's ability to predict target movement, adapt to environmental factors, and ensure successful target engagement.

Overall, the guidance strategy of hypersonic missiles is a sophisticated process that transitions from relatively passive navigation in the mid-course phase to highly active, sensor-driven targeting in the terminal phase. This ensures that even at extremely high speeds and altitudes, the missile can achieve precise impact on its target, whether it is fixed or mobile, heavily defended or evasive.

Survivability in Adverse Conditions

The survivability of hypersonic missiles in adverse conditions is a critical aspect of their design and operational capabilities. Hypersonic missiles, traveling at speeds exceeding Mach 5, face extreme environmental challenges, including intense heat, atmospheric drag, and complex defence mechanisms. Engineers incorporate a variety of strategies to ensure that these missiles maintain functionality and effectiveness, even in hostile and unpredictable environments.

Thermal Protection: One of the most significant challenges for hypersonic missiles is managing the immense heat generated by air friction as they travel through the atmosphere. The surface temperatures of these missiles can exceed 1,000°C, and in some cases, approach 3,000°C, particularly during long-duration flights at low altitudes. This intense heating can lead to material degradation, which compromises structural integrity and aerodynamic performance.

To combat this, hypersonic missiles use advanced thermal protection systems (TPS). These systems often include materials such as Ultra-High Temperature Ceramics (UHTCs) and ablative coatings that can withstand high temperatures without deteriorating. These materials protect critical components, such as electronics and propulsion systems, from extreme thermal stresses.

Additionally, regenerative cooling techniques, where fuel is circulated through channels in the missile's structure to absorb and dissipate heat, are commonly used in scramjet engines. This not only cools the missile's external structure but also preheats the fuel, improving combustion efficiency.

Hypersonic vehicles face significant challenges in terms of materials survivability, as their key subsystems—primary structures, thermal protection systems (TPS), and propulsion systems—must endure extreme conditions such as high heat flux, oxidative environments, and intense mechanical stresses. Developing materials that can withstand these demanding environments is a limiting factor in advancing hypersonic technologies. To design the next generation of resilient materials, an integrated approach that spans from atomistic to macroscopic scales is required. This comprehensive framework enables better understanding and manipulation of materials' properties, particularly through computationally-informed strategies [584].

Metals for Hypersonic Vehicles: Refractory metals, known for their high melting points and thermal stability, are widely used in hypersonic vehicles. However, they exhibit weaknesses in oxidation resistance and strength when subjected to extreme temperatures, particularly in environments with high heat flux. The introduction of high-entropy alloys (HEAs), also referred to as multi-principal element alloys (MPEAs), offers potential improvements. These advanced alloys could reduce density while enhancing oxidation kinetics, making them more suitable for hypersonic applications. The development of these alloys is still in its infancy, and a deeper understanding of their thermophysical properties in operational conditions is needed [584].

Emerging manufacturing methods, such as additive manufacturing, combined with high-throughput characterization techniques and machine learning algorithms, are critical in discovering novel metal compositions rapidly. These approaches can help uncover materials with enhanced high-temperature performance and resistance to oxidation, paving the way for advanced cooling systems and robust metallic structures capable of withstanding the extreme conditions faced by hypersonic vehicles [584].

Composites for Hypersonic Vehicles: Composites offer excellent strength-to-weight ratios and can tolerate high temperatures, especially in inert atmospheres. However, when exposed to extreme heat in oxidative environments, uncoated composites degrade due to oxidation and erosion. One area of focus is enhancing protective high-temperature emissive coatings, which help composites withstand such environments. Improving the bond between coatings and composite matrices, as well as the interfaces between fibres and the matrix, remains a challenge due to differences in material properties [584].

Advances in thermomechanical models and simulations are essential for predicting material behaviour in hypersonic conditions, as current computational techniques struggle to accurately scale properties from microscale fibres to bulk composites. Multi-scale reinforcement strategies, which consider the anisotropic nature of fibre-reinforced materials, are an important area for future development. New reinforcement techniques could improve the overall performance of composites and make them more suitable for hypersonic applications [584].

Ceramics for Hypersonic Vehicles: Refractory ceramics and ultra-high temperature ceramics (UHTCs) are essential for hypersonic applications due to their high melting points and thermal conductivities. However, their use is often limited by low thermal shock resistance and high density when used as monolithic components. The most promising application for these ceramics is in thermal barrier coatings and ceramic matrix composites (CMCs) [584].

One example is transition metal carbides incorporating silicon carbide (SiC), which have gained attention for their ability to endure high thermal and mechanical loads. Despite this, oxidation at temperatures above 1600°C remains a significant challenge. Continued research into the structure-processing-property relationships of ceramics is essential to overcome these issues. This includes exploring transformation toughening (a process that enhances material toughness), improving oxidation resistance, and understanding the fibre-matrix bonding in UHTC composites [584].

Future breakthroughs in ceramics for hypersonic vehicles are likely to come from innovations in additive manufacturing, machine learning-based materials design, and high-entropy UHTC compositions. These methods allow for the creation of complex ceramics with customized properties, tailored for the extreme demands of hypersonic flight. Such materials, formed using these advanced methods, are expected to push the boundaries of what is currently achievable with conventional manufacturing processes, supporting the development of more resilient hypersonic vehicles [584].

Overcoming the materials limitations for hypersonic vehicles requires a multi-disciplinary approach involving advanced materials design, innovative manufacturing techniques, and computational tools. Continued research into metals, composites, and ceramics is essential to meet the aerothermal, oxidative, and mechanical challenges posed by hypersonic flight [584].

Resistance to Atmospheric Conditions: Hypersonic missiles travel through varying atmospheric layers, each posing different challenges. In the upper atmosphere, there is less drag, but as the missile descends into denser air during the terminal phase, the forces acting on it increase significantly. The missile must maintain structural integrity and precise navigation, even as it experiences rapid changes in pressure, air density, and wind conditions.

To enhance survivability, engineers focus on aerodynamic shaping, ensuring that the missile's design minimizes drag and manages heat buildup effectively. Blunted nose cones and wave-rider designs help mitigate shockwave-induced heating and optimize lift-to-drag ratios. This design philosophy ensures that the missile can maintain control, avoid overheating, and remain manoeuvrable even under intense atmospheric stresses.

Manoeuvrability to Evade Defences: A key element of survivability for hypersonic missiles is their ability to evade interception by enemy defence systems. Due to their high speeds, hypersonic missiles are difficult to detect and track, but they are still vulnerable to interceptors and missile defence systems as they approach their target.

To overcome this, hypersonic missiles are equipped with advanced manoeuvring capabilities. During the terminal phase, the missile performs rapid, unpredictable manoeuvres to evade interception attempts. This is crucial when navigating through enemy airspace protected by anti-missile defence systems. The ability to perform evasive manoeuvres at hypersonic speeds, while under thermal and aerodynamic stress, is a significant factor in the missile's survivability.

In addition to aerodynamic controls, reaction control systems (RCS) may be used to adjust the missile's orientation and trajectory. These systems help fine-tune the missile's flight path, making it harder to predict and intercept.

Resistance to Plasma and Electromagnetic Interference: At hypersonic speeds, the intense heat generated by friction with the atmosphere creates a plasma sheath around the missile. This ionized layer of gas can interfere with communication systems, radar guidance, and external sensor readings, potentially affecting the missile's ability to receive updates from command centres or satellites.

To mitigate the effects of plasma interference, advanced hypersonic systems rely on inertial navigation systems (INS) and onboard sensors that function autonomously without requiring external signals. INS, which tracks the missile's position and motion using gyroscopes and accelerometers, provides navigation data even when external signals are blocked. Additionally, plasma-resistant materials and technologies that allow for intermittent or burst transmissions through the plasma sheath are under development, further enhancing the missile's operational capability in GPS-denied or communication-challenged environments.

Enhanced Durability Against Physical Damage: Hypersonic missiles are often subjected to harsh physical environments, including atmospheric turbulence, wind shear, and fragmentation effects from enemy defences like anti-missile missiles or directed-energy weapons. To survive these potential threats, hypersonic missiles must be designed with high structural durability.

This involves using composite materials and refractory metals that can endure extreme heat and stress without degrading. The missile's aerodynamic surfaces, which are critical for control and manoeuvrability, must be resilient enough to handle debris impacts or near-miss explosions without losing functionality. Materials like titanium alloys, carbon-carbon composites, and silicon carbide play key roles in ensuring the missile's ability to withstand physical damage.

Electromagnetic and Electronic Countermeasures: In adversarial environments, electronic warfare (EW) measures, such as jamming or spoofing, can disrupt a missile's guidance systems. To enhance survivability, hypersonic missiles are equipped with electronic countermeasures (ECM) that protect against electronic interference. Additionally, the use of radar-absorbing materials and stealth technology helps reduce the missile's radar signature, making it more difficult for enemy defence systems to detect and track.

COUNTER-HYPERSONIC DEFENCE SYSTEMS

Missile Defence: Current and Emerging Technologies

Several existing missile defence systems, such as the Patriot Advanced Capability-3 (PAC-3), the Aegis Ballistic Missile Defense (BMD) system, and the Terminal High Altitude Area Defense (THAAD), are already showing potential for intercepting hypersonic threats. These systems have been primarily designed for ballistic missile defence, but they also possess capabilities that can be adapted to counter the emerging hypersonic threat. This underscores the growing need for defence systems to evolve as they face new challenges posed by the extreme speeds and manoeuvrability of hypersonic weapons [552].

The Patriot Advanced Capability-3 (PAC-3) missile defence system represents a significant advancement in missile interception technology, particularly in its ability to counter tactical ballistic missiles and emerging hypersonic threats. The PAC-3 builds upon previous iterations of the Patriot missile system, evolving from its original design aimed at intercepting aircraft and cruise missiles to a more sophisticated system that employs a hit-to-kill mechanism. This mechanism allows the PAC-3 to destroy its target through direct impact rather than relying on an explosive warhead, enhancing its effectiveness against fast-moving threats [585, 586].

The operational capabilities of the PAC-3 are centred around its advanced radar and tracking systems, which detect incoming threats during their terminal phase. Upon detection, the system engages its interceptor missile, which utilizes onboard radar and sophisticated targeting algorithms to home in on the target. This process involves several critical steps: detection, tracking, engagement, and the hit-to-kill impact. The agility of the PAC-3 is particularly noteworthy, as it allows the system to effectively engage not only traditional ballistic missiles but also manoeuvring hypersonic vehicles, which pose a significant challenge due to their high speeds and unpredictable trajectories [276, 587].

Figure 39: A PATRIOT Advanced Capability 3 (PAC-3) interceptor is launched from Omelek Island during MDA's historic integrated flight test on Oct. 24, 2012. U.S. Missile Defense Agency, CC BY 2.0, via Flickr.

The PAC-3's design incorporates a two-layer missile defence strategy, which includes both exoatmospheric high-layer interceptors and endoatmospheric low-layer interceptors. This layered approach is essential for addressing the complexities of modern missile threats, particularly given the short time frames available for interception in terminal phases [585]. The system's ability to adapt to various types of threats, including hypersonic vehicles that can exceed speeds of Mach 5, underscores its strategic importance in contemporary defence frameworks [276, 324].

Moreover, recent advancements in guidance and control algorithms have further enhanced the PAC-3's interception capabilities. For instance, the integration of advanced guidance laws, such as those based on sliding mode control and dynamic inversion, has been shown to improve the accuracy and responsiveness of interceptors against high-speed targets [277, 588]. These developments are crucial as they allow the PAC-3 to maintain effectiveness in increasingly complex and dynamic combat environments.

The Aegis Ballistic Missile Defense (BMD) System is a critical component of the United States' defence strategy against ballistic missile threats. Deployed on U.S. Navy cruisers and destroyers, the Aegis BMD system utilizes advanced radar and interceptor technology to detect, track, and neutralize incoming missiles. The system primarily employs the SPY-1 radar, which is capable of detecting and tracking ballistic missiles at high altitudes, and the Standard Missile-3 (SM-3) interceptor, designed to engage threats in their mid-course phase before they re-enter the Earth's atmosphere [589, 590].

The operational effectiveness of the Aegis BMD system is enhanced by its integration of various sensor platforms, including space, sea, and land-based systems. This multi-layered approach allows for comprehensive situational awareness and rapid response capabilities. Upon detection of a missile threat, the Aegis system can launch an SM-3 interceptor, which ascends into space to intercept the missile mid-flight. The SM-3 employs a hit-to-kill mechanism, utilizing kinetic energy to destroy the incoming threat through direct collision [74, 590]. This capability is particularly significant in the context of emerging hypersonic missile threats, as the system's design allows for engagement during the glide phase of such missiles [591].

Figure 40: A Standard Missile-3 (SM-3) Block 1A interceptor is launched from the USS Decatur (DDG -73) during a successful intercept test. USS Decatur with its Aegis Weapon System detected and tracked the target with its on board AN/SPY-1 radar. The test, resulted in the intercept of one medium-range ballistic missile target by THAAD, and one medium-range ballistic missile target by Aegis Ballistic Missile Defense (BMD). Defense Visual Information Distribution Service, Public Domain, via NARA & DVIDS Public Domain Archive Public Domain Archive.

Recent advancements in missile technology have also led to the development of the SM-6 interceptor, which provides additional flexibility by enabling engagement in both the terminal phase and against targets within the atmosphere. This versatility is crucial as missile threats evolve, necessitating adaptive defence strategies (Bradley et al., 2013). The Aegis BMD system's architecture is characterized as a System of Systems (SoS), which emphasizes the complex interactions among its various components. This framework allows for a more nuanced understanding of the system's effectiveness and operational dynamics, particularly in the context of multi-layered kill chains [48, 49.

The Terminal High Altitude Area Defense (THAAD) system is a sophisticated missile defence technology designed to intercept and destroy short, medium, and intermediate-range ballistic missiles during their terminal phase of flight. This system is particularly notable for its land-based configuration and its use of a hit-to-kill mechanism, which relies on kinetic energy to neutralize threats, distinguishing it from other systems like the PAC-3, which is primarily focused on lower-altitude engagements.

Figure 41: A Terminal High Altitude Area Defense (THAAD) interceptor is launched from a THAAD battery located on Wake Island, during Flight Test Operational (FTO)-02 Event 2a, conducted Nov. 1, 2015. During the test, the THAAD system successfully intercepted two air-launched ballistic missile targets. Defense Visual Information Distribution Service, Public Domain, via NARA & DVIDS Public Domain Archive Public Domain Archive.

The operational effectiveness of THAAD is significantly enhanced by its AN/TPY-2 radar, which is capable of long-range detection of incoming missiles. This radar system provides critical early warning and tracking capabilities, allowing THAAD to engage threats with sufficient response time. The interceptors launched by THAAD are designed to be highly manoeuvrable, enabling them to target fast-moving objects in the upper atmosphere, including emerging threats such as hypersonic missiles. The combination of advanced radar technology and agile interceptors positions THAAD as a formidable defence against a variety of missile threats.

The working process of THAAD can be broken down into three main phases: long-range detection, launch, and hit-to-kill interception. Initially, the AN/TPY-2 radar detects and tracks the incoming missile, providing vital data for the engagement. Following this, the THAAD system launches an interceptor missile that ascends to a high altitude, where it can effectively engage the target during its terminal phase. The final phase involves the hit-to-kill mechanism, where the interceptor collides with the target, utilizing kinetic energy to destroy it. This method of interception is critical, as it minimizes the risk of collateral damage from explosive warheads.

Figure 42: A Terminal High Altitude Area Defense (THAAD) interceptor is launched from a THAAD battery located on Wake Island, during Flight Test Operational (FTO)-02 Event 2a, conducted Nov. 1, 2015. During the test, the THAAD system successfully intercepted two air-launched ballistic missile targets. Defense Visual Information Distribution Service, Public Domain, via Picryl.

In addition to its core functionalities, THAAD's interceptors incorporate advanced technologies such as thrust vector control (TVC) and energy-management steering manoeuvres, which enhance their agility and precision during flight. These features are essential for adapting to the dynamic nature of modern missile threats, which may employ evasive manoeuvres or decoys. The system's design reflects a comprehensive approach to missile defence, integrating cutting-edge technology to address the complexities of intercepting ballistic missiles in various flight phases.

Missile Defence System	Primary Function	Key Capabilities	Working Process	Adaptation for Hypersonic Threats
Patriot Advanced Capability-3 (PAC-3)	Intercepting tactical ballistic missiles	- **Hit-to-kill** mechanism - Agile interceptor missile - Advanced radar tracking	- **Detection:** Radar array detects the incoming missile. - **Tracking:** System tracks missile trajectory and velocity. - **Engagement:** PAC-3 missile launched to intercept. - **Hit-to-Kill:** Direct impact with the target.	- PAC-3's agility and precision make it effective for intercepting fast-moving hypersonic missiles in the terminal phase.
Aegis Ballistic Missile Defense (BMD) System	Mid-course ballistic missile interception	- **SPY-1 radar** for long-range detection - **SM-3/SM-6 interceptors** for mid-course and terminal phase interception - Space, sea, and land sensor integration	- **Detection and Tracking**: SPY-1 radar detects a high-altitude missile. - **Mid-Course Interception:** SM-3 interceptor is launched. - **Hit-to-Kill**: The SM-3 uses kinetic energy to destroy the missile in space.	- Aegis's flexibility with **SM-6** interceptors allows it to engage both mid-course and terminal phase hypersonic threats.
Terminal High Altitude Area Defense (THAAD)	Intercepting short, medium, and intermediate-range ballistic missiles	- **Hit-to-kill** mechanism - Long-range **AN/TPY-2 radar** - High-altitude interception	- **Long-Range Detection:** AN/TPY-2 radar detects missile at long ranges. - **Launch:** THAAD interceptor climbs to high altitude. - **Hit-to-Kill:** Missile destroys the target via direct impact.	- THAAD's high-altitude and long-range capabilities make it suitable for engaging high-speed hypersonic missiles in their terminal phase.

Figure 43: Table summarizing the working process and features of the PAC-3, Aegis BMD, and THAAD systems.

The adaptation of existing missile defence systems to counter hypersonic missiles is a topic of increasing relevance due to the unique challenges posed by these advanced threats. Hypersonic missiles, which travel at speeds exceeding Mach 5, exhibit unpredictable manoeuvrability and operate at lower altitudes compared to traditional ballistic missiles, complicating detection and interception efforts [5]. This response synthesizes the capabilities of three missile defence systems—PAC-3, Aegis BMD, and THAAD—and discusses future directions for missile defence technology.

The PAC-3 missile defence system is designed with agility that is crucial for engaging fast-moving, low-altitude hypersonic missiles during their terminal phase. Its advanced guidance and control systems allow for rapid adjustments, which are essential for intercepting targets that can change trajectories unexpectedly [73]. The PAC-3's design emphasizes precision and responsiveness, making it a suitable candidate for addressing the challenges posed by hypersonic threats [592].

Similarly, the Aegis Ballistic Missile Defense (BMD) system demonstrates significant potential in countering hypersonic missiles through its mid-course and terminal phase engagement capabilities. The incorporation of advanced interceptors, such as the SM-6, enhances Aegis BMD's effectiveness against manoeuvring targets. The system's ability to integrate data from various sensors allows for improved tracking and targeting of hypersonic threats, which is critical given their high speed and low flight profiles [19, 593]. The Aegis system's multi-layered defence strategy is particularly advantageous in creating a comprehensive defence against diverse missile threats, including hypersonic missiles [594].

THAAD (Terminal High Altitude Area Defense) also presents a viable option for intercepting hypersonic weapons due to its high-altitude and long-range interception capabilities. By engaging threats as they descend towards their targets, THAAD can leverage its altitude advantage to intercept hypersonic missiles before they reach their intended targets. This capability is further enhanced by ongoing research into cooperative guidance laws that optimize interception strategies against hypersonic vehicles [324, 595]. The integration of advanced sensor technologies and cooperative engagement strategies is essential for maximizing THAAD's effectiveness against these fast-moving threats [596].

Looking ahead, the future of missile defence systems may involve the incorporation of directed energy weapons or innovative interception methods, such as utilizing clouds of shrapnel to disrupt hypersonic missiles during flight. However, the immediate focus remains on improving sensor integration across all defence platforms—land, sea, air, and space—to enhance detection and interception capabilities against hypersonic threats

[597, 598]. This integrated approach is vital for creating a robust defence architecture capable of addressing the complexities introduced by hypersonic technology.

Missile defence, however, is not a solitary endeavour—it is best approached as a "team sport," where the integration of various defence layers and systems across domains is crucial. Hypersonic weapons pose a unique challenge because they operate at different altitudes and speeds compared to traditional ballistic or cruise missiles, which means that effective defence requires cooperation between land-based, sea-based, air-based, and space-based sensors and defence systems. Improving communication and data-sharing between these platforms could enhance the ability to detect, track, and intercept hypersonic missiles at different stages of their flight. For example, a more seamless integration of radar systems, early warning satellites, and interceptor missiles could dramatically increase defence effectiveness [552].

Moreover, directed energy weapons—such as high-energy lasers or microwaves—are being researched as potential solutions for stopping hypersonic missiles. These technologies could destroy missiles by heating them or disrupting their electronic systems mid-flight. Another concept involves deploying clouds of shrapnel to damage or destroy the missile through physical impact. Both ideas represent potential future countermeasures that could complement traditional missile interception strategies [552].

As with many military technologies, the development of hypersonic weapons and their corresponding defence systems is likely to result in a continuous cycle of measure-countermeasure. Each advancement in offensive missile capabilities will likely drive innovation in defence technologies, creating an ongoing arms race between attackers and defenders. While current hypersonic missile technologies still face challenges in terms of performance and mission effectiveness, future developments are expected to significantly improve their capabilities. As a result, the threat posed by hypersonic missiles will likely remain a persistent issue for military strategists in the coming years [552].

Detection, Tracking, and Interception Challenges

The detection and tracking of hypersonic missiles present unique challenges due to the curvature of the Earth and the extreme speeds at which these missiles travel. Ground-based radars, which rely on line-of-sight to detect threats, will encounter limitations in detecting hypersonic missiles compared to other missile types. Specifically, because hypersonic missiles fly at lower altitudes than traditional ballistic missiles but higher than terrain-hugging cruise missiles, ground-based radar systems detect them later than ballistic missiles but earlier than cruise missiles. This makes their detection window narrower, requiring highly responsive detection systems [552].

The rapid speed of hypersonic missiles means that they can cover significant distances in a very short time, reducing the window for detection and response [51, 278]. Moreover, the lower flight altitude of hypersonic missiles compared to ballistic missiles further complicates radar detection, as ground-based systems can only detect these threats later in their flight path due to the Earth's curvature [26, 138]. The curvature of the Earth poses a significant limitation for ground-based radar systems, which can only detect hypersonic missiles at relatively short ranges. This limitation is particularly pronounced when compared to ballistic missiles that ascend to higher altitudes, where they can be detected much earlier in their flight. Hypersonic missiles, while detectable earlier than terrain-hugging cruise missiles, still present a challenge for timely tracking and interception due to their low-altitude flight profile [434]. The ability of hypersonic missiles to manoeuvre also adds complexity to their tracking, as traditional radar systems may not be equipped to handle the rapid changes in trajectory that these missiles can execute [599].

Hypersonic missiles generate intense heat as they move through the atmosphere. This heat creates a plasma sheath around the missile, which poses both detection advantages and challenges. The plasma effect refers to the ionized gas enveloping the missile at hypersonic speeds, making the missile more detectable in certain situations. The plasma plume that forms is more visible to radar than the missile itself, providing an unintended tracking signal to enemy radars. This phenomenon paradoxically makes it easier for radar systems to identify and follow the missile, even though the missile itself is designed to be fast and elusive [552].

The launch location and trajectory of the missile also influence detection times. For example, a hypersonic glide vehicle (HGV) launched from the same point as a ballistic missile of similar range will typically result in a shorter radar warning time. A ballistic missile with a range of approximately 1,860 miles (3,000 km) might allow a ground-based radar system about 12 minutes of warning before impact. In contrast, a hypersonic glide vehicle of the same range could reduce this warning time to around 6 minutes due to its lower altitude and rapid speed. However, the difference in warning time could be mitigated by the presence of space-based sensors. These sensors, orbiting high above Earth, can detect missile launches and flight paths earlier than ground-based radars due to their vantage point, making them an essential part of early-warning systems for hypersonic threats [552].

Thus, while hypersonic missiles pose challenges in terms of detection and tracking, especially for ground-based radars, advancements in space-based detection technologies and plasma visibility offer potential solutions to reduce the time lag between missile launch and detection, improving defence response times [552].

Tracking hypersonic missiles presents significant challenges due to their extreme speed and unpredictable flight paths. Traditional tracking systems, primarily designed for slower aerial threats, often lack the necessary speed and accuracy to maintain a lock on hypersonic missiles once they are detected. The rapid acceleration and manoeuvrability of these missiles complicate the tracking process, as conventional radar systems struggle to keep pace with their dynamics [50, 278]. Moreover, hypersonic missiles generate substantial heat and plasma during flight, creating a detectable heat signature via infrared sensors. However, this signature diminishes rapidly after the boost phase, making it difficult to track the missile effectively beyond initial detection [600].

The limitations of ground-based radar systems, exacerbated by the plasma effect that obscures the missile's trajectory, have led to increased interest in space-based sensor technologies. The U.S. is actively developing a network of space-based infrared sensors, such as the Space-Based Infrared System (SBIRS) and the forthcoming Next Generation Overhead Persistent Infrared (OPIR) satellites, which aim to provide continuous tracking capabilities for hypersonic missiles throughout their flight [600]. However, distinguishing hypersonic missiles from the Earth's background using infrared sensors remains a significant challenge. As a result, researchers are exploring alternative technologies, including hyperspectral and ultraviolet sensors, which could potentially detect the chemical byproducts generated during hypersonic flight, thereby enhancing tracking capabilities [601].

The development of these advanced tracking systems is critical, as hypersonic vehicles represent a substantial shift in military capabilities, necessitating new strategies for detection and interception [278]. The integration of multiple sensor modalities, including infrared and potentially hyperspectral sensors, could provide a more robust solution for tracking these fast-moving threats, ensuring that defence systems can respond effectively to the evolving landscape of aerial warfare [600].

Intercepting hypersonic missiles presents significant challenges primarily due to their remarkable speed and manoeuvrability. Traditional missile defence systems, such as the Patriot Advanced Capability-3 (PAC-3), Aegis Ballistic Missile Defense (BMD), and Terminal High Altitude Area Defense (THAAD), were originally designed to counter ballistic missile threats. However, these systems face limitations when tasked with intercepting hypersonic threats, which can travel at speeds exceeding Mach 5 and execute evasive manoeuvres mid-flight, making them difficult to track and intercept effectively [4, 602]. The rapid changes in trajectory and the short time frame available for interception necessitate enhancements to existing systems to improve their responsiveness and accuracy [73, 603].

The manoeuvrability of hypersonic missiles complicates interception efforts, as these weapons can alter their flight paths significantly during their descent phase, thus challenging the predictive capabilities of interceptor systems [73, 323]. The engagement of such targets requires advanced guidance and control strategies that can accommodate the high speeds and unpredictable trajectories of hypersonic missiles [603, 604]. Research indicates that the existing missile defence systems need to evolve, incorporating new technologies and methodologies to enhance their interception capabilities against these advanced threats [65, 604, 605].

In response to these challenges, innovative interception methods are being explored, including the use of directed energy weapons, such as lasers, and the deployment of clouds of shrapnel to overwhelm hypersonic missiles [598, 606]. These technologies, while still in developmental stages, show promise in providing effective countermeasures against hypersonic threats. For instance, the concept of using fine fragments and particles to exploit the vulnerabilities of hypersonic vehicles during their glide phase has been proposed as a viable interception strategy [598]. This approach leverages the high relative velocities and the thermal and pressure vulnerabilities of hypersonic missiles, potentially increasing the chances of successful interception [598].

The evolution of sensor networks and their technological integration is critical for addressing the challenges posed by hypersonic missiles. The U.S. military's investment in space-based sensor networks, particularly through the deployment of Low Earth Orbit (LEO) and Medium Earth Orbit (MEO) satellites, is a strategic response to enhance missile defence capabilities. These satellite systems are designed to work in conjunction with advanced ground-based radars to create a persistent and integrated missile defence architecture that can effectively detect and track hypersonic threats throughout their flight paths, even in environments where GPS signals are compromised or in contested battle spaces [607].

The integration of diverse sensor technologies—such as infrared, ultraviolet, radar, and hyperspectral sensors—plays a pivotal role in improving the military's tracking capabilities. This multi-layered defence system is designed to ensure redundancy, thereby mitigating risks associated with environmental conditions, enemy interference, or other factors that could impair the functionality of any single sensor type [608]. The collaborative innovation between military and civilian sectors is essential in this context, as it allows for the rapid development and deployment of advanced technologies that enhance situational awareness and operational effectiveness in the battlefield [609].

Moreover, the military's acceptance and integration of the Internet of Things (IoT) into its operations signify a transformative shift in how sensor networks are utilized. The IoT facilitates the connection of various battlefield assets, enabling a more comprehensive and real-time understanding of the operational environment. This interconnectedness not only enhances intelligence, surveillance, and reconnaissance operations but also supports automated supply chain logistics, which is crucial for maintaining operational readiness in dynamic combat scenarios [610, 611]. The evolution of these technologies underscores the importance of a coordinated approach to military-civilian integration, which fosters innovation and enhances the overall effectiveness of defence systems [612, 613].

International Collaboration on Countermeasures

International collaboration on countermeasures against hypersonic weapons has emerged as a critical area of focus due to the significant global security implications associated with the rapid development of hypersonic technologies. As hypersonic missiles possess the capability to evade traditional missile defence systems due to their speed and manoeuvrability, nations are increasingly recognizing the necessity of cooperative defence strategies to mitigate these risks. This collaborative effort is particularly evident among key alliances such as NATO, the Quadrilateral Security Dialogue (Quad), the Five Eyes intelligence-sharing alliance, and various bilateral agreements between countries like Australia and Japan.

The United States, in conjunction with NATO allies, is spearheading initiatives to address the hypersonic threat. NATO has formally acknowledged the development of hypersonic missiles as a substantial security challenge, prompting joint research and development efforts aimed at enhancing missile defence capabilities. The NATO Defense Innovation Accelerator, for instance, is designed to foster innovation and facilitate the sharing of advanced technologies among member states, specifically targeting counter-hypersonic measures [12, 614]. This approach underscores the alliance's commitment to establishing a comprehensive missile defence architecture that integrates various sensor systems to ensure early detection and effective response to hypersonic threats [314].

In the Indo-Pacific region, the Quad—comprising the United States, India, Japan, and Australia—has also prioritized addressing hypersonic threats. While the Quad's primary focus encompasses broader regional security challenges, there is a growing emphasis on technology-sharing agreements related to missile defence capabilities [615]. India, through its Defence Research and Development Organisation (DRDO), is actively developing hypersonic missile technology and is collaborating with the U.S. to enhance its missile defence systems, including shared early warning systems [615]. This partnership exemplifies how nations are pooling resources and expertise to counter the evolving hypersonic threat landscape.

The Five Eyes alliance, consisting of the U.S., UK, Canada, Australia, and New Zealand, has expanded its intelligence-sharing efforts to include hypersonic defence. This collaboration involves coordinating intelligence collection and developing early warning capabilities to track hypersonic missile threats effectively [31]. The United Kingdom, in particular, is investing significantly in hypersonic countermeasures and is working closely with the U.S. to advance early detection and interception technologies, leveraging the intelligence-sharing framework established by the Five Eyes network [31].

Bilateral agreements between Australia and Japan further illustrate the commitment of these nations to enhance their defence capabilities against hypersonic threats. Both countries are investing in missile defence technologies and exploring the integration of space-based sensors in collaboration with the U.S [615]. Japan's enhancement of its Aegis Ballistic Missile Defense system and Australia's Air 6500 Integrated Air and Missile Defense program are indicative of their strategic alignment in addressing the challenges posed by hypersonic weapons [615].

While the European Union's role in hypersonic missile defence may be more limited compared to NATO or the U.S., there is a growing interest in developing European missile defence capabilities. Collaborative research initiatives within the European Defense Fund (EDF) aim to advance missile tracking and detection systems that could eventually integrate into broader NATO or international defence frameworks (Wright, 2022). This reflects an acknowledgment of the need for a unified European response to the hypersonic threat.

Finally, the development of space-based sensor networks is becoming increasingly vital in the collaborative efforts to counter hypersonic missiles. The U.S. Space Development Agency (SDA) is working on satellite constellations capable of detecting and tracking hypersonic missiles, which will be integrated into broader defence systems shared among international partners, including NATO [314]. This technological advancement is crucial for enhancing early warning capabilities and improving the overall effectiveness of missile defence systems.

Area of Collaboration	Details	Initiatives
1. U.S. and NATO Allies	The U.S. and NATO are collaborating to address the threat posed by hypersonic missiles. NATO is focusing on integrating land, sea, and space-based defence systems to detect and track hypersonic missiles early.	**NATO Defense Innovation Accelerator**: Fosters technology-sharing among NATO members to enhance hypersonic defences.
2. The Quadrilateral Security Dialogue (Quad)	Comprising the U.S., India, Japan, and Australia, the Quad focuses on technology sharing and missile defence. Cooperation is increasing to counter the hypersonic threat.	**U.S.-India Cooperation**: Includes India's DRDO efforts to develop hypersonic tech and early warning systems.
3. Five Eyes Intelligence Sharing	The Five Eyes alliance (U.S., UK, Canada, Australia, New Zealand) is focused on intelligence sharing, early warning capabilities, and coordinated hypersonic defence strategies.	**UK Hypersonic Defense**: Collaborative projects between the UK and U.S. for early detection and interception.
4. Australia and Japan Bilateral Agreements	Australia and Japan are upgrading their missile defence systems to better counter hypersonic threats through joint agreements with the U.S.	**Japan's Missile Defense Initiatives**: Enhancing Japan's Aegis BMD and Australia's Air 6500 defence program.
5. The European Union (EU)	The EU is working on missile detection and tracking technologies that could integrate with broader NATO or international hypersonic defence efforts.	**European Defense Fund (EDF)**: Funds research on missile tracking and detection for future integration into NATO.
6. Space-Based Sensor Networks	The U.S., Japan, and NATO allies are collaborating on space-based missile tracking, with the U.S. leading efforts to develop satellite constellations for hypersonic missile detection and tracking in low-Earth orbit (LEO) and medium-Earth orbit (MEO).	**Space Development Agency (SDA)**: Developing LEO and MEO satellite constellations for hypersonic missile tracking.

Figure 44: The key areas of collaboration and initiatives in hypersonic missile defence.

Ethical and Legal Considerations

Compliance with International Laws and Treaties

The technology behind hypersonic weapons has been hailed as a major breakthrough in aerodynamic propulsion and material science. This has raised concerns among U.S. policymakers, especially as near-peer adversaries like Russia and China have made significant advancements in this area. Hypersonic weapons are often compared to revolutionary technologies like nuclear weapons due to their potential impact on warfare and global security. However, there is debate within academic and military circles about how these weapons will influence strategic outcomes. Some argue that hypersonic technology offers strategic advantages for countries aiming to achieve certain military objectives. However, others worry about its destabilizing effects on international security, highlighting concerns about the potential for inadvertent escalation during conflicts and the erosion of strategic stability [616].

One of the main worries associated with hypersonic weapons is their potential to trigger a new arms race. Some scholars argue that the proliferation of hypersonic technology beyond the current leading developers—the U.S., Russia, and China—could destabilize global security. These concerns have spurred calls for arms control and non-proliferation initiatives aimed either at preventing the spread of hypersonic technologies or limiting their deployment once developed. The demise of past arms control agreements, such as the Anti-Ballistic Missile Treaty, has also been seen as a contributing factor to the current race to develop hypersonic capabilities [616].

Despite these concerns, there has been a lack of systematic analysis to identify the specific challenges hypersonic weapons pose to global strategic stability. Many current policy discussions rely on assumptions that the spread of hypersonic weapons is inherently destabilizing or that arms control agreements can mitigate the risks they pose. However, without careful analysis of the various ways these weapons affect nuclear and conventional deterrence dynamics, there is a risk of implementing ineffective or counterproductive solutions. Some aspects of hypersonic weapons may, in fact, be stabilizing in certain strategic contexts, yet little research has explored this possibility. Without a deeper understanding of the nuances, policymakers may fail to address the actual risks these weapons introduce or, worse, exacerbate global instability through poorly conceived arms control efforts [616].

The development of international laws and treaties concerning hypersonic weapons is an evolving field, as existing arms control agreements have not fully adapted to the technological advancements presented by these systems. Hypersonic weapons pose unique challenges to international security and law, necessitating a review of current treaties and the exploration of new frameworks. This overview discusses several key international treaties and agreements that, while not specifically designed for hypersonic weapons, provide relevant legal principles and frameworks that could be applied to their regulation.

The Outer Space Treaty (1967) is a foundational international agreement that governs the activities of states in outer space. It prohibits the placement of nuclear weapons or any weapons of mass destruction (WMD) in space, emphasizing that outer space should be used for peaceful purposes. Although hypersonic weapons are not explicitly mentioned, the treaty's provisions could apply to hypersonic glide vehicles (HGVs) that utilize outer space during their flight. The legal implications of hypersonic weapons traversing or operating in outer space raise questions about compliance with the treaty's restrictions on militarization and the deployment of WMDs. Furthermore, the development of space-based missile defence systems in response to hypersonic threats may also come under scrutiny under this treaty.

The Hague Code of Conduct Against Ballistic Missile Proliferation (2002) is a non-legally binding agreement aimed at curbing the proliferation of ballistic missiles capable of delivering WMDs. While it primarily addresses ballistic missiles, its principles regarding transparency and notification of missile launches could extend to hypersonic systems. The lack of direct references to hypersonic missiles in the HCOC underscores the need for updated international norms to regulate these emerging technologies. As hypersonic weapons can be employed for both conventional and nuclear payloads, they may indirectly fall under the guidelines of the HCOC, particularly concerning transparency and launch notifications.

The Missile Technology Control Regime (MTCR), established in 1987, is an informal export control agreement that seeks to prevent the proliferation of missile technologies capable of delivering WMDs. Although the MTCR is not a treaty, it plays a crucial role in regulating the transfer of technologies related to missiles, including hypersonic systems. Many hypersonic missile technologies, particularly in propulsion and guidance systems, fall within the MTCR's scope. However, the focus on non-proliferation presents challenges, as major powers developing hypersonic capabilities often prioritize national security over compliance with MTCR provisions.

The New START Treaty (2010), signed between the United States and Russia, aims to limit the number of deployed strategic nuclear warheads and delivery systems. Although hypersonic weapons are not explicitly mentioned, there is ongoing debate about whether certain HGVs or hypersonic cruise missiles (HCMs) could be classified as "new types of strategic offensive arms" under the treaty. The development of Russia's Avangard hypersonic glide vehicle has raised concerns regarding its potential inclusion within the warhead limits set by New START. As the treaty approaches its expiration in 2026, discussions about incorporating hypersonic systems into future arms control frameworks are becoming increasingly relevant.

The Non-Proliferation Treaty (NPT) is a landmark agreement aimed at preventing the spread of nuclear weapons and promoting peaceful nuclear energy. While it primarily focuses on nuclear weapons, it implicitly covers delivery systems such as ballistic missiles and potentially hypersonic weapons. Hypersonic missiles capable of delivering nuclear warheads raise significant concerns under the NPT, particularly if non-nuclear states pursue such capabilities. Although existing nuclear states are not directly limited in their development of conventional hypersonic weapons under the NPT, the destabilizing effects of these systems on global security remain a pressing issue.

International Law/Treaty	Description	Relevance to Hypersonic Weapons
The Outer Space Treaty (1967)	Prohibits the placement of nuclear weapons or WMDs in outer space and restricts militarization.	Hypersonic weapons passing through or near space may raise legal questions under the treaty's ban on WMDs in space. Space-based sensor networks for detecting hypersonic threats could also come under scrutiny.
The Hague Code of Conduct (2002)	A voluntary agreement aimed at curbing the proliferation of ballistic missiles capable of delivering WMDs.	Transparency of missile launches and test notifications could be extended to hypersonic weapons. Need for updated norms to regulate transparency of hypersonic systems, especially in terms of conventional and nuclear payloads.
Missile Technology Control Regime (MTCR)	A non-legally binding export control agreement regulating the transfer of missile technologies capable of delivering WMDs.	Hypersonic missile technologies, such as propulsion and guidance systems, are covered under MTCR. However, enforcement is challenging for major powers focused on national security.
New START Treaty (2010)	Limits the number of deployed strategic nuclear warheads and delivery systems between the U.S. and Russia.	Russia's development of hypersonic glide vehicles (HGVs) like *Avangard* raises concerns about counting them under warhead limits. Future arms control frameworks may need to include hypersonic systems.
Non-Proliferation Treaty (NPT)	Aims to prevent the spread of nuclear weapons and promote peaceful nuclear energy.	Hypersonic missiles capable of carrying nuclear warheads could raise non-proliferation concerns. No direct limitations on conventional hypersonic weapons for nuclear states, but concerns about destabilization remain.
Proposed International Norms and Initiatives	Ongoing discussions at the international level about new arms control agreements tailored to hypersonic weapons.	Bilateral agreements between major powers (U.S., Russia, China) to limit testing and deployment. Transparency initiatives for hypersonic missile testing and verification mechanisms. Norms on space weaponization intersecting with hypersonic regulation.
Challenges in Regulating Hypersonic Weapons	Several challenges exist, such as the difficulty in defining hypersonic systems, lack of consensus among major powers, and the rapid pace of technological advancements in hypersonic systems.	Difficulty fitting hypersonic systems into current frameworks (ballistic vs. cruise missiles). Major powers view hypersonic weapons as critical to national defence, complicating arms control talks. Technological gaps in regulation due to rapid advancements.

Figure 45: Key aspects of international laws, treaties, and the challenges posed by regulating hypersonic weapons.

In light of the strategic implications of hypersonic weapons, there have been discussions about creating new arms control agreements specifically tailored to address the challenges posed by these systems. Proposed initiatives include bilateral agreements between major powers to limit the deployment or testing of hypersonic missiles, transparency initiatives to enhance notification and verification mechanisms, and norms regarding the weaponization of space. However, significant challenges remain in regulating hypersonic weapons, including difficulties in defining these systems, a lack of consensus among major powers, and the rapid pace of technological advancements that often outstrip existing legal frameworks.

While existing international treaties such as the Outer Space Treaty and the NPT touch on aspects relevant to hypersonic weapons, there is currently no comprehensive legal framework governing these emerging technologies. The development and deployment of hypersonic weapons by major military powers are driving calls for new arms control agreements or modifications to existing treaties to address the unique challenges they present. As the technology continues to evolve, international collaboration and legal frameworks will likely need to adapt to ensure global security.

Humanitarian and Ethical Implications of Hypersonic Strikes

The development and deployment of hypersonic weapons have raised significant humanitarian and ethical concerns primarily due to their unprecedented speed, manoeuvrability, and capacity for delivering precise and overwhelming force. These characteristics, while advantageous from a military standpoint, pose substantial risks related to civilian harm, global stability, and the evolving nature of warfare.

Hypersonic weapons present unique challenges that complicate existing frameworks for international humanitarian law (IHL). Their ability to manoeuvre mid-flight and evade traditional missile defence systems makes them particularly destabilizing, as they can strike targets with little warning, thereby increasing the likelihood of civilian casualties [2, 5]. The rapid response times associated with hypersonic technology may also encourage pre-emptive strikes, further heightening the risk of escalation in conflict scenarios [7]. This potential for rapid and overwhelming force raises ethical questions regarding the proportionality and necessity of such weapons in warfare, as the principles of IHL demand that military operations must minimize civilian harm [617, 618].

Moreover, the deployment of hypersonic weapons could undermine global stability by altering the strategic balance among nuclear powers. The speed and agility of these weapons could lead to miscalculations during crises, as states may feel pressured to respond quickly to perceived threats, potentially escalating conflicts into nuclear exchanges [4, 8]. This dynamic challenges traditional deterrence strategies and raises concerns about the adequacy of existing arms control agreements, which may not sufficiently address the implications of hypersonic technology [35]. The ethical implications of such technologies extend to the dehumanization of warfare, as the reliance on automated systems for decision-making could diminish accountability and moral responsibility in military operations [619, 620].

The advent of hypersonic weapons has raised significant concerns regarding civilian casualties and infrastructure damage due to their speed and potential for evasion of existing defence systems. These weapons can strike targets at speeds exceeding five times the speed of sound, drastically reducing the time available for civilians in targeted areas to evacuate or seek shelter. Traditional missile systems often allow for early detection and response, but hypersonic missiles can evade many defence systems, increasing the likelihood of civilian casualties even when military targets are the primary focus. The rapid approach of these weapons limits the window for protective measures, leading to a heightened risk of collateral damage in populated areas where civilian infrastructure is often in close proximity to military objectives [7, 621].

The destructive potential of hypersonic strikes in densely populated regions can result in catastrophic damage to essential infrastructure, including hospitals, schools, and water supply systems. Such damage not only leads to immediate casualties but also precipitates long-term humanitarian crises by disrupting access to vital resources. For instance, the ongoing conflict in Ukraine has illustrated how military actions can severely compromise healthcare access, with many medical facilities being attacked or rendered inoperable due to infrastructure damage [622]. The implications of hypersonic weapons extend beyond immediate military objectives; they can destabilize entire communities by creating prolonged periods of instability and deprivation, as seen in various conflict zones where civilian infrastructure has been targeted or collateral damage has occurred [621].

Moreover, the ethical considerations surrounding the use of hypersonic weapons are complex. The potential for high civilian casualties raises questions about the moral implications of deploying such technology in warfare. The increasing precision of these weapons does not necessarily correlate with a reduction in collateral damage; rather, it can lead to a false sense of security regarding their use, as the expectation of precision may not always align with the chaotic realities of combat [623]. This disconnect can exacerbate the challenges faced by civilian populations, particularly in urban environments where military and civilian infrastructures are intermingled. Thus, the deployment of hypersonic weapons necessitates a re-evaluation of military strategies to mitigate civilian harm and protect critical infrastructure [7].

The emergence of hypersonic weapons, particularly those capable of carrying nuclear warheads, introduces significant risks to global security, primarily due to the potential for miscalculation and unintended escalation during military conflicts. The unique characteristics of hypersonic weapons, which travel at speeds exceeding Mach 5, create extremely short reaction times for military and political leaders. This rapid pace can compel decision-makers to act hastily, often based on incomplete or inaccurate information, thereby increasing the likelihood of miscalculations in crisis situations. Such dynamics are particularly concerning in the context of nuclear deterrence, where the fear of a first strike could lead to pre-emptive actions that escalate conflicts to catastrophic levels, as noted by Coetzee, who discusses the dual-use nature of hypersonic weapons and their potential destabilizing effects on nuclear deterrence stability [7].

Moreover, the dual-use nature of hypersonic weapons complicates the strategic landscape. These weapons can be equipped with either conventional or nuclear warheads, which blurs the lines between conventional and nuclear warfare. As highlighted by Raza, the ambiguity surrounding the nature of an incoming hypersonic strike can provoke overreactions from adversaries, who may misinterpret a conventional attack as a nuclear one and respond with nuclear force [5]. This potential for misinterpretation is exacerbated by the rapid advancements in hypersonic technology, which can undermine existing missile defence systems, making it increasingly difficult for states to accurately assess threats [2]. The implications of this ambiguity are profound, as they could lead to a cycle of escalation that heightens the risk of nuclear conflict.

In addition to the immediate risks posed by hypersonic weapons, the broader implications for international security and humanitarian concerns cannot be overlooked. The humanitarian initiative on nuclear weapons has sought to reframe the discourse surrounding these weapons, emphasizing their catastrophic humanitarian consequences rather than their strategic utility [624]. This shift in focus is critical, as it highlights the dire humanitarian implications of any nuclear escalation, which could result in widespread devastation and loss of life on a global scale. The urgency of addressing these risks is underscored by the ongoing arms race among major powers, as they strive for technological superiority in hypersonic capabilities, further complicating the security environment [4].

The ethical concerns surrounding targeting decisions in the context of hypersonic weapons are multifaceted and deeply intertwined with the implications of precision warfare. Hypersonic weapons, characterized by their high speed and manoeuvrability, present a significant shift in military strategy. This capability not only enhances the precision of strikes but also raises critical ethical questions regarding the justification and consequences of their use, particularly in pre-emptive or decapitation strikes against high-value targets, such as political leaders or military command centres.

The potential for hypersonic weapons to facilitate politically motivated assassinations is particularly alarming. The literature indicates that state-directed political assassinations can destabilize regions and exacerbate conflicts, as seen in the context of Israel's political landscape, where such actions are often justified under the guise of national security [625]. Furthermore, research by Chin et al. highlights that successful leader assassinations can lead to intensified conflicts and significant political shifts, particularly in repressive regimes [626]. This suggests that the use of hypersonic weapons for targeted killings could lead to similar destabilizing effects, undermining political stability and democratic processes.

Ethical implications of employing hypersonic weapons extend to the principles of proportionality and necessity in warfare. The ability to conduct precise strikes may create a dangerous precedent where military force is favoured over diplomatic solutions, as noted in the discussions surrounding the targeted killing of Qassem Soleimani [627]. This incident exemplifies how the justification for such strikes can be framed within a legal and ethical context, yet still lead to significant geopolitical tensions and potential escalations of conflict [628]. The argument posited by Pratt emphasizes that military assassination can be subjected to ethical debate, particularly when it aligns with Just War Theory, yet the risks associated with normalizing such actions are profound [629].

The strategic implications of hypersonic weapons also challenge existing international norms aimed at preventing escalatory or pre-emptive military actions. The literature suggests that the rapid response capabilities afforded by these weapons may encourage states to prioritize military solutions over diplomatic engagement, thereby eroding the frameworks established to manage international conflict [7]. This shift could lead to a more militarized global landscape, where the use of force becomes the default response to perceived threats, undermining efforts to resolve disputes through dialogue and negotiation.

The emergence of hypersonic weapons poses significant challenges to existing international legal frameworks, particularly those designed to protect civilians during armed conflict. The principles of distinction and proportionality, which are central to the laws of armed conflict, mandate that military forces must differentiate between combatants and civilians and ensure that civilian harm is not excessive in relation to the military advantage gained. However, the rapid speed and lethality of hypersonic weapons complicate compliance with these principles. In fast-moving conflicts, where intelligence on targets may be incomplete or outdated, the ability to adhere to these legal standards becomes increasingly difficult, potentially leading to violations of international humanitarian law [343, 630].

Moreover, the proliferation of hypersonic weapons among multiple global powers threatens to undermine existing arms control agreements and non-proliferation efforts. Currently, hypersonic weapons are not governed by any specific international treaty, which raises concerns about their impact on international arms control regimes. The absence of regulatory frameworks for these weapons could lead to an arms race, as nations may prioritize the development of hypersonic capabilities over commitments to reduce the number and impact of weapons of mass destruction. This scenario exacerbates global security challenges and increases the likelihood of conflict, as states may feel compelled to enhance their military capabilities in response to perceived threats from hypersonic advancements [343, 631].

The destabilizing influence of new weapons technologies, including hypersonic systems, is further emphasized by the need for updated arms control frameworks. As highlighted in various studies, the current international security environment is characterized by a lack of effective arms control measures, which has led to a resurgence of military modernization efforts among nuclear powers. This situation not only undermines existing treaties but also complicates the enforcement of norms related to nuclear restraint and responsibility [631, 632]. The potential for hypersonic weapons to disrupt strategic stability necessitates a re-evaluation of arms control strategies to address the unique challenges posed by these emerging technologies [343, 631].

The integration of hypersonic weapons into military arsenals presents profound challenges to international law and norms. The principles of distinction and proportionality may be compromised in the face of rapid technological advancements, while the lack of regulatory frameworks for hypersonic systems threatens to destabilize existing arms control efforts. As the international community grapples with these issues, the need for a comprehensive approach to arms control that incorporates the realities of modern warfare becomes increasingly urgent [343, 630, 631].

The emergence of hypersonic weapons significantly impacts global strategic stability, primarily due to their speed, precision, and ability to evade existing missile defence systems. These characteristics challenge the traditional deterrence model based on mutually assured destruction (MAD), which has historically been a cornerstone of nuclear strategy. The ability of hypersonic weapons to potentially disable an adversary's nuclear capabilities before a response can be mounted creates a scenario conducive to pre-emptive strikes, thereby increasing the risk of nuclear conflict in high-stakes situations [7, 633, 634].

The integration of hypersonic technology into military arsenals has led to a re-evaluation of strategic stability, particularly among major powers like the United States, Russia, and China. These nations are actively developing hypersonic systems, which could disrupt the existing balance of power and encourage aggressive military postures. The implications of this arms race are profound, as countries may feel compelled to enhance their own capabilities in response, thereby fostering an environment of insecurity and instability [5, 343, 634]. This dynamic could lead to a situation where the threat of nuclear escalation becomes more pronounced, as states might perceive hypersonic weapons as a means to achieve a first-strike advantage [7, 8].

Moreover, the humanitarian costs of such instability could be catastrophic. The potential for a miscalculation or accidental launch in a high-tension scenario could result in significant casualties and long-term geopolitical ramifications. The strain on international relations and the institutions designed to prevent conflict could further exacerbate the situation, as nations may prioritize military readiness over diplomatic engagement [7, 634]. The destabilizing effects of hypersonic weapons, therefore, extend beyond military considerations, threatening the very foundations of global security and cooperation [8, 633, 634].

The humanitarian and ethical implications of hypersonic strikes are profound and multifaceted. These weapons increase the risk of civilian casualties, complicate international security frameworks, and heighten the potential for global instability. As the development and deployment of hypersonic weapons continue, there is an urgent need for international dialogue, regulation, and cooperation to address the challenges they pose. Without such efforts, the widespread adoption of hypersonic technologies could lead to greater conflict and humanitarian crises in the future.

The Need for Global Arms Control Agreements

The pursuit of hypersonic arms control has become increasingly urgent as these weapons transition from an emerging technology into deployed assets, with more countries developing and fielding them on a variety of delivery platforms. Both China and Russia have already deployed hypersonic glide vehicles (HGVs), and the United States is on the verge of deploying its own hypersonic systems. The rapid proliferation of these weapons poses a significant challenge to global strategic stability, as hypersonic weapons' speed and manoeuvrability make them difficult to intercept. This creates an escalation risk and amplifies the dangers of miscalculation during a conflict. Therefore, it is prudent to pursue arms control before these weapons are fielded in larger numbers by major powers, further increasing the threat they pose to strategic stability [341].

In recent years, both the United States and Russia have expressed interest in addressing hypersonic weapons within their ongoing bilateral dialogue on strategic stability. Resuming this dialogue in 2021, both nations signalled a willingness to engage in initial discussions on future arms control arrangements that could potentially encompass hypersonic systems. This dialogue represents a valuable opportunity to manage the risks posed by hypersonic weapons, and failing to seize it could result in a missed chance to mitigate their destabilizing effects [341].

One proposed method for mitigating these risks involves confidence-building measures (CBMs). These are reciprocal actions aimed at reducing the dangerous consequences of certain weapons systems without the formality of binding treaties. Confidence-building measures can promote transparency and foster a more stable environment, potentially leading to the establishment of more comprehensive arms control agreements. For hypersonic weapons, CBMs could include basing restrictions, information sharing, and military-to-military dialogues between nations [341].

For example, some experts have suggested restrictions on the co-location of nuclear and conventional forces to reduce the ambiguity surrounding a hypersonic missile's intended target. Additionally, it has been proposed that Russia and the United States could voluntarily agree not to base heavy bombers carrying hypersonic weapons, such as the U.S. B-1B bomber equipped with the Air-Launched Rapid Response Weapon (ARRW), outside their national territories. This measure would help prevent misunderstandings during potential conflicts, where such bombers might be misinterpreted as carrying nuclear payloads [341].

Other confidence-building measures could focus on enhancing transparency. One suggestion is the exchange of data on the acquisition and deployment of precision-guided missiles, including hypersonic systems, to provide greater clarity about each nation's capabilities and intentions. Furthermore, mutual demonstrations of hypersonic weapon systems and inviting observers to military exercises could build trust between nations and reduce the risk of accidental escalation [341].

Arms control, export control, and other measures aimed at regulating the development and deployment of hypersonic missiles have been proposed by various analysts as a way to manage the potential risks these advanced weapons pose to global stability. The idea of limiting hypersonic missile proliferation through arms control treaties or export control agreements reflects concerns about the destabilizing effects these weapons could have on the global military balance, particularly among nuclear-armed states. However, achieving consensus on these measures faces significant challenges, including political distrust and the diminishing effectiveness of traditional arms control frameworks [1].

Some analysts advocate for amending existing arms control treaties or creating new ones specifically designed to limit the deployment of hypersonic missiles. For example, the New START treaty, a nuclear arms control agreement between Russia and the United States, originally aimed to limit the number of strategic nuclear warheads and their delivery systems. Some versions of hypersonic missiles, such as Russia's Avangard, could potentially fall under the treaty's purview. However, in February 2023, Russia announced that it would suspend its participation in the New START treaty, further complicating efforts to extend arms control mechanisms to cover hypersonic weapons. This suspension highlights the broader issue of political mistrust that currently hinders arms control discussions. A 2019 report by the UN Office for Disarmament Affairs pointed to a general decline in the belief that arms control agreements could be mutually beneficial, suggesting that the global political climate is less conducive to new disarmament efforts [1].

The New START Treaty, which limits strategic nuclear arms between the U.S. and Russia, currently excludes weapons like hypersonic glide vehicles and hypersonic cruise missiles, which do not follow a ballistic trajectory for more than 50% of their flight. However, Article V of the treaty allows either party to bring up new strategic offensive weapons for consideration under the Bilateral Consultative Commission (BCC). Some legal experts believe that the U.S. could propose the inclusion of hypersonic weapons within New START's limits through this mechanism. However, as New START is set to expire in 2026, this solution would only be temporary [329].

An alternative approach proposed by some experts is the negotiation of a new multilateral arms control agreement specifically addressing hypersonic weapons. These analysts advocate for a moratorium or ban on hypersonic weapon testing, arguing that such a test ban would be verifiable and effective in preventing an arms race and maintaining strategic stability. A test ban would halt the continued development and deployment of these advanced weapons, reducing the potential for escalation among rival states [329].

The creation of new arms control agreements for hypersonic missiles is particularly challenging due to the strategic importance of these weapons to countries like the United States, Russia, and China, which view them as critical to maintaining technological parity and enhancing their military capabilities. Countries with hypersonic missile programs may be reluctant to agree to limitations, fearing that doing so could disadvantage them in future conflicts or undermine their deterrence capabilities. In addition, negotiating arms control agreements specific to hypersonic missiles would require clear definitions of what constitutes a hypersonic weapon, and states would need to agree on mechanisms for verifying compliance, which can be difficult in the context of highly classified weapons programs [1].

International export controls have been suggested as another mechanism to prevent the spread of hypersonic missile technology to countries that do not yet possess it. Export control agreements could limit the transfer of missile components, guidance systems, and propulsion technologies that are crucial to building hypersonic weapons. However, while export controls may slow the proliferation of hypersonic technology to less-developed nations, they are unlikely to be effective in curbing the ambitions of countries that already have advanced missile programs. Nations like Russia, China, and the US have the industrial and technological capacity to continue developing hypersonic missiles independently of any export restrictions, and countries like India and France are rapidly advancing their own capabilities through domestic research and international partnerships [1].

Given the challenges of formal arms control and export control, some analysts have proposed alternative, non-treaty mechanisms to manage the risks associated with hypersonic missiles. These mechanisms could involve informal agreements between states aimed at minimizing ambiguity and reducing the potential for miscalculations that could escalate conflicts. For instance, countries could engage in information exchanges about their hypersonic missile programs, sharing details about their deployment doctrines and operational postures to avoid misunderstandings. By increasing transparency, states could reduce the likelihood that an adversary might misinterpret a missile test or deployment as an aggressive move, thereby lessening the risk of unintended escalation [1].

Additionally, dialogues on military doctrines and postures could help clarify the intended use of hypersonic missiles, particularly whether they are intended for conventional or nuclear missions. One of the major concerns with hypersonic missiles is the possibility of warhead ambiguity—the inability to distinguish whether a missile is carrying a conventional or nuclear payload. Informal agreements that keep conventional and nuclear-armed hypersonic missiles distinct could help prevent a conventional missile from being mistaken for a nuclear one, which could otherwise trigger a nuclear response. Such agreements would not necessarily require formal treaties but could take the form of confidence-building measures between major powers [1].

Other analysts argue that a hypersonic weapons test ban might be impractical. They contend that distinguishing hypersonic weapons from other conventional systems that have shorter ranges and are less prompt, but still pose strategic challenges, is difficult. These analysts believe that rather than pursuing a test ban, transparency and confidence-building measures could be a more feasible solution. For instance, they suggest the exchange of technical data on hypersonic systems, joint technical studies, providing advance notice of missile tests, selecting distinct launch locations for tests, and placing limitations on sea-based hypersonic tests as ways to reduce tensions and enhance trust between nations developing these technologies [329].

While arms control and export controls remain important tools for managing the spread of hypersonic missile technology, they face significant hurdles, including political distrust and the reluctance of advanced missile states to limit their capabilities. Non-treaty mechanisms, such as information exchanges and dialogues on missile postures, offer a more flexible approach that could help reduce the risks of escalation and misunderstanding. However, these informal approaches require strong diplomatic engagement and cooperation between key stakeholders to be effective. Without such measures, the rapid development and deployment of hypersonic missiles could increase tensions, complicate global arms control efforts, and potentially destabilize the strategic balance between major military powers.

PART V - FUTURE OF HYPERSONIC TECHNOLOGY

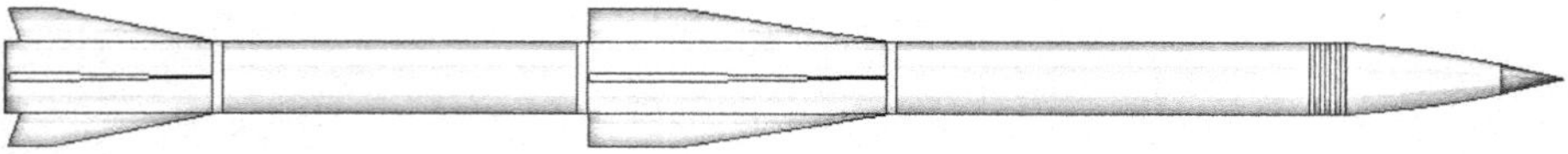

Emerging Technologies Supporting Hypersonic Advancements

Advances in Artificial Intelligence and Autonomous Systems

Advancements in Artificial Intelligence (AI) and autonomous systems have significantly influenced the development and operational capabilities of hypersonic weapons, which are defined as projectiles traveling at speeds exceeding Mach 5. These advancements can be categorized into several key areas: AI-driven flight control systems, autonomous targeting and decision-making, enhanced threat detection and tracking, materials and thermal management, AI-enhanced autonomous launch platforms, improved simulations and testing, autonomous mission planning, and defensive AI systems.

Hypersonic weapons encounter extreme aerodynamic and thermal stresses during flight, necessitating advanced control systems. AI-driven flight control systems are capable of dynamically adjusting flight paths in real-time, which is essential for maintaining stability and precision in navigation under rapidly changing conditions. Research indicates that AI can process vast amounts of sensor data more rapidly than human operators, which enhances manoeuvrability and targeting capabilities in extreme environments [73, 324, 635]. The integration of AI into flight control systems allows for improved adaptability to the unpredictable dynamics of hypersonic travel, thereby enhancing overall mission success rates.

Specific examples of how AI-driven flight control systems enhance the operation of hypersonic weapons include:

- **Dynamic Flight Path Adjustments**: Hypersonic weapons face rapidly changing aerodynamic conditions, such as turbulent airflow and thermal stresses. AI-driven systems can process real-time data from onboard sensors, allowing the weapon to adjust its trajectory dynamically. For instance, the use of neural networks in AI flight control enables real-time corrections to account for atmospheric variations, reducing the chances of trajectory deviation during high-speed manoeuvres.

- **Enhanced Manoeuvrability**: Hypersonic weapons are capable of evasive manoeuvres to avoid detection and interception. AI systems can rapidly analyse sensor inputs to predict incoming threats and execute precise evasive actions. This is particularly useful in hostile environments where human operators may struggle to process information quickly enough to make effective decisions.

- **Thermal Management Optimization**: Hypersonic missiles experience extreme heat due to atmospheric friction at high speeds. AI-based systems can optimize cooling strategies and flight paths to minimize exposure to high thermal loads, adjusting control surfaces to improve thermal dissipation during flight.

- **Targeting Precision**: AI-driven systems can enhance the accuracy of targeting in hypersonic weapons by integrating data from multiple sources, such as GPS, radar, and terrain-matching systems. In situations where targets are mobile or camouflaged, AI algorithms can predict their movements and adjust the flight trajectory accordingly to ensure a successful strike.

These examples demonstrate how AI improves the adaptability, precision, and survivability of hypersonic weapons, ensuring greater mission success rates despite the extreme conditions encountered during flight.

In the context of precision strikes, hypersonic weapons must operate effectively in time-sensitive environments. AI systems facilitate autonomous targeting and decision-making by assessing real-time data feeds and adjusting to target changes dynamically. This capability is crucial for operations in rapidly evolving battlefields, where the ability to reroute or select alternate targets can significantly impact mission outcomes [635]. The implementation of AI in decision-making processes allows for a more agile response to battlefield conditions, ultimately increasing the effectiveness of hypersonic weapons in combat scenarios.

Specific examples of how AI systems enhance precision strikes for hypersonic weapons include:

- **Autonomous Target Adjustment**: AI-enabled hypersonic weapons can autonomously adjust their flight paths in real time based on changes in the target's location or battlefield conditions. For example, during a rapidly evolving combat scenario, such as a mobile launcher changing position, AI can process data from satellites, drones, and ground sensors to reroute the missile to the new target location. This dynamic target tracking ensures the weapon stays effective in fluid combat environments.

- **Target Prioritization**: In time-sensitive strikes, AI systems can assess multiple targets and prioritize them based on threat levels or mission objectives. For instance, AI algorithms could analyse incoming intelligence data and autonomously decide to strike a higher-value target if it becomes visible or exposed, adjusting mid-flight without human intervention. This flexibility is crucial in chaotic combat situations where quick decisions can shift mission success.

- **Adaptive Decision-Making in Real Time**: AI-based systems in hypersonic weapons can adapt to unexpected events on the battlefield, such as enemy air defence activation. For example, if radar systems detect the launch of interceptor missiles, AI can instantly calculate the optimal evasive manoeuvres or alternative routes to avoid interception, ensuring the weapon reaches its target with minimal risk of being neutralized.

- **Battlefield Awareness and Coordination**: AI also enhances coordination between different military assets. For instance, in a precision strike operation, an AI system could integrate data from reconnaissance drones and other hypersonic missiles to create a shared operational picture. This allows for coordinated strikes on multiple targets, improving efficiency and minimizing collateral damage.

These examples highlight the ability of AI systems to enhance hypersonic weapon performance by enabling autonomous decision-making, rapid response to target changes, and real-time adaptation in complex combat scenarios

The rapid speed and manoeuvrability of hypersonic weapons pose significant challenges for detection and interception. AI algorithms have been developed to enhance the prediction of hypersonic flight paths, improving both offensive and defensive capabilities. By analysing multiple data streams from various sources, such as satellites and radar systems, AI systems can provide more accurate trajectory predictions and response plans [635, 636]. This capability is essential for developing effective countermeasures against hypersonic threats, as it allows for timely and precise tracking of fast-moving targets.

Some examples of how AI algorithms improve detection and interception of hypersonic weapons include:

- **Trajectory Prediction in Missile Defense Systems**: AI systems can enhance the capability of missile defence platforms like Aegis and THAAD by analysing data from radar and satellite sources to predict the flight path of incoming hypersonic threats. For instance, the Space-Based Infrared System (SBIRS) gathers thermal data on missile launches, and AI processes this information to forecast the missile's trajectory, even with the rapid speed and manoeuvrability of hypersonic weapons. This allows defence systems to position interceptors effectively in real time.

- **Integration of Multiple Sensor Data**: AI enables the fusion of data from various platforms, such as space-based sensors, ground-based radars, and maritime tracking systems, to build a more accurate picture of hypersonic missile flight paths. For example, AI algorithms in the U.S. Missile Defense Agency's systems analyse inputs from global sensor networks to create a predictive model of a hypersonic missile's behaviour, allowing for faster decision-making and countermeasure deployment.

- **Adaptive Learning in Interception Algorithms**: Machine learning models continuously adapt based on new data from test launches and simulations, refining their ability to predict the evasive manoeuvres of hypersonic missiles. For example, the Pentagon's DARPA program is exploring AI-powered defence systems that can adapt in real time to the unpredictable manoeuvres of hypersonic missiles, increasing the accuracy of interception attempts.

- **Satellite Tracking Enhancement**: AI improves the processing speed of satellite-based early warning systems by automating the detection and tracking of hypersonic weapons' unique signatures, such as heat and plasma plumes. This is critical for intercepting fast-moving targets that would otherwise escape detection due to their high speed and erratic flight paths. AI algorithms help defence systems track the missile continuously despite the challenges posed by atmospheric interference and high velocity.

These examples illustrate how AI-driven solutions are critical for addressing the detection and interception challenges posed by the next generation of hypersonic weapons.

The extreme conditions experienced by hypersonic weapons necessitate the use of advanced materials capable of withstanding high temperatures and stresses. AI is being utilized in the design and testing of these materials, optimizing their performance under hypersonic conditions [637]. Furthermore, autonomous systems can monitor and adjust onboard thermal management systems in real-time, ensuring optimal performance during flight. This integration of AI into materials science and thermal management is critical for enhancing the durability and effectiveness of hypersonic weapons.

Examples of how AI is utilized in the design and testing of materials for hypersonic weapons and in onboard thermal management include:

- **AI-Driven Materials Design for Hypersonic Environments**: Researchers are using AI to develop materials capable of withstanding the extreme temperatures and pressures of hypersonic flight. For example, machine learning algorithms are being employed to identify new ultra-high-temperature ceramics (UHTCs) that have enhanced thermal resistance and durability. These algorithms analyse vast datasets of material properties and predict the behaviour of new compounds, optimizing them for the unique stresses encountered in hypersonic conditions. NASA's research into new materials for spacecraft heat shields, which must endure similar conditions to hypersonic vehicles, utilizes AI for materials discovery and testing, showcasing the broader application of this technology.

- **Real-Time Thermal Management Systems**: Hypersonic weapons generate extreme heat due to air friction, particularly during sustained flight in the atmosphere. AI is used in advanced thermal management systems that autonomously adjust cooling mechanisms to prevent structural failure. For instance, thermal sensors onboard hypersonic vehicles feed data into AI systems that can regulate heat distribution across the vehicle, adjusting airflow or coolant circulation in real time. This approach has been used in experimental hypersonic vehicles like the Boeing X-51 Waverider, where AI-managed heat dissipation is crucial for the vehicle's performance during sustained Mach 5+ speeds.

- **Autonomous Testing in Hypersonic Wind Tunnels**: AI systems are increasingly being integrated into hypersonic wind tunnel testing to simulate the stresses and conditions materials face during flight. For example, the U.S. Department of Defense has used AI models to run simulations that predict how new materials will perform under extreme temperatures and pressures before physical prototypes are built. This reduces the time and cost of testing and accelerates the development of more resilient materials for hypersonic applications.

The development of mobile launch platforms for hypersonic weapons has been significantly advanced through the integration of AI and autonomous systems. These platforms can operate with minimal human intervention, autonomously identifying optimal launch conditions and adapting to environmental changes or enemy defences [638]. This capability not only increases operational flexibility but also enhances the survivability of launch platforms in contested environments.

The following are some examples of how AI and autonomous systems have enhanced the development and operation of mobile launch platforms for hypersonic weapons:

- **AI-Enhanced Targeting and Launch Optimization**: Hypersonic weapon platforms, such as mobile truck-mounted systems or naval vessels, increasingly integrate AI algorithms to autonomously identify optimal launch conditions. These algorithms process real-time data from weather sensors, satellite feeds, and radar systems to determine the most favourable launch windows. For instance, Russia's Avangard hypersonic missile system benefits from AI-driven decision-making, allowing it to autonomously adapt to environmental changes and select the optimal launch trajectory, ensuring precise and timely engagement of targets.

- **Autonomous Mobile Launch Platforms**: Modern hypersonic missile systems, such as China's DF-ZF and Russia's Kinzhal, are designed to be deployed from mobile platforms like aircraft, trucks, or submarines. AI systems play a crucial role in enabling these platforms to navigate and position themselves autonomously in unpredictable, contested environments. For example, AI onboard mobile launch platforms can detect and avoid enemy surveillance systems or defences, enhancing the survivability of the platform by minimizing detection risk. This is particularly valuable in environments where human operators may not have real-time data or fast enough reaction times to make critical adjustments.

- **Adaptation to Enemy Defences**: In addition to environmental factors, mobile launch platforms equipped with AI can react to dynamic enemy defence measures. Autonomous systems can modify launch parameters, such as flight trajectory and timing, based on incoming threats like anti-ballistic missile defences or jamming attempts. For example, the U.S. is exploring autonomous battlefield management systems that coordinate mobile hypersonic platforms, ensuring seamless adjustment to evolving enemy defences and allowing the launch system to remain agile and responsive in high-threat zones.

AI-powered simulations are revolutionizing the testing of hypersonic weapons by providing virtual environments to evaluate performance across various scenarios. This reduces the need for costly physical tests and allows for rapid iteration of designs based on real-time feedback from simulations [638]. Autonomous testing systems can further enhance the accuracy and efficiency of hypersonic weapons development, ensuring that systems are thoroughly vetted before deployment.

AI-powered simulations are playing a transformative role in the testing and development of hypersonic weapons by providing virtual environments that can assess their performance in various conditions. Implementation examples include:

- **Reduced Physical Testing Costs and Risks**: Hypersonic weapons testing, traditionally reliant on expensive and high-risk physical trials, is now increasingly augmented by AI-driven simulations. For example, the U.S. Department of Defense has been leveraging AI-powered tools like digital twins—virtual models of physical systems—to simulate the behaviour of hypersonic weapons in different scenarios. These simulations allow for stress testing in various environmental conditions such as extreme heat, atmospheric turbulence, and varying altitudes, drastically reducing the number of physical prototypes required and the associated costs.

- **Rapid Iteration of Design**: AI-powered simulations provide real-time feedback on weapon designs, allowing engineers to make adjustments quickly. For instance, AI systems can evaluate aerodynamics, material stresses, and thermal management in a variety of simulated flight conditions. This has been critical for the development of systems like the Hypersonic Air-breathing Weapon Concept (HAWC), where design iterations can be tested virtually before physical prototypes are constructed. These simulations enable thousands of design iterations, each tested under different parameters, improving the overall robustness of the weapon system.

- **Autonomous Testing Systems**: Autonomous systems integrated into AI-powered simulations help refine testing efficiency and accuracy. In some cases, these systems automatically adjust the virtual environment based on real-world data inputs from past trials. An example of this is Lockheed Martin's X-59 Quiet Supersonic Transport, where AI simulations play a crucial role in predicting performance outcomes across varying terrains and atmospheres. Such systems ensure that by the time the physical hypersonic weapon is tested, the majority of critical performance issues have been identified and addressed through virtual trials.

- **Scenario-Based Testing for Real-Time Adjustments**: AI simulations allow for the modelling of complex, time-sensitive combat scenarios that hypersonic weapons are likely to encounter. For instance, the U.S. military is incorporating AI simulations to test how hypersonic weapons respond in counter-defence scenarios where real-time flight path adjustments are critical. This approach helps developers to refine targeting and evasion algorithms before actual deployment.

AI-powered simulations offer significant benefits, including reduced testing costs, rapid design iterations, and enhanced accuracy in predicting real-world performance. These tools are becoming integral to the safe and efficient development of next-generation hypersonic weapons systems.

AI facilitates autonomous mission planning for hypersonic weapons, enabling them to select and adjust their routes and tactics in real-time. By analysing vast datasets, these systems can identify the most effective pathways to targets while avoiding detection and countermeasures [635, 638]. This capability is vital for maintaining the strategic advantage of hypersonic weapons in complex operational environments.

AI has revolutionized mission planning for hypersonic weapons by enabling real-time autonomous adjustments to routes and tactics, including:

- **Real-Time Route Optimization**: AI-driven systems can autonomously select and modify flight paths based on real-time data analysis. For instance, during a mission, AI can evaluate satellite and radar feeds to identify potential threats, such as enemy radar or missile defence systems, and reroute the weapon to avoid detection. This capability was demonstrated in tests with advanced hypersonic systems where AI used input from various sensors to ensure the missile followed the optimal path, bypassing hostile zones and minimizing risk. AI was critical in adapting routes as new threats emerged during the mission.

- **Target Prioritization**: In scenarios involving multiple targets or mobile targets, AI allows hypersonic weapons to prioritize based on mission objectives. For example, the U.S. Air Force's experimentation with hypersonic platforms integrated AI to autonomously determine which targets were most vulnerable or strategically significant. The AI system analysed battlefield data and recommended tactical changes mid-flight, enhancing mission effectiveness by dynamically adjusting the focus based on new intelligence.

- **Countermeasure Avoidance**: AI-driven systems in hypersonic weapons are equipped with the ability to detect and adapt to enemy countermeasures such as jamming or electronic warfare. A notable example is the use of AI in DARPA's Hypersonic Air-breathing Weapon Concept (HAWC), where the AI system autonomously adjusted flight tactics to avoid radar detection by constantly recalculating the missile's radar cross-section and adjusting its trajectory to evade tracking. The system evaluated both terrain and environmental conditions to maintain stealth while navigating hostile territories.

- **Enhanced Decision-Making in Complex Environments**: During complex operations involving multiple threats and changing environmental conditions, AI systems can continuously analyse vast datasets from satellites, ground sensors, and other intelligence sources. In the Russian Avangard hypersonic missile system, AI is used to assess dynamic factors like wind patterns, thermal gradients, and enemy defences to continuously update flight paths. This ensures that the missile can maintain a high degree of manoeuvrability, crucial for surviving against sophisticated defence systems.

In response to the emerging threat of hypersonic weapons, AI-driven defensive systems are being developed to detect, track, and neutralize incoming missiles. These autonomous systems can react significantly faster than human-operated systems, providing critical response time for intercepting fast-moving targets [604, 636]. The integration of AI into defensive strategies is essential for enhancing national security and countering the challenges posed by hypersonic technologies.

The following are examples of how AI is being integrated into these defensive strategies:

- **Lockheed Martin's Missile Defense Systems**: Lockheed Martin has been actively integrating AI into its missile defence technologies to enhance real-time tracking and interception capabilities. AI algorithms process data from multiple sensors, such as radar and satellites, to track hypersonic missiles. For example, in their Aegis Ballistic Missile Defense (BMD) system, AI assists in rapidly analysing vast amounts of data from the SPY-1 radar to track fast-moving hypersonic missiles and recommend interception strategies. This significantly improves reaction times, allowing defensive systems to adjust trajectories and deploy countermeasures faster than manual control could manage.

- **DARPA's Glide Breaker Program**: The Defense Advanced Research Projects Agency (DARPA) is using AI in its Glide Breaker program, which is designed to intercept hypersonic threats. AI plays a crucial role in the decision-making processes by using predictive algorithms to assess and adapt to the unpredictable flight paths of hypersonic glide vehicles. By leveraging real-time data from various tracking systems, AI can predict missile trajectories and guide interception systems with unprecedented accuracy, enabling faster and more efficient defensive responses.

- **Northrop Grumman's Missile Defense Solutions**: Northrop Grumman is incorporating AI in its missile defence platforms, such as the Integrated Air and Missile Defense Battle Command System (IBCS). AI-driven systems within IBCS fuse data from sensors across land, sea, air, and space to create a comprehensive picture of the battlespace, allowing for real-time tracking of hypersonic weapons. The AI system can autonomously allocate resources to intercept threats by selecting the most appropriate weaponry and launch sites, drastically reducing the time required for human intervention and decision-making.

- **Raytheon's Hypersonic Defense Program**: Raytheon Technologies is developing AI-enabled systems that improve tracking and intercepting capabilities for hypersonic weapons. For instance, Raytheon's Missile Defense Agency (MDA) contracts involve AI technologies that enhance radar and sensor fusion to track hypersonic missiles more effectively. AI-driven algorithms process high-speed data from various tracking systems to accurately predict the missile's flight path, allowing defensive systems to react and neutralize the threat in real time.

Next-Generation Propulsion and Material Science

Next-generation propulsion systems are pivotal for the advancement of hypersonic weapons, which require sustained speeds exceeding Mach 5 while ensuring control and stability. The development of these systems is focused on several innovative technologies, including scramjet improvements, dual-mode propulsion, hybrid rocket-scramjet systems, and pulse detonation engines (PDEs).

Scramjet Improvements: Scramjet engines are integral to hypersonic propulsion due to their ability to compress incoming air at supersonic speeds without rotating machinery. Recent studies highlight ongoing research aimed at enhancing scramjet efficiency through improved fuel injection techniques, combustion stability, and drag reduction. For instance, Lee and Lee [639] discuss the fluid and combustion dynamics in dual-mode scramjets, emphasizing their enhanced specific impulse across various hypersonic flight Mach numbers, which is crucial for extending operational capabilities. Additionally, Zhu [70] emphasizes the importance of scramjet engines in hypersonic air-breathing vehicles, noting their role in compressing air and optimizing propulsion performance. These advancements are critical for achieving longer flight durations and operational effectiveness in military applications.

Dual-Mode Propulsion: The concept of dual-mode propulsion, which allows engines to operate in both ramjet and scramjet modes, is gaining traction for hypersonic cruise missiles. This adaptability enables missiles to utilize ramjet propulsion for initial acceleration before transitioning to scramjet mode at hypersonic speeds. Xue et al. [249] provide a comprehensive survey on the conceptual design of hypersonic aircraft powered by RBCC (Rocket-Based Combined Cycle) engines, which effectively combine the strengths of both ramjet and scramjet technologies to optimize performance across a wide range of flight conditions. Furthermore, the work on fluid dynamics in dual-mode scramjets reinforces the potential for these systems to enhance the operational envelope of hypersonic vehicles [639].

Rocket-Scramjet Hybrid Systems: The integration of rocket and scramjet propulsion systems represents a significant advancement in hypersonic missile technology. This hybrid approach allows for rapid acceleration via rockets, followed by sustained high-speed flight using scramjets. Research indicates that such systems can provide strategic advantages in time-sensitive missions. For example, the study by Isono et al. [640] discusses the arrangement of rocket engine locations in combined cycle engines, which is essential for optimizing performance in hypersonic flight scenarios. Additionally, the hybrid rocket-based combined cycle engine (HRBCC) concept explored by Hillion [199] suggests that this technology could facilitate both intra- and extra-atmospheric hypersonic flight, thereby enhancing mission versatility.

Pulse Detonation Engines (PDEs): PDEs are emerging as a promising propulsion technology for hypersonic applications. Unlike traditional engines that maintain continuous combustion, PDEs operate by detonating fuel-air mixtures in pulses, which can yield higher efficiency and thrust-to-weight ratios at hypersonic speeds. Research by Rosato et al. [267] highlights the potential of stabilized detonation for hypersonic propulsion, indicating that this technology could revolutionize propulsion systems for both terrestrial and interplanetary travel. Furthermore, Zhao discusses the combustion characteristics of detonation cycle engines (DCR), which have shown promise in achieving hypersonic flight capabilities [641].

The future of hypersonic propulsion systems is characterized by innovative technologies such as scramjets, dual-mode engines, hybrid systems, and PDEs. Each of these advancements plays a crucial role in enhancing the performance, efficiency, and operational capabilities of hypersonic weapons, thereby shaping the landscape of modern military aviation.

The development of materials for extreme environments, particularly in the context of hypersonic flight, is a critical area of research in material science. Hypersonic vehicles experience extreme aerodynamic heating and mechanical stresses, necessitating advanced materials that can maintain structural integrity and thermal resistance under such conditions. This synthesis will explore various material categories, including Ultra-High-Temperature Ceramics (UHTCs), Carbon-Carbon Composites, Metal Matrix Composites (MMCs), Active Thermal Protection Systems, and Self-Healing Materials, all of which are pivotal for the advancement of hypersonic technology.

Ultra-High-Temperature Ceramics (UHTCs) are essential for components exposed to extreme heat, such as the leading edges and nose cones of hypersonic vehicles. UHTCs, including zirconium diboride and hafnium carbide, can withstand temperatures exceeding 2,000°C, making them suitable for thermal structures in hypersonic applications. Research indicates that these materials exhibit excellent long-term ablation resistance, crucial for maintaining performance in the harsh conditions of hypersonic flight [642]. The ability of UHTCs to maintain structural stability under sustained aerodynamic heating is a significant advantage, addressing one of the major challenges in hypersonic weapon systems [347, 642].

Carbon-Carbon Composites are another critical material in hypersonic technology. These composites are known for their lightweight and high-temperature resistance, retaining structural properties even at elevated temperatures. They are commonly used in heat shields and thermal protection systems, effectively preventing heat from degrading critical components of missiles [347, 642]. The mechanical properties of these composites, particularly their compressive strength at high temperatures, make them suitable for structural applications in extreme environments [643].

Metal Matrix Composites (MMCs) represent a promising area of research for future hypersonic missiles. By combining metals with reinforcing ceramic or fibre materials, MMCs achieve superior strength-to-weight ratios, which are essential for performance under hypersonic conditions [644]. The incorporation of materials such as silicon carbide into aluminium matrices enhances the mechanical and thermal properties, making MMCs ideal candidates for components that must endure high temperatures and mechanical stresses [644]. Recent studies have highlighted the effectiveness of MMCs in providing improved wear resistance and mechanical stability compared to conventional materials [644].

Active Thermal Protection Systems are being explored to dynamically manage heat during hypersonic flight. These systems utilize circulating coolant fluids to dissipate heat, thereby prolonging the endurance of missiles and protecting internal components from thermal degradation [642]. This innovative approach to thermal management is crucial for enhancing the reliability and performance of hypersonic systems, as it addresses the challenges posed by extreme thermal environments [347, 642].

Self-Healing Materials are an emerging field that holds significant promise for extending the lifespan and durability of hypersonic weapons. These materials can autonomously repair themselves when subjected to mechanical stress or damage, mitigating wear and tear caused by extreme flight conditions [645, 646]. The development of self-healing polymers and composites has gained traction, with research focusing on their application in aerospace structures to enhance resilience and reduce maintenance needs [645, 646]. The integration of self-healing capabilities into materials used in hypersonic vehicles could revolutionize their operational longevity and reliability [645, 646].

The advancement of material science for extreme environments, particularly in hypersonic applications, is characterized by the development of UHTCs, carbon-carbon composites, MMCs, active thermal protection systems, and self-healing materials. Each of these material categories plays a vital role in addressing the challenges posed by hypersonic flight, ensuring that systems can withstand the intense thermal and mechanical stresses encountered during operation.

Artificial intelligence (AI) is increasingly recognized as a transformative force in the development of materials and systems for hypersonic applications, where extreme speeds and temperatures impose significant demands on structural components. The application of AI algorithms in this domain is multifaceted, encompassing material design, simulation, thermal management, and structural optimization.

In the realm of material design and simulation, AI facilitates the rapid development of advanced materials capable of withstanding the harsh conditions encountered during hypersonic flight. Traditional methods of material design often involve extensive experimental testing, which can be both time-consuming and costly. However, AI-driven approaches, particularly those utilizing machine learning (ML) algorithms, allow researchers to simulate material behaviours under hypersonic conditions before physical testing occurs. For instance, Chiang et al. [647] highlight the development of a toolkit that expedites particle-based simulations, significantly reducing the computational costs associated with predicting material properties. Similarly, Sun et al. [648] discuss a knowledge transfer framework that enhances the prediction of alloy properties, demonstrating the efficacy of machine learning in material characterization. These AI models can predict failure points in materials, such as cracking or deformation, by analysing large datasets from previous experiments, thus optimizing material compositions for hypersonic applications [648].

The ability of AI to simulate various scenarios enables researchers to identify the ideal balance between weight, strength, and heat resistance in materials used for critical components like nose cones and leading edges of hypersonic vehicles. For example, the work by Zhao et al. [649] emphasizes the importance of understanding heat and drag reduction mechanisms, which are crucial for developing materials that can endure extreme thermal loads. The predictive capabilities of AI not only streamline the design process but also reduce the reliance on costly physical testing, thereby accelerating the development of hypersonic systems.

In addition to material design, AI plays a vital role in real-time thermal management during hypersonic flight. The extreme heat generated by friction necessitates sophisticated cooling systems to maintain the structural integrity of the vehicle. AI-powered thermal management systems can autonomously adjust cooling mechanisms based on real-time flight data, as discussed by Zheng et al., cited in Sun et al. [648], who explore the integration of machine learning in managing thermal loads effectively. These systems monitor temperature fluctuations and make dynamic adjustments to prevent overheating, thereby enhancing the overall efficiency and safety of hypersonic vehicles.

Furthermore, AI contributes to the optimization of structural layouts in hypersonic systems. Through techniques such as topology optimization, AI algorithms can refine structural configurations by removing unnecessary material while reinforcing areas subjected to high loads. This approach leads to lighter and stronger designs, which are essential for withstanding the combined forces of heat, pressure, and mechanical stress during hypersonic flight [650]. The integration of AI in structural optimization not only improves performance but also facilitates the exploration of complex design spaces that would be challenging to navigate using traditional methods.

AI is fundamentally reshaping the landscape of hypersonic technology by enabling more efficient material design, testing, and real-time management of critical systems. The ability to simulate material behaviour, predict failure points, optimize structural layouts, and autonomously manage thermal loads positions AI as an integral component in overcoming the challenges posed by high-speed atmospheric travel. As research in this field continues to advance, the role of AI in developing durable, reliable, and efficient hypersonic vehicles will undoubtedly expand.

Additive manufacturing (AM), commonly referred to as 3D printing, is significantly transforming the aerospace and defence sectors, particularly in the development of hypersonic missile systems. This technology enables the production of complex, high-performance components that are often impractical to manufacture using traditional methods such as casting, forging, and machining. The ability to fabricate intricate geometries is crucial for hypersonic missiles, where lightweight and durable components are essential for optimal performance in extreme environments [651, 652].

One of the key advantages of additive manufacturing is its capability to create complex geometries and provide precise control over material properties. Hypersonic missiles encounter extreme conditions, including high temperatures and intense aerodynamic loads, necessitating components like rocket nozzles and thermal shields to be designed with intricate internal structures. For instance, additive manufacturing allows for the integration of internal cooling channels within rocket nozzles, which are vital for managing heat during high-speed flight. This capability reduces the risk of overheating and enhances the reliability of missile components by minimizing the number of joints and potential failure points [651, 652]. The design freedom afforded by AM enables engineers to optimize these components directly from 3D models, overcoming many constraints associated with traditional manufacturing processes [652].

Additive manufacturing facilitates rapid prototyping and customization, which are invaluable in the fast-paced development of hypersonic systems. Engineers can quickly iterate designs based on testing feedback without the need for extensive tooling changes, significantly reducing development time and costs [651, 652]. This flexibility allows for the fine-tuning of material properties, such as thermal resistance and structural strength, tailored to specific mission requirements [651]. The ability to customize parts also enhances the performance of hypersonic missiles, enabling the integration of advanced materials that can withstand extreme conditions while maintaining lightweight characteristics [652].

In terms of material usage, additive manufacturing allows for the creation of lightweight yet highly durable components. The technology supports the use of advanced materials such as titanium alloys and nickel-based superalloys, which are essential for high-performance applications in hypersonic systems. By optimizing part geometries, AM can reduce material usage in non-critical areas while reinforcing high-stress regions, thereby enhancing overall structural integrity [651, 652]. For example, thermal shields designed for hypersonic missiles must endure temperatures exceeding 3,000°C; additive manufacturing enables the incorporation of sophisticated cooling features that would be challenging to achieve with conventional methods [651, 652].

Furthermore, efficient cooling systems are critical for hypersonic applications due to the extreme heat generated at speeds above Mach 5. Additive manufacturing allows for the design of intricate internal cooling channels within missile components, such as fuel injectors and thermal shields. This capability not only improves heat management but also simplifies the manufacturing process by consolidating multiple components into a single part, thereby enhancing efficiency and reducing assembly time [651, 652].

Additive manufacturing is revolutionizing the production of hypersonic missile systems by offering unparalleled design freedom, rapid prototyping capabilities, and the ability to produce lightweight yet durable components. The technology significantly enhances the performance of critical parts, including rocket nozzles, thermal shields, and cooling systems, thereby pushing the boundaries of what is achievable in hypersonic flight. As the competition in hypersonic technology intensifies, the role of additive manufacturing in developing next-generation missile systems is expected to grow even more prominent [651, 652].

Integrating Hypersonic Systems with Cyber Warfare and Space Operations

Integrating hypersonic systems with cyber warfare and space operations represents a major shift in modern military capabilities, where advanced weaponry, cyber domains, and space-based infrastructure converge to create new forms of strategic dominance. Hypersonic weapons, characterized by their high speed and manoeuvrability, introduce complex challenges and opportunities when integrated with these other domains.

Cyber warfare involves the use of digital attacks by one nation or entity against another, with the intent to compromise critical infrastructure, military systems, government networks, or financial institutions. These cyberattacks aim to disrupt, disable, or destroy digital networks, gain unauthorized access to sensitive information, or destabilize a target politically or socially. The tactics used in cyber warfare vary widely and include hacking, malware deployment, phishing, Distributed Denial of Service (DDoS) attacks, and the spread of misinformation or disinformation to create unrest or confusion.

Cyber warfare strategies often include offensive measures designed to infiltrate or damage an adversary's digital systems. Hacking and penetration are key techniques in which attackers gain unauthorized access to networks to steal or manipulate data, or disrupt essential functions. Malware and ransomware attacks cripple systems by embedding malicious code that can lock out users or siphon data, with ransomware demanding payment to restore functionality. DDoS attacks, on the other hand, overload networks by bombarding them with traffic, making systems inoperable, particularly targeting critical services such as government websites, banks, or hospitals. Data manipulation and espionage involve altering or stealing sensitive data, often for intelligence gathering or to undermine the security of a nation.

Defensive strategies in cyber warfare are crucial for maintaining the integrity of national networks and systems. Cybersecurity infrastructure includes firewalls, encryption, and intrusion detection systems that form the first line of defence against potential attacks. Nations also develop robust incident response teams trained to detect, respond to, and mitigate the effects of cyberattacks, ensuring that systems can be restored quickly. Intelligence and monitoring activities help in continuously tracking threats, assessing vulnerabilities, and securing a nation's critical digital infrastructure against potential cyber intrusions.

The primary objective of cyber warfare is often to disrupt the target nation's vital services and infrastructure. Disruption of critical services, such as power grids or transportation networks, can halt essential civilian and military functions. Economic warfare attacks financial systems, stock markets, and banking institutions to destabilize the economy and spread panic. Additionally, psychological warfare through cyber means involves using misinformation and disinformation campaigns to influence public opinion or create social unrest, further weakening the target's stability.

Cyber warfare has increasingly become a pivotal aspect of global conflicts, particularly in the last decade. Notable incidents such as the Stuxnet attack in 2010, the Ukraine power grid attack in 2015, and the SolarWinds hack in 2020 exemplify the evolving nature of warfare in the digital age.

The Stuxnet worm, widely attributed to the United States and Israel, specifically targeted Iran's nuclear facilities, causing significant disruptions to its uranium enrichment processes. This incident is often cited as one of the first known uses of cyber warfare to achieve strategic military objectives, marking a shift in how states engage in conflict [653]. The sophistication of Stuxnet, which was designed to manipulate industrial control systems, underscores the potential of cyber operations to inflict damage without traditional kinetic warfare [654].

Similarly, the 2015 cyber attack on Ukraine's power grid, attributed to Russian actors, left hundreds of thousands without electricity and highlighted the vulnerabilities of critical infrastructure to cyber threats. This incident not only demonstrated the destructive potential of cyber warfare but also raised concerns about the implications for national security and civilian safety in the context of international conflicts [655]. The attack exemplified how cyber operations can blur the lines between military and civilian targets, complicating the legal and ethical frameworks surrounding warfare [656].

The SolarWinds hack in 2020 further illustrated the complexities of cyber espionage and warfare. By embedding malicious code in a widely used software update, Russian-backed actors gained long-term access to sensitive networks of U.S. government agencies and corporations. This incident revealed the vulnerabilities inherent in supply chain security and the potential for cyber operations to compromise national security on a massive scale [657]. The implications of such breaches extend beyond immediate damage, as they can undermine public trust in governmental institutions and disrupt international relations [658].

These incidents collectively highlight the transformative impact of cyber warfare on global conflicts. The integration of cyber capabilities into military strategies necessitates a re-evaluation of traditional concepts of warfare, sovereignty, and international law. As cyber warfare continues to evolve, it is imperative for nations to develop robust frameworks to address the challenges posed by this new domain of conflict [659].

As digital threats continue to grow in complexity, cyber warfare has become a central component of modern military and political strategies. Nations around the world are building both offensive and defensive cyber capabilities to gain strategic advantages, signalling a shift from traditional battlefield engagements to the digital domain, where conflicts are fought over control of data, networks, and critical infrastructure.

Hypersonic Systems and Cyber Warfare

The integration of hypersonic systems with cyber warfare represents a significant advancement in military capabilities, intertwining the rapid deployment of hypersonic missiles with the complexities of cybersecurity. Hypersonic systems, characterized by their ability to travel at speeds exceeding Mach 5, rely heavily on sophisticated digital infrastructures, including command and control (C2) networks and real-time data links for navigation and targeting. These digital systems are vulnerable to cyber threats, which can exploit weaknesses in communication networks to disrupt operations, intercept data, or launch attacks such as Distributed Denial of Service (DDoS) to incapacitate hypersonic capabilities during critical missions [660].

Cyber vulnerabilities in hypersonic systems can be exploited by adversaries aiming to sabotage missile guidance systems or disrupt supply chains. For instance, a cyberattack targeting the production line of missile components could lead to the deployment of compromised or malfunctioning weapons, significantly undermining operational effectiveness. The potential for cyber operations to disrupt hypersonic systems highlights the necessity for robust cybersecurity measures that can protect against such threats. As noted in the literature, the increasing reliance on interconnected systems necessitates a proactive approach to cybersecurity, particularly in military applications where the stakes are exceptionally high [661].

To counter these threats, military forces are increasingly leveraging artificial intelligence (AI) and machine learning (ML) technologies to enhance their defensive cyber capabilities. AI can facilitate real-time monitoring and detection of cyber threats, allowing for rapid responses to potential breaches. The integration of AI into cybersecurity frameworks enables the automation of threat detection and response, significantly improving the resilience of hypersonic systems against cyberattacks [662]. Furthermore, AI-driven solutions can analyse vast amounts of data to identify patterns indicative of malicious activities, thereby enhancing the overall security posture of military operations involving hypersonic technologies.

Moreover, the application of AI in cybersecurity extends beyond mere defence; it also encompasses offensive strategies that can be employed to disrupt an adversary's hypersonic capabilities. For example, AI can be utilized to identify vulnerabilities in enemy systems, allowing for targeted cyber operations that could incapacitate their hypersonic missile infrastructure. This dual-use nature of AI in cyber warfare underscores its critical role in modern military strategy, where the intersection of hypersonic systems and cybersecurity is increasingly becoming a focal point of national defence initiatives [663].

Integration with Space Operations

The integration of hypersonic weapons with space operations is a critical aspect of modern military strategy, particularly in enhancing their operational effectiveness through satellite-based navigation, missile tracking, and C5ISR systems. This integration not only improves the capabilities of hypersonic systems but also introduces new vulnerabilities that must be addressed.

Satellite-Based Navigation and Targeting: Hypersonic missiles depend heavily on advanced guidance systems for precision targeting, particularly during the terminal phase of flight. The Global Positioning System (GPS) and emerging Low Earth Orbit (LEO) navigation systems are vital for ensuring accurate missile guidance [39]. However, this reliance on space-based systems exposes hypersonic weapons to potential threats from anti-satellite (ASAT) weapons, jamming, and cyberattacks aimed at disrupting GPS signals. To counter these vulnerabilities, military strategies are increasingly focused on diversifying satellite constellations, which can provide redundancy and resilience against adversarial actions [39]. The development of LEO constellations, for instance, enhances the robustness of navigation systems, making it more challenging for adversaries to disrupt missile guidance [39].

Missile Tracking and Detection: Space-based missile detection systems are crucial for modern missile defence architectures. The U.S . Space-Based Infrared System (SBIRS) and other upcoming space-tracking networks are designed to detect and track missile launches, including those of hypersonic glide vehicles (HGVs) [50]. The unique flight profiles of hypersonic missiles, characterized by their manoeuvrability and low-altitude trajectories, present significant challenges to traditional missile defence systems [5]. Integrating advanced space-based sensors, such as hyperspectral and infrared sensors, can significantly enhance the detection and tracking capabilities for hypersonic threats, allowing for quicker defensive responses [50]. Countries like the U.S., China, and Russia are actively developing tailored space-based early-warning systems to address these challenges [5].

C5ISR Integration: The successful integration of hypersonic weapons into the broader C5ISR framework is essential for effective military operations. Space-based assets provide real-time data that enable decision-makers to monitor missile trajectories, adjust targeting, and respond to emerging threats [664]. This interconnected approach ensures that hypersonic systems can operate cohesively, receiving continuous updates from satellites, ground-based radar, and airborne assets [664]. Furthermore, the integration of space-based Tracking, Telemetry, and Command (TT&C) infrastructure is critical for maintaining operational control over hypersonic glide vehicles, especially in contested environments where communication links may be compromised [664].

Offensive and Defensive Implications

The integration of offensive hypersonic operations with cyber and space capabilities presents significant implications for modern warfare. Offensive hypersonic weapons can be utilized in conjunction with cyber warfare tactics to disrupt an adversary's critical infrastructure. For example, disabling or jamming satellite-based tracking systems can create opportunities for hypersonic strikes to occur undetected. This approach aligns with the concept of "defending forward," which emphasizes proactive measures to undermine enemy capabilities before they can be effectively deployed [665, 666]. The ability to conduct cyberattacks that disable missile defence systems further enhances the effectiveness of hypersonic strikes, as it leaves adversary forces vulnerable to precision attacks [665, 666].

Moreover, the development of space-based early warning systems is crucial for detecting and tracking hypersonic glide vehicles (HGVs). These systems offer advantages over traditional ground-based systems due to their broader spatial coverage and improved tracking capabilities [39, 667]. The integration of artificial intelligence (AI) into these systems can facilitate real-time analysis of multiple data streams, enabling rapid decision-making and effective countermeasures against hypersonic threats [39, 668]. This coordinated effort between cyber defence, space tracking, and missile interception is essential for a comprehensive defence strategy against hypersonic threats.

Defensive measures may also involve targeting an adversary's space assets to degrade their command-and-control capabilities related to hypersonic missile operations. By disrupting the communication and coordination of these systems, a nation can significantly reduce the effectiveness of an adversary's hypersonic capabilities [669]. The use of AI and machine learning in these defensive strategies can enhance the ability to predict and respond to hypersonic threats, further solidifying the importance of integrating cyber and space operations in national defence strategies [668].

Hypersonic Weapons in Space

Hypersonics in Near-Earth Orbit: Strategic Potential

The strategic potential of hypersonic weapons operating in near-Earth orbit (NEO) represents a transformative shift in military capabilities, significantly impacting global security and warfare. Hypersonic systems pose unique challenges to traditional defence mechanisms and military operations in space. The integration of these weapons into near-Earth orbit offers both advantages and challenges for defence planners and policymakers, making them a focal point in the modern arms race.

Hypersonic missiles do not traditionally operate in near-Earth orbit in the same manner as satellites or spacecraft. However, there are advanced hypersonic systems, such as hypersonic glide vehicles (HGVs), that exploit the boundary between space and the Earth's atmosphere to enhance their operational capabilities. Here's a detailed breakdown of how hypersonic missiles interact with near-Earth orbit:

Hypersonic glide vehicles are launched into space aboard a ballistic missile, similar to an intercontinental ballistic missile (ICBM). However, unlike traditional ICBMs that follow a predictable ballistic trajectory and re-enter the atmosphere in a steep descent, HGVs re-enter the Earth's atmosphere much sooner and glide through the upper layers of the atmosphere at hypersonic speeds. This gliding phase allows them to travel at near-orbital altitudes for a longer portion of their journey. During this time, they can take advantage of:

- **Reduced Air Resistance:** The thinner atmosphere in the upper layers reduces drag, allowing the missile to maintain high speeds while conserving energy.

- **Unpredictable Flight Paths:** HGVs can manoeuvre within the upper atmosphere, making their trajectories difficult to predict, complicating interception by missile defence systems.

- **Prolonged Flight Time:** By staying in the near-Earth orbit region longer, hypersonic systems can extend their range and delay detection by ground-based sensors.

While traditional ballistic missiles leave the atmosphere entirely and re-enter it on a steep, predictable trajectory, hypersonic missiles—particularly HGVs—skim the boundary of space. They re-enter the atmosphere at shallower angles, gliding at high speeds while using the Earth's curvature to alter their path. This near-Earth glide phase allows the missile to operate in a region where both gravity and atmospheric forces can be balanced to achieve long-range flight and high-speed manoeuvres, making them difficult to track and intercept.

Hypersonic systems exploit near-Earth altitudes (typically around 60 to 100 km above the Earth's surface), which are still within the Earth's atmosphere but above the altitude where most traditional missile defence systems are optimized to operate. By gliding in this region, these missiles can:

- **Avoid Early Detection:** Most early warning systems are designed to detect objects that leave the atmosphere (like ICBMs) or stay well within it (like cruise missiles). Hypersonic weapons that glide through the upper atmosphere can evade detection longer.

- **Maximize Manoeuvrability:** The thin atmosphere in near-Earth orbit allows for precise adjustments to the missile's trajectory. These lateral movements make it harder for ground-based interceptors to predict the missile's path.

Operating near the edge of space allows hypersonic missiles to integrate with space-based systems, such as satellites, for real-time adjustments and coordination. Command, control, communications, computers, intelligence, surveillance, reconnaissance, and targeting (C5ISRT) networks—often involving space-based sensors—can communicate with hypersonic missiles for improved targeting accuracy, especially in the terminal phase of flight when precision becomes critical.

A future concept, which some nations are rumoured to be exploring, is a type of weapon system that operates in low-Earth orbit before re-entering the atmosphere to strike its target. This would involve launching a missile into orbit and releasing a hypersonic glide vehicle (HGV) to descend at extremely high speeds onto a target, enabling strikes from nearly anywhere around the globe. China's 2021 test of a hypersonic missile with a fractional orbital bombardment system (FOBS) raised significant international concerns, as it suggested that a hypersonic glide vehicle could be placed in low-Earth orbit and then directed toward a target, adding further complexity to defence strategies.

Operational complexities in operating hypersonic missiles in Near-Earth Orbit:

- **Heat and Material Stresses:** The intense heat generated by friction in the upper atmosphere requires advanced materials capable of withstanding extreme temperatures. Traditional missile materials are not sufficient to handle these conditions.

- **Plasma Effects:** At hypersonic speeds, the missile generates a sheath of plasma around itself due to the intense heat of atmospheric friction. This plasma can block communication with satellites and ground stations, complicating guidance and control.

- **Detection and Tracking:** While the altitude of these missiles can reduce drag and extend their flight time, it also creates unique challenges for radar and infrared tracking systems. The high speed and low radar cross-section of hypersonic missiles make them difficult to detect and track accurately.

Hypersonic missiles that operate in near-Earth orbit represent a significant evolution in missile technology. They combine speed, manoeuvrability, and altitude advantages, making them extremely difficult to detect and intercept. Their ability to exploit the boundary between space and the atmosphere allows them to evade traditional missile defences and extend their reach globally. However, the development of countermeasures, such as space-based sensors and enhanced missile defence systems, is critical to addressing this growing strategic threat. As hypersonic missile technology continues to advance, so too will the complexity of maintaining global security in the face of these new weapons.

One of the key strategic advantages of hypersonic weapons in NEO is their unpredictable flight paths and enhanced manoeuvrability. Unlike traditional intercontinental ballistic missiles (ICBMs), which follow predictable trajectories, hypersonic glide vehicles (HGVs) launched from orbit can alter their flight paths mid-course. This capability significantly complicates interception efforts by missile defence systems, allowing these weapons to bypass fortified defence networks and strike critical targets with minimal warning [2]. For instance, Russia's Avangard hypersonic glide vehicle, which can reportedly fly at speeds around Mach 27 while performing rapid manoeuvres, exemplifies how hypersonic systems can exploit altitude and speed to evade detection and enhance strike precision [2].

Additionally, the global reach and rapid response capabilities of hypersonic weapons in NEO are noteworthy. Once positioned in low-Earth orbit (LEO), these weapons can strike targets anywhere on the globe within minutes, thereby providing military forces with unprecedented rapid-response capabilities. This "prompt global strike" capacity enables timely responses to emergent threats, such as time-sensitive targets or crises, effectively reducing traditional warning systems and reaction times [35]. The U.S. military's pursuit of hypersonic technologies through programs like the Conventional Prompt Global Strike (CPGS) underscores the strategic imperative of these capabilities [2, 35].

Hypersonic systems operating in near-Earth orbit can effectively counter Anti-Access/Area Denial (A2/AD) strategies employed by nations such as China and Russia. These countries have developed sophisticated A2/AD systems designed to restrict adversarial access to critical military or economic zones. Hypersonic weapons can penetrate these heavily defended areas by leveraging their speed and manoeuvrability, thus bypassing ground-based missile defences and electronic warfare systems [2, 35]. For example, China's DF-ZF hypersonic glide vehicle and Russia's Kinzhal are specifically designed to challenge missile defence systems through unpredictable flight paths and high speeds, further enhancing their effectiveness against A2/AD strategies [2].

The integration of hypersonic weapons with existing space-based infrastructure presents additional strategic advantages. These systems can utilize satellites for real-time updates on battlefield conditions, target adjustments, and enhanced precision. The advancements in tracking, telemetry, and command (TT&C) systems further augment the capabilities of hypersonic weapons, enabling comprehensive reconnaissance and strike capabilities [2]. This integration allows for more accurate and timely military interventions, making hypersonic systems more effective in dynamic combat environments. The deployment of hypersonic weapons in near-Earth orbit, however, presents significant challenges and risks, particularly concerning the weaponization of space, increased vulnerability of space assets, and the implications for arms control and international law.

The introduction of hypersonic weapons into the space domain raises critical concerns about the militarization of space. The Outer Space Treaty of 1967 explicitly prohibits the placement of weapons of mass destruction in space, yet the advent of hypersonic technology complicates this framework. Hypersonic glide vehicles (HGVs), capable of traveling at speeds exceeding Mach 5, blur the lines between offensive and defensive capabilities, potentially leading to an arms race among major powers such as the United States, Russia, and China [4, 670]. The dual-use nature of these technologies increases the risk of misinterpretation and unintended escalation during crises, as nations may perceive the deployment of such systems as aggressive posturing [7]. Furthermore, the integration of autonomous systems into hypersonic weapons could exacerbate strategic instability, as these systems may operate with a degree of unpredictability, increasing the likelihood of miscalculations [671].

Hypersonic weapons heavily rely on space-based assets for navigation, communication, and targeting, which introduces significant vulnerabilities. Adversaries could target these critical space infrastructures to disrupt hypersonic operations, thereby transforming space into a contested domain [672, 673]. The reliance on satellites for guidance and command-and-control functions means that any degradation of these assets could severely undermine the effectiveness of hypersonic systems [674]. The fragility of space assets, combined with the potential for anti-satellite weapons (ASATs) to target them, creates a precarious situation where the loss of space superiority could have dire consequences for military operations on Earth [673]. This scenario underscores the need for robust protective measures for space assets to maintain strategic advantages.

The deployment of hypersonic weapons poses significant challenges to existing arms control frameworks. Current international agreements do not adequately address the unique capabilities and risks associated with hypersonic systems, leading to concerns about a potential arms race devoid of regulatory oversight [675, 676]. The lack of clear legal boundaries regarding the military use of space could foster an environment of instability, as nations may feel compelled to develop advanced space-based weapon systems without restraint [677]. The absence of comprehensive arms control measures could exacerbate tensions and lead to increased military competition in space, further complicating international relations

[8]. As such, there is an urgent need for dialogue and negotiation among nations to establish new frameworks that address the complexities introduced by hypersonic technologies and their implications for global security.

The integration of hypersonic missile systems with near-Earth orbit capabilities represents a transformative development in modern warfare. The speed, manoeuvrability, and global reach of these systems give them a strategic advantage in overcoming traditional defences and executing rapid strikes. However, the challenges associated with space weaponization, increased vulnerability of space assets, and the lack of comprehensive arms control measures raise significant concerns about the long-term implications for global security. As nations continue to develop these technologies, it is critical to establish regulatory frameworks and collaborative defence strategies to manage the risks posed by hypersonic weapons in near-Earth orbit.

Space-Based Launch Platforms

A significant and evolving aspect of hypersonic weapons development is the concept of space-based launch platforms, which introduces the idea of launching these weapons from space, significantly enhancing their tactical flexibility and effectiveness.

The Missile Defense Agency (MDA) and the Space Development Agency (SDA) are advancing U.S. missile defence capabilities through the integration of space-based systems, as demonstrated by their collaboration on the upcoming launch of six satellites into low-Earth orbit. This launch, conducted by SpaceX's Falcon 9 rocket from Space Launch Complex 40 in Cape Canaveral, marks a significant step in enhancing missile detection, tracking, and defence [678].

Among the six satellites, two belong to MDA's Hypersonic and Ballistic Tracking Space Sensor (HBTSS) system, and the remaining four are part of SDA's Tranche 0 (T0) Tracking Layer. These satellite systems will form part of a larger missile defence network, designed to counter increasingly sophisticated threats, including hypersonic and ballistic missiles [678].

Figure 46: SpaceX's Falcon 9 rocket and Dragon spacecraft lift off from Launch Complex-40 at Cape Canaveral Air Force Station, Fla. Original from NASA. Digitally enhanced by rawpixel. Rawpixel, CC BY 2.0, via Flickr.

The Hypersonic and Ballistic Tracking Space Sensor (HBTSS) system and the Space Development Agency's (SDA) Tranche 0 (T0) Tracking Layer are pivotal elements in the United States' missile defence architecture, particularly in addressing the challenges posed by advanced missile threats such as hypersonic glide vehicles (HGVs) and ballistic missiles. These systems are designed to enhance the detection, tracking, and interception capabilities of missile defence networks, which are increasingly necessary in a security environment characterized by rapid technological advancements and evolving threats.

The U.S. Space Force's Space Development Agency (SDA) is constructing the Proliferated Warfighter Space Architecture (PWSA), a transformative satellite network aimed at enhancing military capabilities in space. This architecture, previously known as the National Defense Space Architecture, is designed to provide low-latency surveillance, missile tracking, and global data transport. It marks a significant leap in how the U.S. military plans to utilize space for both defence and communication [679].

Established in 2019, the SDA was tasked with rapidly advancing the U.S.'s space-based defence infrastructure to maintain superiority in an increasingly competitive space domain. The PWSA is the SDA's flagship initiative aimed at overcoming the limitations of older military satellite systems that rely on large, costly satellites stationed in geosynchronous orbit (GEO). These older systems are highly vulnerable due to their small numbers and predictable orbits [679].

The PWSA addresses these shortcomings by deploying hundreds of smaller, cost-efficient satellites in low Earth orbit (LEO), operating as a resilient network capable of continuous tracking and data relay. This "mesh network" concept allows for improved resilience against attacks and ensures continuous monitoring of critical targets on Earth [679].

The PWSA is organized into multiple layers, each fulfilling specific roles [679]:

1. **Transport Layer**: Acts as the backbone, enabling global data transfer via laser-based inter-satellite links.

2. **Tracking Layer**: Equipped with infrared sensors to detect and track ballistic and hypersonic missile threats.

3. **Custody Layer**: Provides real-time tracking of mobile, time-sensitive targets on the ground.

4. **Navigation Layer**: Offers alternative positioning, navigation, and timing (PNT) capabilities in case GPS becomes unavailable.

5. **Battle Management Layer**: Manages command, control, and data processing for efficient, time-sensitive operations.

6. **Support Layer**: Provides logistical support and ground station connectivity.

These layers will be built and deployed in "tranches" every two years, with each tranche introducing new capabilities and expanding global coverage. Tranche 0 launched in 2023 as a proof-of-concept, while Tranche 1 (2024) will offer initial warfighting capabilities, followed by Tranche 2 (2026), which aims to provide global missile tracking and communication coverage.

The PWSA will provide several critical benefits to military operations [679]:

- **Beyond-Line-of-Sight (BLOS) Communications**: The Transport Layer enables global, real-time data links, extending tactical communication networks. A recent demonstration in November 2023 successfully transmitted Link-16 messages from space, proving the system's ability to support battlefield communications anywhere on the globe.

- **Missile Warning and Tracking**: The Tracking Layer provides advanced missile detection and tracking capabilities, especially for hypersonic threats, which are difficult to detect using traditional radar systems.

- **Tactical Intelligence, Surveillance, and Reconnaissance (ISR)**: Persistent surveillance of ground targets in real-time allows military commanders to maintain "custody" of mobile or time-sensitive targets, ensuring quicker response times during conflicts.

- **Resilient Global Connectivity**: The Transport Layer, utilizing laser links between satellites, creates a high-bandwidth, secure communication network that is resistant to jamming and difficult to disrupt.

To accelerate deployment, the SDA is using a spiral development model, which leverages commercially available satellite technology rather than investing years in designing complex custom-built satellites. This approach allows the SDA to field new capabilities faster while incorporating the latest technological advancements [679].

Each tranche of satellites will be launched every two years, expanding the capabilities and coverage of the PWSA. Tranche 0 (2023) demonstrated key functions of the Transport and Tracking Layers, while Tranche 1 (2024) will deliver initial operational capabilities in missile tracking and beyond-line-of-sight communications [679].

Building the PWSA presents several technical challenges, including the integration of diverse satellite systems into a unified network. Moreover, developing new technologies such as optical inter-satellite links is complex and requires continuous testing. However, the SDA's incremental approach has already shown promising progress. By the 2030s, the PWSA could include thousands of satellites, offering unprecedented global coverage, advanced threat tracking, and secure communication capabilities [679].

The Proliferated Warfighter Space Architecture is poised to revolutionize the U.S. military's space operations. Its resilient, proliferated satellite network offers advanced missile tracking, real-time surveillance, and secure global communications. As the SDA continues to expand the architecture with future tranches, the PWSA will be crucial for deterring and countering adversarial threats, helping maintain U.S. dominance in an increasingly contested space domain [679].

The HBTSS is a sophisticated space-based sensor platform that provides "birth-to-death" tracking of missile threats, meaning it monitors these threats from launch through to impact. This capability is crucial for tracking hypersonic glide vehicles, which can manoeuvre unpredictably and fly at low altitudes, thereby evading traditional ground-based radar systems [39, 40]. The system's ability to deliver fire-control quality data allows for precise tracking, which is essential for guiding missile defence systems to intercept these fast-moving threats effectively [50]. The integration of HBTSS with terrestrial radar and interceptor systems forms a multi-layered defence strategy that enhances the overall efficacy of U.S. missile defence [680].

Together, the HBTSS and the T0 Tracking Layer exemplify advancements in missile defence capabilities, particularly in countering sophisticated missile technologies developed by potential adversaries. The integration of these systems into a cohesive missile defence network underscores the U.S. commitment to maintaining a strategic advantage in an increasingly complex security landscape where traditional radar systems struggle to keep pace with fast-moving and manoeuvrable targets [680]. As such, these systems are critical for ensuring national security and responding effectively to emerging missile threats.

In parallel, the SDA's Tranche 0 Tracking Layer is a foundational component of the Proliferated Warfighter Space Architecture (PWSA), representing the initial phase of a broader network of space-based sensors aimed at missile detection and tracking [39, 681]. The T0 satellites, deployed in Low-Earth Orbit (LEO), are designed to provide wide-field-of-view sensors that facilitate early missile warning and tracking, complementing the more detailed tracking capabilities of the HBTSS [39, 681]. This layered approach not only improves the detection of missile launches but also enhances tracking during flight, thereby increasing the likelihood of successful interception [39].

The integration of the Hypersonic and Ballistic Tracking Space Sensor (HBTSS) satellites with the Proliferated Warfighter Space Architecture (PWSA) marks a significant advancement in missile defence. The HBTSS satellites introduce a "birth-to-death" tracking system, enabling continuous monitoring of missile threats from their launch to interception. This real-time tracking is particularly crucial for countering hypersonic glide-phase weapons, which are notoriously difficult to detect and track due to their high speeds and unpredictable flight paths [678].

Collaborative testing between the Missile Defense Agency (MDA) and Space Development Agency (SDA) involves testing the combined efficacy of missile tracking and fire-control systems in orbit. By synchronizing data between their satellite systems, both agencies aim to enhance the accuracy of missile defence responses. This collaboration will result in faster detection and hand-offs of missile threat data, enabling more effective engagements from defence systems on the ground [678].

The SDA's Proliferated Warfighter Space Architecture seeks to establish a multi-layered satellite network capable of monitoring advanced threats, such as hypersonic missiles. The architecture integrates wide-field-of-view satellites for initial threat detection and warning, while medium-field-of-view sensors are used for more precise fire control, providing a comprehensive and layered defence system [678].

These advanced missile defence systems represent an evolution in defence strategy. The HBTSS satellites, which focus on detailed missile tracking, and SDA's satellites, offering broader surveillance, work together to create a faster, more reliable missile defence response network. This integrated system will ensure that the U.S. can track and engage missile threats with greater accuracy [678].

After the satellites are launched, on-orbit testing will be conducted over two years. This testing phase is essential for validating the performance of the satellites and ensuring they integrate seamlessly into the existing U.S. missile defence infrastructure. These tests will help ensure that the systems can operate effectively in real-time scenarios, strengthening the nation's overall missile defence capability [678].

These developments reflect the increasing importance of space-based missile defence systems in countering threats posed by adversaries like

China and Russia, who are also advancing hypersonic missile technologies. By deploying these satellites, the U.S. aims to maintain a strategic advantage in missile defence, especially in contested environments where traditional radar systems might struggle [678].

The MDA's continued investment in HBTSS and SDA's role in expanding satellite networks indicate the growing reliance on space to enhance national security and global defence capabilities. These efforts are essential to staying ahead of missile threats in an era where hypersonic weapons can evade terrestrial defences due to their high speeds and unpredictable trajectories. This launch is part of the National Security Space Launch program, further demonstrating the U.S.'s commitment to leveraging space for defence purposes [678].

L3Harris, a defence contractor, announced in December 2023 that it has received approval from the U.S. Space Development Agency (SDA) to proceed with the production of 16 satellites designed to detect and monitor hypersonic missiles [682]. These satellites are crucial for identifying and tracking hypersonic missile threats, which are known for their high speeds, manoeuvrability, and unpredictable flight paths. Hypersonic weapons are considered destabilizing because they are designed to evade traditional missile defence systems, making early detection critical to national security [682].

The approval for production followed the successful completion of both a critical design review and a production readiness review, signifying that the design is solid and ready for manufacturing. The SDA, part of the U.S. Space Force, is developing the Proliferated Warfighter Space Architecture (PWSA), a layered network of satellites that includes both a Transport Layer and a Tracking Layer. The Transport Layer consists of interconnected communication satellites that relay data, while the Tracking Layer contains sensor satellites tasked with identifying missile threats, including hypersonic weapons [682].

L3Harris was awarded a $700 million contract in 2022 to design and build 16 of the Tracking Layer satellites under Tranche 1 of the SDA's program. The company has partnered with Maxar Technologies, which is supplying the satellite buses. These satellites will play a critical role in enhancing the U.S.'s ability to track hypersonic missiles, which are difficult to monitor due to their high manoeuvrability and low flight profiles [682].

These Tracking Layer Tranche 1 satellites are expected to be launched in 2025, further bolstering the U.S.'s defence capabilities. Additionally, L3Harris has previously delivered four prototype tracking satellites as part of Tranche 0, although delays have affected the planned launch schedule for these initial units. The SDA's long-term goal is to create a proliferated low-Earth orbit satellite network that provides continuous global coverage for hypersonic and ballistic missile tracking, integrating both defence and communications systems for a robust national security infrastructure. Alongside L3Harris, companies such as Northrop Grumman and Raytheon RTX are also involved in producing satellites for the Tracking Layer under separate contracts [682].

Space-based launch platforms represent a significant advancement in military technology, particularly in the context of hypersonic missiles. These platforms, which can be positioned in Low Earth Orbit (LEO) or beyond, are designed to launch missiles with extreme speed and precision, thereby enhancing global strike capabilities. The strategic advantage of such systems lies in their ability to reduce the time required for a missile to reach its target compared to traditional ground or air-based systems, thus altering the dynamics of modern warfare.

The development of hypersonic technologies is crucial for the efficacy of space-based launch platforms. Hypersonic airbreathing vehicles, including scramjet-powered systems, are being explored for their potential to serve as reusable launch vehicles. These vehicles can significantly lower transport costs and improve payload delivery to orbit, thus enhancing the striking capacity of military forces [683]. The integration of advanced propulsion systems, such as scramjets, is essential for achieving the necessary speeds and altitudes for effective missile deployment from space [542, 683].

Moreover, the concept of launching missiles from space aligns with the ongoing research into rapid response launch systems. Studies have demonstrated the feasibility of utilizing data link technologies to enhance command and control capabilities between space and ground platforms, facilitating a more integrated approach to military operations [684]. This integration is vital for ensuring that space-based launch platforms can operate effectively within the broader context of military strategy and operations.

Additionally, the geopolitical implications of space-based launch systems cannot be overlooked. As hypersonic technologies advance, there are critical questions regarding territorial claims and military airspace restrictions in near-space environments. The potential for hypersonic missiles to operate in these zones necessitates a re-evaluation of existing legal frameworks governing airspace and military operations [685]. This evolving landscape underscores the importance of governance and international cooperation in managing the implications of such technologies.

The advantages of space-based platforms, particularly in the context of hypersonic missile systems, are multifaceted and significant. These advantages can be categorized into global reach and rapid deployment, increased stealth and ambiguity, flexible trajectories and enhanced manoeuvrability, and deterrence and strategic posturing.

Space-based hypersonic missiles benefit from their elevated launch position, which allows for significantly reduced flight times compared to traditional ground-based systems. The proximity to potential targets enables these missiles to reach hypersonic speeds rapidly, thereby shortening the response window for adversarial defence systems. This capability is underscored by the challenges faced by missile defence systems in intercepting such fast-moving threats, as highlighted by He et al. [73], who note that the engagement of hypersonic targets presents new challenges due to their speed and the limited time available for defence systems to react. Furthermore, the development of hypersonic vehicles is gaining global attention, as evidenced by the advancements made by countries like the United States, which has successfully tested hypersonic flight at speeds exceeding Mach 6, as demonstrated by the X-51 flight [642].

The launch of hypersonic missiles from space can occur with minimal detectable signatures, contrasting sharply with ground or air-based launches that are more easily tracked. The rapid descent from orbit complicates detection efforts, as noted by Law et al. [50], who discuss the difficulties in tracking hypersonic glide vehicles (HGVs) due to their near-space trajectories and superior manoeuvrability. This stealth capability is further enhanced by the potential for multi-mission functionalities of space-based platforms, which can obscure the intent and timing of missile launches, thereby overwhelming existing missile defence systems.

Unlike traditional ballistic missiles that follow predictable parabolic paths, hypersonic missiles launched from space can alter their trajectories dynamically. This flexibility allows them to respond to changing battlefield conditions, complicating interception efforts for missile defence systems that rely on trajectory prediction. The nonlinear dynamics of hypersonic flight necessitate advanced guidance and control systems, as discussed by Mehta et al. , who emphasize the importance of adaptive control mechanisms to enhance missile penetration capabilities [26]. The unpredictability of these trajectories poses additional challenges for adversaries, making effective defence increasingly difficult.

The mere existence of space-based hypersonic missile platforms can serve as a significant deterrent, reshaping global military strategies. The potential for immediate strikes from virtually anywhere in the world forces nations to reconsider their defence postures and rapid-response capabilities. As highlighted by Raza, the emergence of hypersonic weapon systems is viewed as a game-changer in modern warfare, potentially undermining conventional military capabilities and altering the balance of power [5]. The strategic implications of such systems extend beyond mere military applications, influencing diplomatic relations and international security dynamics.

The development and deployment of space-based launch platforms for hypersonic missiles present numerous challenges that can be categorized into technological and logistical barriers, sustainability and maintenance issues, legal and ethical concerns, and vulnerabilities to countermeasures.

The complexity of launching missiles from space necessitates sophisticated technologies. Precise launch mechanisms and accurate targeting systems are essential to ensure effectiveness under the extreme conditions of space. The integration of advanced propulsion and guidance systems is critical, as hypersonic vehicles require robust navigation capabilities to operate effectively in near-space environments [686]. Moreover, the high costs associated with developing and maintaining these technologies can deter investment and complicate logistics, particularly when considering international regulations that may restrict the deployment of offensive capabilities in space [50].

The operational longevity of space-based platforms hinges on their maintenance and reliability. Regular monitoring and servicing of satellites or other space assets pose significant logistical challenges, particularly given their remote positioning in orbit. The costs associated with such maintenance can be substantial, and the technical difficulties of replacing or repairing systems in space further complicate the operational framework of these platforms [687]. Continuous operational readiness is crucial, as any downtime could compromise strategic advantages [684].

The Outer Space Treaty of 1967 imposes restrictions on the placement of nuclear weapons or weapons of mass destruction in orbit, raising significant legal and ethical questions regarding the militarization of space [688]. While hypersonic missiles may not directly fall under the treaty's prohibitions, the dual-use nature of space technologies could lead to perceptions of escalation and provoke international tensions. The development of space-based platforms could be interpreted as a move towards militarization, prompting concerns from other nations and potentially leading to an arms race in space [689].

Space-based assets are increasingly vulnerable to anti-satellite weapons (ASAT), which are being developed by several nations, including the U.S., China, and Russia. The potential for these platforms to become targets for ASAT systems introduces a new dimension to space warfare, as nations may prioritize the development of technologies specifically aimed at neutralizing space-based capabilities [688]. This vulnerability could lead to an escalation of military capabilities in space, further complicating the strategic landscape and increasing the risks associated with deploying hypersonic missile platforms from orbit [19].

The integration of space-based launch platforms with hypersonic missile technology is poised to significantly alter the landscape of international military strategy. The pursuit of hypersonic capabilities by major powers such as the United States, China, and Russia reflects a growing emphasis on strategic superiority in warfare. Hypersonic weapons, characterized by their ability to travel at speeds exceeding five times the speed of sound, present unique advantages in terms of rapid strike capabilities and evasion of traditional missile defence systems [5, 62]. However, this technological evolution raises profound international concerns, particularly regarding the potential for an arms race and the violation of existing space treaties.

The development of hypersonic missile systems is often viewed as destabilizing, as these weapons could be mistaken for ballistic missiles, thereby increasing the risk of miscalculation and unintended escalation during conflicts [62, 690]. The potential for hypersonic missiles to be equipped with nuclear warheads further complicates this issue, as it blurs the lines between conventional and nuclear warfare [5, 62]. As nations continue to advance their hypersonic capabilities, the international community faces the challenge of adapting existing regulatory frameworks to address the implications of deploying offensive military assets in space. The Missile Technology Control Regime (MTCR) and other arms control agreements may need to be reevaluated to prevent the proliferation of these advanced systems [691].

Moreover, the integration of space-based platforms with hypersonic technology could lead to increased tensions among global powers, as nations vie for technological superiority. The emergence of hypersonic weapon systems is already perceived as a new wave of arms competition, particularly in strategic regions such as the Indian Ocean [5]. The potential for an arms race is further exacerbated by the changing nature of the space industry, which is witnessing the rise of NewSpace entrants that may not adhere to traditional export controls and regulatory measures [691]. This situation necessitates a coordinated international response to mitigate the risks associated with hypersonic missile proliferation and to ensure compliance with existing treaties governing the use of outer space for military purposes [690].

The Militarization of Space and Future Wars

The militarization of space has emerged as a significant concern in contemporary international relations, as global powers increasingly recognize space as a critical domain for military operations and strategic dominance. Historically, space was primarily utilized for peaceful endeavours, such as scientific exploration and communication; however, recent decades have witnessed a transformation in its role, with nations viewing it as essential for national security [677, 692]. The integration of advanced satellite technology, missile defence systems, and the potential deployment of space-based weapons has profound implications for the nature of future conflicts [56, 693].

Space infrastructure is vital for military operations on Earth, encompassing satellite communications, global positioning systems (GPS), reconnaissance, and surveillance capabilities. Military forces rely heavily on space-based assets to coordinate operations, navigate precision-guided munitions, and gather intelligence [673, 694]. The U.S. Global Positioning System (GPS) and Russia's GLONASS are pivotal for missile guidance, logistics, and troop movements, underscoring the strategic importance of space in modern warfare [692, 695]. As such, the militarization of space is not merely a theoretical concern but a practical reality that shapes military strategies worldwide.

The development of anti-satellite (ASAT) weapons represents a critical aspect of space militarization. Nations including the United States, China, Russia, and India have either developed or tested ASAT capabilities, which can disrupt an adversary's communication and intelligence networks during conflicts [676, 693]. The potential for kinetic energy weapons, often referred to as "rods from God," illustrates the growing interest in offensive military capabilities in space, despite these concepts remaining largely theoretical at present [696]. Additionally, directed energy weapons, such as laser systems mounted on satellites, are being explored as means to neutralize threats without kinetic force, further highlighting the evolving landscape of space warfare [692, 697].

Countries are increasingly formulating military doctrines that integrate space operations into their broader strategies. The establishment of the U.S. Space Force in 2019 exemplifies this trend, recognizing space as a distinct warfighting domain [677, 698]. Similarly, China and Russia have developed their own space-oriented military branches, reflecting a shared belief that future conflicts will significantly involve space operations [56, 693]. These doctrines emphasize the necessity of maintaining space superiority and safeguarding space assets from hostile actions, which is crucial for national defence.

The emergence of hypersonic missiles, which may operate within the upper atmosphere and utilize space during their flight trajectories, introduces new challenges for missile defence systems [673, 696]. The development of space-based sensors, such as the Hypersonic and Ballistic Tracking Space Sensor (HBTSS), aims to enhance the detection and tracking of hypersonic missiles, thereby improving defensive capabilities [694, 697]. This intersection of hypersonic technology and space operations complicates the existing landscape of military strategy and defence.

International governance regarding the militarization of space remains limited, primarily due to existing treaties such as the Outer Space Treaty of 1967, which prohibits the placement of weapons of mass destruction in space but does not prevent the deployment of conventional weapons or ASAT capabilities [677, 699]. The lack of enforceable regulations on space-based weapons exacerbates tensions among space-faring nations and complicates efforts to prevent the militarization of space [692, 698]. The potential for satellite disruption, orbital debris generation, and cyber warfare in space poses significant risks to global security [673, 700].

The growing militarization of space could reshape global power dynamics, as nations vie for control over strategic assets and technologies. Countries like the U.S., Russia, and China are heavily investing in space capabilities, while other nations, including India, Japan, and members of the European Union, are also pursuing militarization efforts [56, 677]. This arms race in space raises the likelihood of conflicts, necessitating robust international cooperation and governance to ensure space remains a peaceful domain [692, 699].

The Future of Global Military Strategy

How Hypersonics could Change Warfare

Hypersonic weapons travel at speeds greater than Mach 5 thus dramatically reducing the time available for an adversary to respond. Traditional ballistic missiles can be detected and tracked early in their trajectory due to predictable flight paths, but hypersonic weapons, particularly boost-glide vehicles and hypersonic cruise missiles, can alter their course mid-flight and reach their targets much faster. This means defence systems need to detect and react in minutes or even seconds, adding immense pressure on military decision-making processes and reducing the window for countermeasures.

This transformation necessitates a comprehensive re-evaluation of existing military frameworks to effectively counter the unique challenges posed by these advanced weapon systems.

The rapid speed and manoeuvrability of hypersonic weapons drastically reduce the time available for detection and response, compelling military planners to transition from traditional early-warning systems to real-time detection frameworks. Conventional missile defence systems, which are primarily designed for slower ballistic missiles, are inadequate against hypersonic threats due to their limited reaction times [2, 5]. As a result, there is a pressing need for automated decision-making systems that leverage artificial intelligence (AI) and machine learning to analyse vast amounts of sensor data swiftly and initiate defensive measures [671]. Furthermore, the integration of space-based sensors is becoming increasingly critical, as ground-based radar systems struggle to detect low-altitude hypersonic missiles effectively. The U.S., Russia, and China are already investing in advanced space-based early-warning systems to enhance their detection capabilities [50].

In light of the rapid speeds of hypersonic missiles, military strategies are shifting towards distributed and layered defence systems. This approach involves creating multiple interception opportunities through overlapping defence mechanisms deployed across various geographic locations [5]. Such a strategy necessitates the deployment of interceptors at different altitudes and locations, including land, sea, and space, to ensure comprehensive coverage [2]. Additionally, integrating missile defence systems with cyber and electronic warfare capabilities will be essential to disrupt the guidance systems and communication networks of incoming hypersonic threats [671]. This layered defence strategy aims to enhance the resilience of military assets against the swift and unpredictable nature of hypersonic attacks.

The reduced response time associated with hypersonic missiles may lead to a heightened emphasis on pre-emptive strikes within military strategies. Given that intercepting a launched hypersonic missile is nearly impossible, adversaries may prioritize striking first to neutralize enemy launch platforms before they can be activated [5, 7]. This shift towards a more aggressive posture could escalate the risks of conflict, as the window for deterrence and diplomatic resolution narrows significantly [2]. The strategic implications of this trend necessitate careful consideration of the balance between offensive capabilities and the potential for unintended escalation in military engagements.

The capability of hypersonic missiles to strike with little warning will compel militaries to harden and disperse critical infrastructure, including command-and-control centres, power grids, and airbases. Concentrating military assets in a single location increases vulnerability to rapid destruction by hypersonic strikes, making decentralization a priority [671]. Strategies may involve relocating missile launchers and command centres to mobile platforms or constructing underground facilities designed to withstand high-velocity impacts [2]. This approach aims to enhance the survivability of essential military infrastructure in the face of emerging hypersonic threats.

Naval and aerial assets, such as aircraft carriers and bombers, are particularly susceptible to hypersonic missile attacks due to their size and strategic importance. As hypersonic threats evolve, military strategies will likely adapt to enhance the survivability of these assets through increased investment in hypersonic interceptors and defensive countermeasures, including lasers and electronic jamming systems [671]. The importance of distributed naval operations will also rise, with fleets operating in more dispersed formations rather than traditional carrier strike groups to mitigate the risk of being targeted by hypersonic weapons [2].

The reliance of hypersonic missiles on advanced navigation and communication systems underscores the critical role of cyber and electronic warfare in future military strategies. Efforts will focus on jamming communication links, disrupting satellite guidance systems, and employing cyberattacks to neutralize missile control systems (Masood, 2023). Investments in electronic countermeasures designed to interfere with hypersonic missile systems will become essential components of defence strategies, enhancing the ability to counter these advanced threats [2].

Space-based assets will play a pivotal role in detecting and tracking hypersonic missile launches. The rise of hypersonic threats necessitates a strategic shift towards protecting space assets and developing resilient space networks capable of monitoring missile activity in real-time [50]. Militaries will invest in space-based early-warning systems that utilize a combination of infrared sensors, tracking satellites, and autonomous systems to coordinate responses effectively [671]. This focus on space operations reflects the growing recognition of the strategic importance of space in modern warfare.

The advent of hypersonic weapons necessitates a fundamental transformation in military operations, demanding faster detection, decision-making, and interception capabilities. The integration of AI-driven systems, distributed defences, and cyber warfare strategies will be crucial in adapting to this next-generation threat. As military forces evolve to meet these challenges, the incorporation of space operations into traditional military doctrines will be essential for maintaining strategic advantage in an increasingly complex and dynamic security environment.

Hypersonic missiles, which can glide or manoeuvre at lower altitudes and fly unpredictably, operate in a unique altitude range that complicates detection and interception efforts. They are too high for traditional short-range air defence systems and too low for long-range radar systems, which are typically designed to intercept ballistic missiles on predictable trajectories. This necessitates a re-evaluation of existing missile defence infrastructures, as many current systems are rendered ineffective in this new operational environment [3, 604].

One of the primary strategic shifts will be the development of integrated multi-layered defence systems capable of tracking and engaging hypersonic threats across various altitudes and phases of flight. This approach will likely include space-based sensors, which can provide early detection of hypersonic missile launches. For instance, the U.S. Space Development Agency's Proliferated Warfighter Space Architecture (PWSA) and the Hypersonic and Ballistic Tracking Space Sensor (HBTSS) are designed to enhance detection capabilities throughout the missile's flight path, addressing the limitations of ground-based radar systems [604].

The integration of low and medium-altitude interceptors will be crucial. Current ground-based air defence systems must evolve to incorporate advanced interceptors that can engage fast-moving hypersonic targets at lower altitudes. This may involve enhancing existing systems or developing new interceptor technologies specifically tailored for the unique challenges posed by hypersonic speeds [604].

The unpredictability of hypersonic missile flight paths underscores the necessity for improved early detection and tracking systems. Traditional missile defence systems, which are optimized for ballistic trajectories, must adapt to the new realities of hypersonic threats. This adaptation will likely involve the implementation of artificial intelligence (AI) and machine learning technologies to enhance tracking capabilities. AI can process vast amounts of data in real time, facilitating quicker decision-making and more accurate targeting of hypersonic missiles [604].

Furthermore, the integration of faster data processing and sharing mechanisms will be essential. Early detection systems must be closely linked with command-and-control infrastructures to ensure that threat information is processed and acted upon swiftly, thereby improving the overall responsiveness of defence systems [604].

Given the extreme speeds and manoeuvrability of hypersonic missiles, traditional kinetic interceptors may struggle to keep pace. Consequently, there is likely to be increased investment in directed energy weapons, such as high-energy lasers and microwave systems. These systems offer rapid engagement times and can potentially neutralize hypersonic threats more effectively by targeting missile components mid-flight [604].

Countries like the U.S. are already exploring laser-based missile defence systems, which could provide near-instantaneous targeting capabilities against hypersonic weapons. The ability to engage missiles at various points in their flight trajectory makes directed energy systems a promising avenue for future defence strategies [604].

The rapid speed and unpredictability of hypersonic weapons necessitate significant upgrades to command and control systems. These systems must be capable of facilitating instantaneous responses to emerging threats. One potential solution is the establishment of distributed command centres, which can mitigate the risks associated with a single point of failure. This decentralized approach will enhance resilience against hypersonic threats that could disable key infrastructure [604].

Additionally, the incorporation of autonomous and semi-autonomous decision-making systems will be vital. Given the short response windows associated with hypersonic threats, automated algorithms will be essential for real-time threat identification and mitigation, allowing human operators to oversee but not be burdened by the decision-making process [604].

To maintain effective defences, nations are expected to increase their investments in counter-hypersonic technologies. This includes the development of specialized hypersonic interceptors capable of adapting to the erratic flight paths of hypersonic missiles. The creation of manoeuvrable interceptors that can keep pace with hypersonic threats will be crucial for effective defence [604].

The limitations of ground-based radars in tracking hypersonic weapons highlight the need for expanded space-based military infrastructure. Systems like HBTSS will enable continuous surveillance, allowing defence networks to monitor missiles from launch to impact. Additionally, space-based interceptors could be positioned to target hypersonic missiles during their upper atmosphere or gliding phases of flight, further enhancing defence capabilities [604].

Given the challenges associated with defending against hypersonic missiles once they are launched, some military doctrines may shift towards pre-emptive strikes aimed at disabling launch platforms or hypersonic weapon systems before they can be deployed. This shift could lead to more aggressive military postures, emphasizing the neutralization of threats prior to engagement [604].

The combination of speed, manoeuvrability, and unpredictability associated with these weapons requires the evolution of traditional radar systems and missile defences into more sophisticated, multi-layered, and space-integrated systems. By incorporating advanced tracking technologies, AI-driven defences, directed energy weapons, and rapid decision-making infrastructures, militaries will be better equipped to counter these emerging threats. This evolution will also drive greater investments in space-based defences and new technologies to address the limitations of current missile defence infrastructures [604].

Hypersonic weapons are uniquely suited for first-strike operations due to their speed and ability to evade traditional detection systems. This capability allows for the targeting of high-value military assets, such as command centres and missile silos, with little to no warning. As noted by Coetzee, the speed and agility of hypersonic missiles could render existing missile defences obsolete, prompting nations to adopt pre-emptive strategies to neutralize threats before they can be effectively countered [7]. The integration of hypersonic capabilities into military planning will likely emphasize rapid deployment and high-value targeting, compelling military planners to adapt their strategies to incorporate these advanced systems alongside conventional forces [5].

While hypersonic weapons enhance first-strike capabilities, they also bolster second-strike potential, which is critical for maintaining deterrence. The ability of hypersonic missiles to penetrate defences and strike key infrastructure means that nations equipped with such technology can threaten retaliatory strikes against any adversary that initiates a first attack. This dynamic complicates the deterrence calculus, as adversaries may hesitate to launch a first strike, fearing a swift and devastating response [7]. The need for strategic stability in this context will likely drive countries to improve the survivability and redundancy of their command and control structures, ensuring that they can still launch a retaliatory strike even if their primary military assets are compromised [5].

The introduction of hypersonic weapons necessitates significant upgrades to existing missile defence systems, which are currently ill-equipped to handle the speed and manoeuvrability of these advanced threats. Current systems often fail to detect hypersonic missiles due to their low-altitude flight paths and rapid approach speeds [5]. Consequently, nations will need to invest in new technologies, such as space-based tracking systems and directed energy weapons, to enhance their defensive capabilities [7]. The development of layered defence systems that can detect and intercept hypersonic threats across multiple domains will become a priority for military strategists [5].

The capability of hypersonic weapons to target critical military command centres with minimal warning will encourage a shift towards decentralized military leadership structures. To mitigate the risk of losing command and control capabilities in a hypersonic strike, military strategies will increasingly emphasize distributed operations with built-in redundancies [7]. This approach will ensure that military forces can maintain operational effectiveness even in the face of targeted attacks on leadership and communication systems, allowing for autonomous or semi-autonomous operations when necessary [5].

Hypersonic weapons provide a significant advantage for regional power projection, allowing nations to threaten neighbouring countries with rapid strikes that can alter the balance of power in regional conflicts. This capability enables countries to deploy hypersonic weapons to quickly target military bases or naval assets, thereby limiting an adversary's defensive options [5]. Additionally, smaller or less powerful nations may adopt hypersonic technology as part of an asymmetric warfare strategy, leveraging these advanced systems to counter more technologically advanced adversaries [5].

The development and deployment of hypersonic weapons are likely to spur a new arms race, as nations strive to develop their own hypersonic capabilities or countermeasures. This competition will drive military strategies to focus on forming bilateral and multilateral alliances to share resources and technologies related to missile defence [5]. Countries will seek to collaborate on joint efforts to develop space-based tracking and early-warning systems, as no single nation can effectively counter the hypersonic threat alone [7].

The introduction of hypersonic weapons is poised to significantly alter military doctrines and strategies due to their unique capabilities, which include precision strikes, minimal warning, and the ability to evade conventional defences. This transformation will necessitate a comprehensive re-evaluation of military strategies across various domains.

The threat posed by hypersonic weapons, which can strike high-value targets with little to no warning, will compel nations to rethink their defensive strategies regarding critical infrastructure. The dispersal of military assets is a primary strategy to mitigate the risk of a single strike incapacitating essential facilities. For instance, command centres and airbases may be relocated to more secure and geographically dispersed locations to enhance survivability against hypersonic strikes [5]. Additionally, there will be an increased emphasis on hardening facilities through reinforced construction and the development of deeper bunkers or underground bases. Such investments are crucial to ensure that vital military installations can withstand direct hypersonic attacks, thereby maintaining operational capabilities in the face of emerging threats [5, 8].

The rapid response times associated with hypersonic weapons may lead military leaders to adopt pre-emptive strike doctrines. The pressure to act swiftly to neutralize potential threats before they can launch an attack could escalate conflicts rapidly. This shift towards a first-strike doctrine is particularly concerning, as it may prioritize the neutralization of enemy forces at the onset of a conflict, especially those that are critical for nuclear deterrence [5, 8]. The reduced decision-making window necessitates that military strategies incorporate high-alert readiness, which could further exacerbate tensions between adversaries and increase the likelihood of miscalculation during crises [5, 8].

To counter the challenges posed by hypersonic threats, military strategies will increasingly focus on the development of advanced detection and tracking systems. Conventional missile defence systems are ill-equipped to handle the speed and manoeuvrability of hypersonic missiles, necessitating the integration of space-based sensors and AI-powered early warning systems (Law et al., 2023). Investments in satellite constellations will enable real-time monitoring of hypersonic threats, while AI will enhance the processing and analysis of data from various sources, improving the ability to predict missile trajectories and provide timely alerts (Law et al., 2023).

The inadequacies of existing missile defence systems against hypersonic threats will drive militaries to invest in next-generation technologies. This includes the development of directed energy weapons, such as lasers, capable of engaging fast-moving hypersonic targets [5, 8]. Furthermore, a multi-layered defence approach will likely be adopted, combining space-based sensors, advanced ground-based interceptors, and potentially autonomous drone interceptors to create a robust defence against hypersonic weapons [5, 8].

To mitigate the risks associated with hypersonic weapons targeting centralized command and control structures, militaries will need to adopt decentralized command systems. This strategy involves creating redundant communication networks and decentralized decision-making hubs to ensure operational continuity even after a hypersonic strike [5, 8]. The focus on autonomous operations will also increase, allowing military units to function independently of central command, thereby enhancing resilience against potential disruptions [5, 8].

The advent of hypersonic weapons will likely alter the landscape of nuclear deterrence. The potential for no-warning strikes against nuclear deterrents increases the vulnerability of nations relying on traditional deterrence strategies. Consequently, countries may disperse their nuclear forces across various platforms to mitigate the risk of being neutralized by a single hypersonic strike [5, 8]. Additionally, nations may adopt higher alert levels, shortening the time required to launch retaliatory strikes and reducing decision-making timelines in high-pressure scenarios [5, 8].

The deployment and use of hypersonic weapons are reshaping military strategies significantly, particularly through the integration of space and cyber warfare. This transformation is evident in several critical areas, including increased reliance on space-based sensors, the integration of cyber warfare, the enhancement of C5ISR capabilities, and the necessity for rapid decision-making and automation.

Hypersonic weapons, capable of traveling at speeds exceeding Mach 5, present unique challenges for detection and tracking due to their high velocity and unpredictable flight paths. Traditional radar systems struggle to detect these weapons, especially when they operate at lower altitudes. Consequently, advanced space-based sensor systems, such as the Hypersonic and Ballistic Tracking Space Sensor (HBTSS), are becoming essential for monitoring these threats from launch to intercept [2, 5]. The global and persistent observation capabilities provided by these sensors will be crucial for nations aiming to maintain situational awareness in the face of hypersonic threats.

The strategic implications of this reliance on space-based sensors are profound. Nations will prioritize the protection and control of their space assets, recognizing that satellite constellations are vital for tracking hypersonic threats and guiding responses [4, 7]. This shift towards space dominance necessitates robust space defence strategies to safeguard these assets from adversarial actions, including anti-satellite weapons [35]. Furthermore, military strategies will increasingly adopt layered defence systems that integrate terrestrial and space-based detection and tracking capabilities, creating a comprehensive defence architecture capable of engaging hypersonic missiles throughout their flight paths [20].

The operational effectiveness of hypersonic systems is heavily reliant on advanced digital networks for real-time tracking, guidance, and communication. This dependence introduces vulnerabilities, as these networks can be targeted by cyberattacks, potentially disrupting the communication and guidance systems essential for hypersonic operations [671, 701]. As a result, militaries must prioritize cyber defence to protect the integrity of the digital infrastructure supporting hypersonic weapons [50]. Developing cyber-resilient communication networks will be critical to ensure that data links and guidance systems remain secure against adversarial threats.

Moreover, military strategies may increasingly incorporate offensive cyber capabilities to neutralize hypersonic threats. Disabling or disrupting an adversary's command and control systems, satellite links, or guidance systems through cyberattacks could significantly diminish the effectiveness of hypersonic strikes [8, 702]. This dual approach—strengthening cyber defences while developing offensive capabilities—will be essential in the evolving landscape of military conflict characterized by hypersonic technologies.

Hypersonic systems rely on integrated Command, Control, Communications, Computers, Cyber, Intelligence, Surveillance, and Reconnaissance (C5ISR) capabilities to function effectively. These systems are crucial not only for guiding hypersonic weapons to their targets but also for intercepting them [703, 704]. As hypersonic technologies advance, militaries will likely invest more in real-time data sharing across platforms, necessitating secure communication links and cyber-protected networks [262].

The integration of C5ISR capabilities will lead to a shift towards cross-domain operations, where cyber, space, air, and missile defence systems collaborate seamlessly. For instance, a hypersonic missile might be detected by space-based sensors, tracked by ground radars, and intercepted by cyber-secure missile defence systems [276]. However, the complexity of C5ISR systems also introduces potential vulnerabilities, necessitating comprehensive efforts to identify and secure weak points within these networks to ensure operational continuity during cyberattacks [598, 606].

The speed of hypersonic weapons drastically reduces the reaction time available to military decision-makers, necessitating an increased reliance on automated systems capable of processing data and responding to threats without human intervention [79, 198]. Militaries will integrate artificial intelligence (AI) and machine learning into defence systems to automate threat detection, tracking, and engagement, enabling rapid responses to hypersonic missile trajectories [671, 705].

AI-enhanced command systems will also play a critical role in facilitating swift decision-making during hypersonic missile strike scenarios. By assessing risks, prioritizing threats, and recommending counteractions, AI can significantly reduce the cognitive burden on military leaders, allowing for more informed and timely responses to emerging threats [701, 702]. This shift towards automation and AI integration will be essential for maintaining strategic advantages in the rapidly evolving landscape of modern warfare.

The global proliferation of hypersonic weapons is reshaping military strategies and intensifying an arms race primarily among the United States, Russia, and China. Each of these nations is heavily investing in the development of hypersonic systems, including boost-glide vehicles and hypersonic cruise missiles, which offer significant strategic advantages such as speed, precision, and evasion of traditional missile defence systems. This technological race is compelling other nations, including India and Japan, to enhance their military capabilities to maintain parity and deterrence, leading to increased military spending across the globe [5, 77, 706].

In 2021, for instance, the U.S. allocated over $3 billion for hypersonic weapons research, reflecting a marked increase from previous years. Concurrently, Russia has prioritized systems like the Avangard and Tsirkon missiles, while China has made significant strides with its DF-ZF hypersonic vehicle [5, 77]. This escalation in military budgets is not merely a reaction to perceived threats but also a strategic necessity for nations aiming to counterbalance the advancements made by their rivals. The implications of such spending are profound, potentially diverting funds from other critical areas such as social programs and conventional defence systems, thereby straining national budgets [5, 706].

The unregulated development of hypersonic weapons poses a serious threat to global security. These systems can undermine existing missile deterrence frameworks by drastically reducing response times and complicating counter-strike capabilities, which heightens the risk of miscalculations during conflicts. The unpredictability associated with hypersonic weapons, combined with their ability to strike well-defended targets with little warning, renders traditional arms control measures inadequate [35]. As nations prioritize pre-emptive strike capabilities and escalation dominance, the potential for rapid conflict escalation increases, raising the spectre of nuclear confrontation [5, 35].

Moreover, current arms control agreements, such as the New START Treaty, do not adequately address hypersonic technologies, leaving a regulatory gap that could exacerbate the arms race. Experts argue for the urgent need to establish new international treaties that specifically regulate hypersonic weapons to prevent further destabilization of global security [35]. The competitive atmosphere surrounding hypersonic development complicates diplomatic efforts, as nations express mutual concerns over each other's advancements, making arms control negotiations increasingly challenging [35].

The emergence of hypersonic weapons also necessitates a re-evaluation of existing deterrence models. Their ability to bypass traditional missile defences and strike swiftly could lead to a shift towards new deterrence strategies that emphasize speed and flexibility. This shift may include integrating hypersonic capabilities with cyber warfare and space-based systems, fundamentally altering the dynamics of nuclear deterrence and increasing the risk of strategic instability [5, 35]. Nations equipped with hypersonic capabilities may adopt a first-strike posture, further complicating the security landscape and increasing the likelihood of conflict [5].

The introduction of hypersonic weapons, especially those launched from naval platforms, is likely to disrupt traditional naval warfare paradigms. Historically, aircraft carriers and naval task groups have served as the backbone of naval power, projecting force across vast distances. However, hypersonic missiles can engage these assets from hundreds or thousands of kilometres away, striking before effective defensive measures can be mobilized [707]. The ability of these missiles to travel at speeds exceeding Mach 5 and to manoeuvre mid-flight complicates interception efforts, rendering existing naval defence systems inadequate [19].

As a result, navies may need to rethink their operational strategies, moving away from centralized carrier groups towards more dispersed and multi-domain operations. This shift will necessitate increased reliance on stealth technologies, the dispersal of naval assets, and the enhancement of underwater platforms to mitigate vulnerability to hypersonic strikes [33]. Furthermore, investments in advanced detection systems, such as space-based sensors, will become essential for early threat identification and response [664].

Hypersonic weapons also pose significant challenges to current air defence systems, which are primarily designed to intercept slower, more predictable threats. The manoeuvrability and low-altitude flight profiles of hypersonic missiles enable them to evade traditional radar and interception systems [708]. This necessitates a re-evaluation of air defence strategies, emphasizing the need for faster, more integrated systems that can respond to these advanced threats in real-time [709].

The vulnerability of airspace to hypersonic attacks could compromise traditional air superiority doctrines, as these missiles may target critical infrastructure, including airbases, before defensive measures can be implemented [710]. Consequently, nations will need to prioritize mobile airfields, hardened shelters, and rapid-reaction forces capable of relocating swiftly to avoid destruction [711]. The integration of space-based sensors and hypersonic interceptors will be crucial to countering these emerging threats effectively.

The unpredictable nature of hypersonic weapons will require a more layered and integrated approach to defence across multiple domains, including space, cyber, and electromagnetic warfare. Enhanced detection capabilities, such as those provided by the Hypersonic and Ballistic Tracking Space Sensor (HBTSS), will be vital for tracking missiles from launch to impact [712]. Additionally, cyber defences will play a critical role in safeguarding the digital networks that support hypersonic weapon guidance and tracking systems [713].

To counter hypersonic threats, nations must invest in next-generation missile defence systems, including advanced lasers and directed energy weapons, capable of engaging these weapons during their glide phase or at terminal stages [714]. This evolution in defence technology will necessitate rapid advancements in military capabilities focused on intercepting hypersonic missiles during their early phases of flight.

The deployment of hypersonic weapons introduces substantial challenges for military strategies and international arms control frameworks. Existing treaties, such as the Non-Proliferation Treaty (NPT) and the now-defunct Intermediate-Range Nuclear Forces (INF) Treaty, were primarily designed to regulate traditional ballistic and cruise missiles, leaving significant gaps in the regulation of hypersonic systems. These weapons, which can carry both nuclear and conventional warheads, present dual-use capabilities that complicate arms control efforts and increase the risk of unregulated proliferation. Oelrich [4] highlights the rapid development of hypersonic glide weapons by major powers like the U.S., Russia, and China, emphasizing the inadequacy of current arms control frameworks to address these emerging threats. This oversight necessitates a re-evaluation of military doctrines to facilitate bilateral or multilateral discussions aimed at establishing new treaties that specifically address hypersonic technologies [5].

The ambiguity surrounding the intent behind hypersonic weapon launches poses another significant challenge. The ability of these weapons to carry either nuclear or conventional payloads complicates the strategic landscape, as adversaries may misinterpret a conventional strike as a nuclear attack, leading to unintended escalation. Coetzee [7] emphasizes that the speed and agility of hypersonic missiles drastically reduce response times, which could encourage pre-emptive strikes and undermine traditional deterrence strategies. This situation necessitates the development of enhanced crisis management protocols and communication channels to mitigate the risks of misinterpretation and accidental conflict [8]. Improved confidence-building measures, such as real-time notifications of missile launches, could be vital in preventing misunderstandings during tense situations [633].

Moreover, the introduction of hypersonic weapons compresses decision-making timelines in conflict scenarios, which poses risks to strategic stability. Traditional deterrence strategies, which rely on the ability to launch a retaliatory strike, may be rendered less effective due to the rapidity of hypersonic attacks [2]. As highlighted by Raza, the pursuit of hypersonic capabilities by major powers could lead to an arms race, as nations feel compelled to develop similar technologies to maintain military parity [5]. This dynamic may result in a significant shift in military spending and national defence priorities, diverting resources from conventional and nuclear capabilities to hypersonic systems, thereby exacerbating tensions and complicating arms control efforts [715].

Hypersonic weapons could fundamentally alter the nature of warfare by reducing reaction times, challenging traditional defence systems, and increasing the unpredictability of military engagements. Their speed and manoeuvrability make them a game-changing capability for nations that possess them, while also raising the stakes in terms of global security and arms control. As more nations develop these technologies, the military, political, and strategic landscape of future conflicts will likely undergo significant changes, necessitating new approaches to defence, diplomacy, and deterrence.

Perceptions on Global Strategic Dynamics

The analysis provided below represents speculative perceptions of various researchers and presents specific worldviews related to the development of hypersonic missile technology and space infrastructure, especially with respect to the strategic influence of China, Russia, and the U.S. The narrative reflects concerns voiced by some researchers and defence analysts about the rapid evolution of these technologies, particularly in the context of geopolitical competition.

Hypersonic missile technology, which is characterized by its high speed and manoeuvrability, is perceived as a major shift in the global strategic landscape. Researchers focusing on U.S. security concerns highlight how these advancements, particularly by China and Russia, could disrupt existing military balances and introduce strategic ambiguities. For example, the ability of hypersonic missiles to fly within the Earth's atmosphere and change course mid-flight makes them harder to detect and defend against than traditional ballistic missiles. Some researchers view this as a new era of warfare that challenges the effectiveness of existing defence systems, while others see it as a potential destabilizing force that could provoke an arms race or miscalculations between global powers of South America, researchers emphasize the dual-use nature of China's expanding space infrastructure, which can be used for both civilian and military purposes. Some analysts interpret China's growing influence in the region, particularly through its space facilities, as a strategic move to extend its military reach. These facilities are seen as supporting China's military objectives, including the command and control of hypersonic missiles, thus potentially altering regional power dynamics and raising sovereignty concerns in host countries .

Others argue that although hypersonic weapons and military posturing may raise concerns, they do not automatically indicate aggressive military intent. Some scholars view China's activities as part of a broader effort to increase its global standing in space exploration and technology, while others raise alarm over the lack of transparency and the potential for these sites to support offensive military operations . The debate around these issues reflects of the geopolitical implications of China's investments in space infrastructure and hypersonic missile technology.

Ultimately, these perspectives reflect the evolving discourse surrounding hypersonic missile technology and international security, with different analysts emphasizing different potential outcomes based on their interpretations of current trends.

The evolution of hypersonic missile technology has caused significant shifts in global strategic dynamics, as highlighted in recent discussions, particularly in a U.S. congressional hearing focused on the advancements by China and Russia. Hypersonic weapons differ from traditional ballistic and cruise missiles, creating unique challenges for detection and defence. Unlike ballistic missiles, which follow predictable trajectories through space and are relatively easy to identify and intercept, hypersonic missiles stay within the atmosphere and can alter their flight paths mid-course. This combination of high speed and unpredictability presents new strategic ambiguities and disrupts existing power balances [716].

The intersection of technological advancements and geopolitical concerns in the Southern Cone of South America is a complex issue that has evolved significantly over the years. Countries in this region have historically developed national space programs through partnerships that emphasize technology transfer and investment. This collaborative approach has fostered a degree of regional integration, particularly within the Southern Cone, which has been recognized as a transnational subregion with enhanced economic ties and shared technological initiatives [716, 717].

However, the recent expansion of foreign-operated space infrastructure, particularly by China, has raised critical questions regarding the intentions behind such investments. China has established a significant number of space facilities in South America, which has prompted concerns from the United States regarding national security [718]. The dual-use capabilities of Chinese technology, which can serve both civilian and military purposes, are particularly alarming as they may influence command and control systems for advanced military technologies [719]. The implications of this dual-use technology extend beyond mere economic interests, as they intertwine with broader security dynamics in the region.

China's strategic approach to space infrastructure in the Southern Cone can be viewed through the lens of its broader national security strategy, which has shifted towards a more proactive stance in recent years [720]. This shift reflects China's desire to enhance its influence in global governance and security frameworks, particularly in regions where it seeks to expand its geopolitical footprint [721]. The establishment of space facilities in South America is not merely an economic endeavour; it is also a strategic move to secure China's interests in a region that is increasingly significant in the context of U.S.-China relations [722].

Moreover, the implications of China's investments in space technology are compounded by the geopolitical tensions that characterize the current international landscape. The U.S. perceives China's activities as a potential threat to its national security, particularly in light of the dual-use nature of the technologies involved [723]. This perception has led to increased scrutiny and a re-evaluation of security policies in the Southern Cone, as countries in the region navigate the delicate balance between fostering technological partnerships and safeguarding their sovereignty against foreign influence [724].

The role of tracking, telemetry, and command (TT&C) stations is pivotal in enhancing China's capabilities to control its hypersonic missile systems remotely. These stations are integral components of China's command, control, communications, cyber, intelligence, surveillance, reconnaissance, and targeting (C5ISRT) network, which has been expanding significantly, particularly in South America [716]. This expansion raises serious concerns regarding international and regional security, as it facilitates direct military links between China and South American nations, thereby altering the strategic balance in the region [725, 726].

The strategic implications of China's military advancements were starkly illustrated in 2021 when China successfully tested a hypersonic glide vehicle (HGV) that demonstrated the capability to circumnavigate the Earth before striking a target. This test involved an intercontinental ballistic missile (ICBM) and underscored the potential risks associated with hypersonic weapons, which can deliver both nuclear and conventional payloads with unprecedented speed and manoeuvrability [19, 727]. The development of such technologies poses significant challenges to existing missile defence systems and raises alarms about the future of global security dynamics [728].

Moreover, the infrastructure that supports these military capabilities is not limited to China itself but extends into regions like South America, where investments in infrastructure are increasingly linked to military objectives. China's Belt and Road Initiative (BRI) has facilitated the establishment of various projects that not only enhance economic ties but also serve strategic military purposes, thereby embedding Chinese influence in the region [729].

Beyond the security concerns, the integration of these dual-use facilities into China's global TT&C network could support the deployment and precision guidance of hypersonic missiles in conflict scenarios. The United States has voiced concerns about these developments, with security analysts noting the potential for these sites to support offensive military operations, including hypersonic missile launches aimed at U.S. space assets. Given the strategic ambiguity and limited oversight of Chinese facilities in the region, host nations may unknowingly become complicit in military activities that could undermine their sovereignty [716].

Additionally, the geopolitical landscape is complicated by the dual-use nature of ground stations, which play an essential role in both civilian and defence applications. Ground stations in the Southern Hemisphere are particularly sought after due to their strategic positioning, providing continuous communication with spacecraft and enabling various functions such as personnel training and data processing. As a result, the proliferation of space infrastructure in South America, while ostensibly focused on scientific exploration, has far-reaching implications for global security, particularly in the command and control of hypersonic weapon systems [716].

In conclusion, the expansion of hypersonic missile technology and its intersection with space infrastructure in the Southern Hemisphere pose significant challenges to global strategic stability. The potential for these facilities to support offensive operations, combined with limited oversight from host nations, raises concerns about sovereignty, defence, and the peaceful use of space. This situation highlights the urgent need for robust international cooperation, defence alliances, and improved governance mechanisms to mitigate the risks posed by the dual-use capabilities of foreign-operated space infrastructure [716].

As a contrary view, Wright and Tracy [730] outline that the United States, Russia, and China are the leading developers of hypersonic systems. The U.S. is focusing on non-nuclear hypersonic missiles such as the Army's Dark Eagle and the Navy's Conventional Prompt Strike. These are expected to be fielded within the next few years. Russia has already claimed operational deployment of systems like the Avangard (a long-range hypersonic glide vehicle) and Tsirkon (a ship-launched missile), with the Kinzhal used in combat during the war in Ukraine. China, similarly, has developed systems like the DF-ZF and Xingkong-2 that are operational or nearing deployment.

Most of these systems are categorized as boost-glide vehicles, which are launched via rocket boosters and glide unpowered over long distances. Hypersonic cruise missiles, which use scramjet engines for propulsion, are also being developed by these countries. These weapons' ability to manoeuvre mid-flight and avoid detection has generated heightened interest, though their true strategic value remains debatable [730].

Hypersonic weapons' military utility has been questioned by several analysts. One key claim is that their low-altitude flight makes them stealthy and difficult to track with existing radar and missile defence systems. The theory is that their speed and altitude enable them to evade interception. However, these purported advantages are balanced by significant limitations [730].

- **High-Speed Atmospheric Drag**: The low-altitude glide through the atmosphere creates substantial drag, which can slow the hypersonic vehicle, making it no faster than traditional ballistic missiles. For example, while a ballistic missile might follow a depressed trajectory, reaching its target more quickly, a hypersonic glide vehicle might actually take longer due to drag.

- **Intense Heating**: The sustained high speeds within the atmosphere subject hypersonic weapons to extreme heating, particularly during prolonged glide phases. This heating effect poses a significant technical challenge, as it requires advanced thermal protection systems. The U.S.'s **HTV-2** test vehicle, designed to reach Mach 20, experienced catastrophic failure in 2011 due to heating issues. Hypersonic vehicles must balance these heat stresses with maintaining structural integrity, further complicating their operational viability.

- **Detectability**: Although often considered stealthy, hypersonic weapons are detectable at various stages of their flight. Large rocket boosters are used to accelerate these weapons, producing bright exhaust plumes that can be easily tracked by early-warning satellites. During their glide phase, hypersonic vehicles generate significant infrared signatures from the heat they produce, making them visible to infrared-based detection systems already in use by many nations.

Hypersonic missiles are touted for their ability to manoeuvre during flight, potentially outmanoeuvring missile defence systems. However, the degree of manoeuvrability is often overstated. The forces required to change direction at such high speeds introduce significant drag, reducing the missile's overall range and speed. Additionally, manoeuvring introduces further challenges in guidance and targeting, limiting the practical military benefits these weapons offer [730].

Furthermore, while hypersonic systems are believed to provide enhanced accuracy, they rely on guidance technologies that are similar to those found in traditional manoeuvrable warheads. Systems like MaRVs (Maneuverable Reentry Vehicles), which were developed decades ago, offer comparable levels of precision without the same level of technological complexity or cost [730].

The significant investment in hypersonic weapons by countries like the U.S., Russia, and China has raised concerns about the return on investment. The United States alone has spent over $10 billion on hypersonic weapons development, with more expected as these systems move into full-scale production. Despite this expenditure, many analysts argue that the costs outweigh the benefits, especially given that conventional missile systems like MaRVs can achieve similar results at a lower cost. The Congressional Budget Office has suggested that hypersonic weapons might cost up to one-third more than comparable ballistic missile systems equipped with manoeuvrable warheads [730].

While hypersonic weapons offer an alluring combination of speed, manoeuvrability, and long-range strike capability, their actual strategic utility remains unclear. They are beset by challenges related to heat management, drag, and detectability. As research continues, the balance between cost, capability, and practicality will shape the future of these systems. The global hypersonic arms race may ultimately contribute more to escalating military tensions than to providing a decisive advantage on the battlefield [730].

Projected Hypersonic Missile Costs

The Congressional Budget Office (CBO) estimated costs for various hypersonic weapon systems in its analysis, focusing on several key categories of expenses. These include procurement, platform integration, and sustainment. The analysis targets systems that could counteract anti-access and area-denial (A2/AD) strategies employed by adversaries like China and Russia. Specifically, the CBO examined U.S. military projects such as the Army's Long-Range Hypersonic Weapon (LRHW), the Navy's Intermediate-Range Conventional Prompt Strike (IR-CPS) missile, and the Air Force's Air-Launched Rapid Response Weapon (ARRW), as well as hypothetical intermediate-range ballistic and hypersonic cruise missiles [731].

CBO estimated the procurement costs for each missile system, considering components like boosters, glide bodies, and canisters. For the LRHW and IR-CPS missiles, CBO based its cost estimates on analogies with existing missile technologies such as Northrop Grumman's Orion 32 rocket boosters. CBO used cost-estimating models to calculate the costs of boosters, thermal protection, and the guidance systems necessary for hypersonic flight. The procurement of a single LRHW/IR-CPS missile, including its canister, is estimated to cost $81 million initially, with the average cost decreasing to $41 million per missile for a total procurement of 300 units, following a learning curve that accounts for production efficiencies over time [731].

For the Air Force's ARRW missile, which is a smaller hypersonic boost-glide system launched from aircraft, CBO estimated an average procurement cost of $15 million per missile over the purchase of 300 units. This figure is based on comparisons with systems like the Army's MGM-140

Army Tactical Missile System (ATACMS), adjusting for the ARRW's greater length and complexity [731].

CBO's analysis also includes platform integration costs, which are the expenses associated with modifying existing military platforms (such as ships, submarines, and aircraft) to carry and launch hypersonic missiles. For example, the Army plans to launch LRHW missiles from M870 trailers, requiring integration with ground vehicles, while the Navy plans to use Zumwalt-class destroyers and Virginia-class submarines as platforms for the IR-CPS missile. Platform integration costs for the Navy include $900 million to replace the guns on Zumwalt-class destroyers with launch tubes for IR-CPS missiles, as well as costs for integrating the advanced payload modules needed to accommodate these larger missiles [731].

For the Air Force's ARRW system, the primary platform is the B-52 bomber. Although the Air Force has already tested ARRW launches from B-52s, CBO estimates an additional $200 million to fully integrate these missiles into operational B-52 systems. Integrating ARRWs onto other platforms, such as the F-15E or other bombers, could increase costs further [731].

CBO's estimates include costs for sustaining the missile systems after deployment, covering activities such as maintenance, repair, and upgrades over a 20-year period. Sustainment costs for the LRHW/IR-CPS system are estimated at $0.5 million per missile annually, based on DoD's sustainment data for Trident missiles. However, because the Trident system is older and nuclear-armed, CBO adjusted these figures to reflect the more modern and conventional nature of the LRHW/IR-CPS. For the Air Force's ARRW missiles, sustainment costs are estimated at $0.1 million per missile annually, reflecting the smaller size and simpler design of the ARRW compared to LRHW/IR-CPS [731].

CBO notes that it did not estimate development costs for these systems, citing the difficulty of tracking funds due to nontraditional funding mechanisms used by the services. Hypersonic missile technologies are still in a relatively immature stage, and cost estimates from the Department of Defense may be optimistic. This uncertainty, combined with the high risks of cost and schedule overruns associated with emerging technologies, adds complexity to CBO's analysis [731].

CBO's analysis provides a framework for understanding the potential costs of hypersonic missile systems based on current development projects by the Army, Navy, and Air Force. These estimates help illustrate the broader financial implications of deploying hypersonic capabilities in response to A2/AD environments, while acknowledging significant uncertainties due to the immaturity of the technology and the evolving nature of defence procurement and sustainment [731].

To estimate how much the U.S. would need to budget over the next 10 years for hypersonic missile procurement, we can draw from the Congressional Budget Office's (CBO) cost analysis for different hypersonic missile systems, including procurement, platform integration, and sustainment.

Key Estimates:

- **Long-Range Hypersonic Weapon (LRHW)/Intermediate-Range Conventional Prompt Strike (IR-CPS)**
  - **Procurement cost**: $81 million for the first missile, with the average cost decreasing to $41 million per missile over a total procurement of 300 missiles.
  - **Total Procurement for 300 missiles**: $12.3 billion.
- **Air-Launched Rapid Response Weapon (ARRW)**
  - **Procurement cost**: $15 million per missile for 300 missiles.
  - **Total Procurement for 300 missiles**: $4.5 billion.
- **Platform Integration Costs**:
  - **Army's LRHW**: Estimated at $40 million per battery for new ground equipment.
  - **Navy's IR-CPS**: $900 million for integrating launch tubes on Zumwalt-class destroyers, with additional integration for Virginia-class submarines, amounting to $1.0 billion.
  - **Air Force's ARRW**: $200 million for integrating ARRWs onto B-52 bombers.
- **Sustainment Costs**:
  - **LRHW/IR-CPS**: $0.5 million per missile annually for sustainment. Over 20 years, this equates to $30 million per missile.
  - **ARRW**: $0.1 million per missile annually for sustainment, or $2 million per missile over 20 years.

10-Year Budget Estimate:

Given the complexity of these programs and various factors, a rough budget estimate can be structured as follows:

- **Procurement**:
  - LRHW/IR-CPS: $12.3 billion
  - ARRW: $4.5 billion
- **Platform Integration**:
  - Army LRHW: $200 million (for five batteries)
  - Navy IR-CPS: $1.0 billion
  - Air Force ARRW: $200 million
- **Sustainment Costs** (for 10 years):
  - LRHW/IR-CPS: 300 missiles x $0.5 million x 10 years = $1.5 billion
  - ARRW: 300 missiles x $0.1 million x 10 years = $300 million

Total 10-Year Budget Estimate:

- **Total Procurement**: $12.3 billion (LRHW/IR-CPS) + $4.5 billion (ARRW) = $16.8 billion.
- **Total Platform Integration**: $1.4 billion.
- **Total Sustainment (10 years)**: $1.8 billion.

Grand Total: **$20 billion over 10 years** (rounded estimate).

This estimate reflects the need for substantial investment in hypersonic missile systems, platform modifications, and sustainment to maintain a strategic advantage, particularly against adversaries like China and Russia. It does not account for additional variables such as potential technological breakthroughs or geopolitical shifts.

Estimating the procurement costs for hypersonic missiles for countries like China, Russia, and India involves examining available data on their defence spending and hypersonic weapons programs, although precise figures are often difficult to ascertain due to security and proprietary concerns. Here's a general breakdown based on available information:

China: China has made significant advancements in hypersonic technology, including systems like the DF-ZF hypersonic glide vehicle and the CM-401 anti-ship missile. Their investments are linked to broader defence goals, particularly to counter U.S. influence in the Indo-Pacific region. Although exact costs are not publicly disclosed, China's defence budget has grown annually, reaching over $230 billion in 2023 [732]. It is assumed that a considerable portion is allocated to hypersonic weapons development, potentially reaching billions over the next decade.

Russia: Russia has operational hypersonic missiles, including the Avangard and Kinzhal systems, which are integrated into its strategic forces. Like China, Russia sees hypersonic weapons as a crucial part of its military doctrine, especially given its relatively limited economic resources compared to the U.S. Russia's defence budget stands at approximately $86 billion [733], with a substantial portion going to the development and procurement of advanced missile systems. Given its economic constraints, Russia likely allocates several billion dollars to hypersonic technologies over the next decade, though it may not match U.S. or Chinese spending levels.

India: India is in the early stages of hypersonic missile development, with projects like the HSTDV (Hypersonic Technology Demonstrator Vehicle). India's defence budget is about $74 billion annually [734], with hypersonic research forming part of its broader strategic initiatives. India's budget for hypersonic missile development may be relatively smaller compared to China and Russia, but it still represents a strategic priority in its efforts to match regional rivals.

While the U.S. is projected to spend tens of billions on hypersonic missile procurement over the next decade, China and Russia are likely investing comparable amounts, albeit tailored to their strategic priorities. India is emerging in this space and will likely increase its hypersonic defence expenditure in the coming years.

Balancing Offense and Defence in a Hypersonic Age

Balancing offense and defence in the hypersonic age requires nations to rethink their strategic and military frameworks due to the unprecedented speed, manoeuvrability, and unpredictability of hypersonic weapons. These capabilities pose both offensive advantages and defensive challenges, disrupting traditional military doctrines and demanding new approaches to maintaining strategic stability.

Offense: Leveraging Hypersonic Capabilities

Hypersonic weapons give countries an edge in terms of speed and surprise in military engagements. Their ability to travel at speeds greater than Mach 5, often while manoeuvring unpredictably, makes them extremely difficult to detect and intercept with existing defence systems. Countries such as the United States, Russia, and China, which are leading the development of hypersonic missile technologies, are exploring ways to use these weapons for both strategic and tactical advantages.

On the offensive front, hypersonic weapons allow for precision strikes on high-value targets deep within enemy territory, bypassing traditional missile defence systems. Their agility allows them to avoid detection by conventional radar and intercept systems, making them suitable for neutralizing critical infrastructure such as command centres, military bases, and aircraft carriers. This gives rise to a new form of first-strike capability, where adversaries could cripple key military assets with little warning. For countries like China and Russia, which have already demonstrated advancements in hypersonic technology, the ability to project power over long distances without detection gives them a significant offensive advantage.

In terms of strategic nuclear deterrence, hypersonic weapons could also enhance second-strike capabilities. The speed and unpredictability of these systems increase the likelihood that a retaliatory strike will succeed even if a country's primary nuclear forces are targeted in a first-strike scenario. This alters the dynamics of nuclear deterrence, as countries may feel greater pressure to act pre-emptively to protect their strategic assets.

Defense: Overcoming Hypersonic Threats

On the defensive side, the development of effective countermeasures against hypersonic weapons remains a pressing challenge. Traditional missile defence systems are designed to track and intercept ballistic missiles that follow predictable parabolic trajectories, while hypersonic missiles can alter their flight paths and operate at lower altitudes, where conventional radar systems are less effective. This calls for a multi-layered defence approach that integrates advanced technologies capable of addressing the unique challenges posed by hypersonic systems.

Key to this defence strategy is the development of space-based sensors and tracking systems, such as the Hypersonic and Ballistic Tracking Space Sensor (HBTSS). These sensors are designed to detect hypersonic missiles during all phases of flight, including their glide phase, providing real-time data for interceptors to engage. Nations will also need to invest in directed energy weapons (such as lasers) and other non-kinetic solutions that can neutralize hypersonic threats in mid-flight.

The rapid response times required to defend against hypersonic missiles mean that artificial intelligence (AI) and automation will play a key role in future defence systems. AI-powered algorithms can process vast amounts of data from space-based and terrestrial sensors, predicting missile trajectories and launching countermeasures faster than human operators can react. This integration of cyber warfare and AI into defence frameworks is crucial, as disrupting the communications or guidance systems of hypersonic missiles could provide another layer of protection.

Strategic Balancing: Avoiding Escalation

In the broader context of international security, balancing offense and defence in the hypersonic age must also address the risk of escalation. Hypersonic weapons' dual-use nature (conventional and nuclear) increases the risk of misinterpretation during conflicts, as an adversary may perceive a conventional hypersonic strike as a nuclear attack. This increases the likelihood of rapid escalation and underscores the need for clear communication channels and new arms control measures.

Countries must work toward establishing international norms governing the use of hypersonic weapons to reduce the risks of inadvertent escalation. This could include agreements on testing transparency, limits on deployment, or restrictions on the use of nuclear-capable hypersonic systems. Diplomatic efforts, particularly among the U.S., Russia, and China, will be crucial in maintaining strategic stability in the face of these emerging technologies.

Preparing for Offense and Defense

In the era of hypersonic weapons, countries worldwide must adopt comprehensive strategies across several domains—military, technological, diplomatic, and infrastructural—if they hope to prepare effectively. Here's how nations, from advanced militaries to emerging powers, can adapt their defence posture in response to the challenges posed by hypersonic systems:

Hypersonic weapons' ability to manoeuvre at extremely high speeds renders traditional radar systems insufficient. To address this, countries like the United States, Russia, and China are advancing space-based sensor systems such as the Hypersonic and Ballistic Tracking Space Sensor (HBTSS). Nations like India and European countries must follow suit by investing in low-Earth orbit (LEO) satellite constellations to enable real-time tracking and enhance response capabilities. Programs like the Space Development Agency's (SDA) Tracking Layer satellites offer a model for comprehensive monitoring systems designed to counter hypersonic threats by providing critical early detection during the weapons' glide phase.

Most missile defence systems currently focus on intercepting ballistic missiles, but hypersonic weapons present new challenges with their lower-altitude, faster-moving flight paths. Countries such as the United States and Israel are leading the charge in laser defence systems and directed energy weapons as countermeasures for hypersonic threats. Russia is also reportedly working on systems to intercept hypersonic missiles, while Japan and South Korea are upgrading their missile defence infrastructure in response to hypersonic advancements by regional powers.

By incorporating these new systems, countries can build layered missile defences that combine space and ground-based assets for comprehensive coverage, similar to the evolving Aegis Ballistic Missile Defense (BMD) and PAC-3 systems in the United States.

Given that hypersonic weapons rely heavily on advanced guidance and digital networks, countries need to enhance their cybersecurity infrastructure to protect communications and tracking systems. Cyber warfare will play a key role in neutralizing hypersonic systems by disrupting adversaries' digital networks. Countries like Israel and the United States are already integrating cyber defence measures, while nations like China are investing heavily in cyber offense, capable of disabling enemy systems.

In the hypersonic era, nations need to bolster offensive and defensive cyber capabilities not just to protect their own assets, but to potentially exploit vulnerabilities in adversarial hypersonic networks.

While the U.S., Russia, and China are spearheading hypersonic development, smaller powers like France, India, and Australia will need to push for international cooperation to establish new arms control frameworks. Existing treaties like the Non-Proliferation Treaty (NPT) and Intermediate-Range Nuclear Forces (INF) Treaty do not cover hypersonic systems, creating a regulatory vacuum.

Arms control dialogues between nuclear powers, especially involving the U.S., Russia, and China, should focus on transparency in hypersonic missile testing and deployment. Meanwhile, non-nuclear nations should advocate for non-proliferation efforts that restrict the transfer of hypersonic technology to non-state actors or hostile countries.

The compressed response time presented by hypersonic weapons will necessitate real-time decision-making systems. Nations must invest in artificial intelligence (AI)-driven systems capable of processing massive amounts of real-time data, predicting hypersonic missile trajectories, and launching countermeasures in seconds.

Countries like the United States are already integrating AI-based systems into their missile defence architecture, while China is advancing AI for autonomous military decision-making. Other nations should follow by incorporating automated response systems and AI-powered crisis management protocols that can reduce the likelihood of miscalculation or escalation during high-stakes conflicts.

The arms race in hypersonic weapons requires all countries to invest heavily in research and development (R&D). For instance, France, India, and Japan are actively researching hypersonic propulsion systems like scramjets to ensure they are not left behind technologically. Nations should also explore additive manufacturing (3D printing) to produce lightweight but durable components for hypersonic missiles and advanced materials that can withstand the extreme temperatures encountered during hypersonic flight.

AI-driven material science is another area where countries can optimize the design and performance of hypersonic systems by predicting failure points and enhancing structural resilience.

Hypersonic weapons fundamentally challenge existing military doctrines, especially with their ability to perform surprise precision strikes. Countries with nuclear capabilities, such as the United States, Russia, and India, will need to adapt their deterrence strategies to reflect the reduced warning times hypersonic systems introduce. The risk of accidental nuclear escalation must be addressed, with nations developing new rules of engagement for hypersonic weapon use.

Countries should also focus on dispersing and hardening critical infrastructure to protect it from potential hypersonic strikes, similar to how NATO has dispersed assets to avoid surprise missile strikes.

The global hypersonic arms race requires nations to adopt a multi-dimensional approach. By investing in space-based tracking, missile defence, cyber warfare integration, and AI-driven decision-making systems, countries can mitigate the risks posed by hypersonic weapons. Diplomatic efforts must also focus on closing the regulatory gap left by existing arms control treaties, and nations must invest in R&D to maintain technological parity. Ultimately, a coordinated global strategy is essential for navigating the challenges of the hypersonic era while maintaining strategic stability.

References

1.Harriss, L. and E. Noone, *Hypersonic missiles*. 2023.

2.Cernat, R., *The Potential of Hypersonic Weapons and the Great Military Powers Strategy Regarding Their Production. Case Study.* Romanian Military Thinking, 2021. **2021**(4): p. 230-245.

3.None, N., E. Noone, and L. Harriss, *Hypersonic Missiles.* 2023.

4.Oelrich, I., *Cool Your Jets: Some Perspective on the Hyping of Hypersonic Weapons.* Bulletin of the Atomic Scientists, 2020. **76**(1): p. 37-45.

5.Raza, I. and N. Mehmood, *Hypersonic Weapon Systems – A New Wave of Arms Race in the Indian Ocean Region.* Margalla Papers, 2023. **27**(1): p. 28-39.

6.Kraska, J. and R. Pedrozo, *Missile Warfare and Nuclear Weapons.* 2022: p. 203-240.

7.Coetzee, E., *Hypersonic Weapons and the Future of Nuclear Deterrence.* Scientia Militaria South African Journal of Military Studies, 2021. **49**(1).

8.Chyba, C.F., *New Technologies &Amp; Strategic Stability.* Daedalus, 2020. **149**(2): p. 150-170.

9.None, N., et al., *The Economic Justification for the Use of Expensive Weapons on the Example of Hypersonic Weapons.* Ekonomika I Upravlenie Problemy Resheniya, 2024. **3/4**(144): p. 71-76.

10.Rogov, S.M., *Global and Regional Stability in the Nuclear World.* Herald of the Russian Academy of Sciences, 2021. **91**(3): p. 374-385.

11.Huang, Z., Y. Zhang, and Y. Liu, *Research on State Estimation of Hypersonic Glide Vehicle.* Journal of Physics Conference Series, 2018. **1060**: p. 012088.

12.Malinowski, P., *Hypersonic Weapon as a New Challenge for the Anti-Aircraft Defense Command and Control System.* Safety & Defense, 2020. **6**(2): p. 89-99.

13.Xue, X., S. Huang, and D. Wei, *An Adaptive Multivariate Student's T process Recursive Method for Hypersonic Glide Vehicle Trajectory Prediction.* Iet Radar Sonar & Navigation, 2023. **17**(6): p. 1061-1077.

14.Jia, S., et al., *Distributed Analytical Formation Control and Cooperative Guidance for Gliding Vehicles.* International Journal of Aerospace Engineering, 2020. **2020**: p. 1-14.

15.Ma, S., et al., *Multidisciplinary Design Optimization of Reentry-Powered Hypersonic Vehicles Based on Surrogate Model.* International Journal of Aerospace Engineering, 2024. **2024**: p. 1-17.

16.Xie, Z., et al., *An Analytical Reentry Solution Based Online Time-Coordinated A* Path Planning Method for Hypersonic Gliding Vehicles Considering No-Fly-Zone Constraint.* Aerospace, 2024. **11**(6): p. 499.

17.Shen, Z., et al., *Penetration Trajectory Optimization for the Hypersonic Gliding Vehicle Encountering Two Interceptors.* 2022.

18.Tang, S. and J. Guo, *Entry Guidance Command Generation for Hypersonic Glide Vehicles Under Threats and Multiple Constraints.* Ieee Access, 2022. **10**: p. 1-15.

19.Zhu, C., et al., *Impact-Time-Control Guidance Law for Hypersonic Missiles in Terminal Phase.* Ieee Access, 2020. **8**: p. 44611-44621.

20.Fan, Y.G., et al., *A Hybrid Model Algorithm for Hypersonic Glide Vehicle Maneuver Tracking Based on the Aerodynamic Model.* Applied Sciences, 2017. **7**(2): p. 159.

21.Chen, S. and Q. Xia, *A Multiconstrained Ascent Guidance Method for Solid Rocket-Powered Launch Vehicles.* International Journal of Aerospace Engineering, 2016. **2016**: p. 1-11.

22.Fu, W., et al., *Guidance Law and Neural Control for Hypersonic Missile to Track Targets.* Discrete Dynamics in Nature and Society, 2016. **2016**: p. 1-10.

23.Rizvi, S.T.I., et al., *Trajectory Optimisation for a Rocket-Assisted Hypersonic Boost-Glide Vehicle.* The Aeronautical Journal, 2017. **121**(1238): p. 469-487.

24.Chen, K., et al., *Tightly Coupled Integrated Navigation Algorithm for Hypersonic Boost-Glide Vehicles in the LCEF Frame.* Aerospace, 2021. **8**(5): p. 124.

25.Gao, J.-z., X. Chen, and J.-y. Song, *Gauss Pseudospectral Method Based Trajectory Optimization for Hypersonic Glide Vehicles.* Destech Transactions on Computer Science and Engineering, 2018(pcmm).

26. Mehta, S.S., et al., *Nonlinear Control of Hypersonic Missiles for Maximum Target Penetration.* 2012.

27. Gao, Y., R. Zhou, and J. Chen, *Integrated Entry Guidance With No-Fly Zone Constraint Using Reinforcement Learning and Predictor-Corrector Technique.* Proceedings of the Institution of Mechanical Engineers Part G Journal of Aerospace Engineering, 2024. **238**(7): p. 728-741.

28. Chekov, A. and S. Babkina, *Hypersonic Weapons: Evolution or Revolution?* International Trends / Mezhdunarodnye Protsessy, 2023. **21**(2): p. 83-102.

29. Chen, J., et al., *Shape Optimization of the Cross-Section for Noncircular Hypersonic Missile Forebody.* International Journal of Aerospace Engineering, 2020. **2020**: p. 1-9.

30. McFarland, J., *The Development of Hypersonic Weapons in the US, China and Russia.* The Rusi Journal, 2023. **168**(1-2): p. 10-18.

31. Wright, T.M., *Hypersonic Missile Proliferation: An Emerging European Problem.* 2022.

32. Brockmann, K. and D. Stefanovich, *Hypersonic Boost-Glide Systems and Hypersonic Cruise Missiles: Challenges for the Missile Technology Control Regime.* 2022.

33. Materak, W., *The Evolution of Air Threats in Future Conflicts.* Safety & Defense, 2023. **9**(1): p. 24-30.

34. Idris, A.C., M.R. Saad, and K. Kontis, *Potential of Micro-Vortex Generators in Enhancing the Quality of Flow in a Hypersonic Inlet-Isolator.* Journal of Advanced Research in Fluid Mechanics and Thermal Sciences, 2020. **77**(1): p. 1-10.

35. Borrie, J., A. Dowler, and П. Подвиг, *Hypersonic Weapons: A Challenge and Opportunity for Strategic Arms Control.* 2019.

36. Teeple, N., *Offensive Weapons and the Future of Arms Control.* The Canadian Journal of European and Russian Studies, 2021. **14**(1): p. 79-102.

37. Zhang, Z., Y. Luo, and Y. Qu, *Distributed Formation Control With Obstacle and Collision Avoidance for Hypersonic Gliding Vehicles Subject to Multiple Constraints.* International Journal of Aerospace Engineering, 2023. **2023**: p. 1-23.

38. Xie, Y., et al., *Dual-Channel and Bidirectional Neural Network for Hypersonic Glide Vehicle Trajectory Prediction.* Ieee Access, 2021. **9**: p. 92913-92924.

39. Cheng, Y., et al., *An LEO Constellation Early Warning System Decision-Making Method Based on Hierarchical Reinforcement Learning.* Sensors, 2023. **23**(4): p. 2225.

40.Li, S., et al., *Multiple Model Tracking for Hypersonic Gliding Vehicles With Aerodynamic Modeling and Analysis.* Ieee Access, 2019. **7**: p. 28011-28018.
41.Li, Z., Y. Wang, and W. Zheng, *Accurately Tracking Hypersonic Gliding Vehicles via an LEO Mega-Constellation in Relay Tracking Mode.* Journal of Systems Engineering and Electronics, 2024. **35**(1): p. 211-221.
42.Zhang, J., et al., *A Search Method for a Hypersonic Gliding Vehicle Based on Early Warning Information Guidance.* Iet Radar Sonar & Navigation, 2022. **16**(7): p. 1162-1175.
43.Meng, F. and K. Tian, *Phased-Array Radar Task Scheduling Method for Hypersonic-Glide Vehicles.* Ieee Access, 2020. **8**: p. 221288-221298.
44.Zhang, B., et al., *Free Form Deformation Method Applied to Modeling and Design of Hypersonic Glide Vehicles.* Ieee Access, 2019. **7**: p. 61400-61413.
45.Patoz, A., et al., *Duty Factor Reflects Lower Limb Kinematics of Running.* Applied Sciences, 2020. **10**(24): p. 8818.
46.Hamed, M., *Knowledge, Attitude, and Practices Toward Dengue Fever Among the Public: A Cross-Sectional Study in the Western Region of Saudi Arabia.* Frontiers in Public Health, 2024. **12**.
47.Carreon, L.Y., et al., *Clinical Outcomes of Decompression Alone Versus and Decompression and Fusion for First Episode Recurrent Disc Herniation.* Global Spine Journal, 2019. **10**(7): p. 832-836.
48.Liu, Y., et al., *Carcinoma Microsatellite Instability Status as a Predictor of Benefit From Fluorouracil-Based Adjuvant Chemotherapy for Stage II Rectal Cancer.* Asian Pacific Journal of Cancer Prevention, 2015. **16**(4): p. 1545-1551.
49.Phuyal, P., et al., *The Knowledge, Attitude and Practice of Community People on Dengue Fever in Central Nepal: A Cross-Sectional Study.* BMC Infectious Diseases, 2022. **22**(1).
50.Law, Y.W., et al., *Detecting and Tracking Hypersonic Glide Vehicles: a Cybersecurity-Engineering Analysis of Academic Literature.* International Conference on Cyber Warfare and Security, 2023. **18**(1): p. 189-198.
51.Wan, S., et al., *Suboptimal Midcourse Guidance With Terminal-Angle Constraint for Hypersonic Target Interception.* International Journal of Aerospace Engineering, 2019. **2019**: p. 1-13.
52.Chen, J., Y. Han, and Y. Ren, *Robust Guidance Algorithm Against Hypersonic Targets.* 2020.
53.Hu, M. and Y. Jiang, *Research on the High Speed of Piston Pumps Based on Rapid Erecting of Launch Vehicles.* Applied Sciences, 2023. **13**(12): p. 7178.
54.Ji, C.-H., C.-S. Kim, and B.S. Kim, *Flight Control Law for Stabilizing Transient Response of the Aircraft During Gun Firing.* Aerospace, 2023. **10**(4): p. 365.

55.None, N., et al., *Simulation Model of a Missile Strike.* Electrical and Computer Systems, 2024(40(116)): p. 63-76.

56.Fontana, S. and F.D. Lauro, *An Overview of Sensors for Long Range Missile Defense.* Sensors, 2022. **22**(24): p. 9871.

57.Kristensen, H.M. and M. Korda, *United States Nuclear Forces, 2020.* Bulletin of the Atomic Scientists, 2020. **76**(1): p. 46-60.

58.Goswami, M., et al., *Potential of Multi Constellation Global Navigation Satellite System in Indian Missile Test Range Applications.* Defence Science Journal, 2020. **70**(6): p. 682-691.

59.Han, Y., W. Meng, and X. Wang, *Design and Experimental Analyses of an Accuracy Verification System for Airborne Target Tracking via Radar Guidance Systems.* Applied Sciences, 2022. **12**(14): p. 6838.

60.Fan, Y., et al., *Lateral Control Strategy for a Hypersonic Cruise Missile.* International Journal of Advanced Robotic Systems, 2017. **14**(2).

61.Stefanovich, D., *Nuclear Posture and Technology Trends in South Asia and Ways Ahead.* National Security Journal, 2022. **3**(4).

62.Жao, T., *Going Too Fast: Time to Ban Hypersonic Missile Tests?* Bulletin of the Atomic Scientists, 2015. **71**(5): p. 5-8.

63.Wang, Y., et al., *Research on Key Technologies of Hypersonic Vehicle.* Journal of Physics Conference Series, 2022. **2183**(1): p. 012011.

64.Brockmann, K. and M. Schiller, *A matter of speed? Understanding hypersonic missile systems.* 2022, Stockholm International Peace Research Institute.

65.Huang, J., et al., *Constraints Maintaining Mesh Refinement Method for Solving Whole Trajectory of Discontinuous Aerodynamics.* Ieee Access, 2023. **11**: p. 110086-110101.

66.Knight, D., *Flight Control of the Kinzhal Missile Using Energy Deposition.* 2023.

67.Yu, S., X. Ni, and F. Chen, *CFD Simulation Strategy for Hypersonic Aerodynamic Heating Around a Blunt Biconic.* International Journal of Aerospace Engineering, 2021. **2021**: p. 1-11.

68.Dong, M. and H. Huang, *Effect of Rotating Channel Turning Section Clearance Size on Heat Transfer Characteristics of Supercritical Pressure Hydrocarbon Fuel.* Energies, 2023. **16**(16): p. 6051.

69.Yang, P., et al., *Fuzzy Control Strategy for Hypersonic Missile Autopilot With Blended Aero-Fin and Lateral Thrust.* Proceedings of the Institution of Mechanical Engineers Part I Journal of Systems and Control Engineering, 2015. **230**(1): p. 72-81.

70.Zhu, M., et al., *Control-Volume-Based Exergy Method of Truncated Busemann Inlets in Off-Design Conditions.* Processes, 2024. **12**(3): p. 535.

71.Hwang, S.-r. and H.J. Lee, *Comparison and Evaluation of Transport Property Prediction Performance of Supercritical Hydrocarbon Aviation Fuels and Their Pyrolyzed Products via Endothermic Reactions.* Energies, 2023. **16**(13): p. 5195.

72.Xinchun Li, Z.W.a.G.D., *Energy and Exergy Analysis of a Closed Brayton Cycle for Power Generation from the Scramjet Cooling Heat* International Journal of Astronautics and Aeronautical Engineering, 2021. **6**(2).

73.He, F., W. Chen, and Y. Bao, *Novel Particle Swarm Optimization Guidance for Hypersonic Target Interception With Impact Angle Constraint.* International Journal of Aerospace Engineering, 2022. **2022**: p. 1-10.

74.Holland, O.T. and S.E. Wallace, *Using Agents to Model the Kill Chain of the Ballistic Missile Defense System.* Naval Engineers Journal, 2011. **123**(3): p. 141-151.

75.Li, H., et al., *Multi-Objective Fault-Tolerant Output Tracking Control of a Flexible Air-Breathing Hypersonic Vehicle.* Proceedings of the Institution of Mechanical Engineers Part I Journal of Systems and Control Engineering, 2010. **224**(6): p. 647-667.

76.Analytica, O., *Hypersonic weapons could jeopardise global security.* Emerald Expert Briefings, 2015(oxan-db).

77.Stone, R., *'National Pride Is at Stake.' Russia, China, United States Race to Build Hypersonic Weapons.* 2021.

78.Peng, W., Z. Feng, and T. Yang, *Rapid Aerodynamic Shape Optimization With Payload Size Constraints for Hypersonic Vehicle.* Ieee Access, 2019. **7**: p. 84429-84447.

79.Yan, B., et al., *A Rapid Penetration Trajectory Optimization Method for Hypersonic Vehicles.* International Journal of Aerospace Engineering, 2019. **2019**: p. 1-11.

80.Mogavero, A. and R.E. Brown, *An Improved Engine Analysis and Optimization Tool for Hypersonic Combined Cycle Engines.* 2015.

81.Cheng, K., et al., *Performance Assessment of Closed-Brayton-Cycle and Thermoelectric Generator Combined Power Generation System Coupled With Hydrocarbon-Fueled Scramjet.* Energies, 2023. **16**(21): p. 7385.

82.Liang, Z., et al., *Decoupled Three-Dimensional Entry Trajectory Planning Based on Maneuver Coefficient.* Proceedings of the Institution of Mechanical Engineers Part G Journal of Aerospace Engineering, 2016. **231**(7): p. 1281-1292.

83.Liu, X., et al., *Backstepping Attitude Control for Hypersonic Gliding Vehicle Based on a Robust Dynamic Inversion Approach.* Proceedings of the Institution of Mechanical Engineers Part I Journal of Systems and Control Engineering, 2014. **228**(8): p. 543-552.
84.Zhu, J., et al., *Highly Constrained Optimal Gliding Guidance.* Proceedings of the Institution of Mechanical Engineers Part G Journal of Aerospace Engineering, 2015. **229**(12): p. 2321-2335.
85.Chen, X., et al., *Phugoid Dynamic Characteristic of Hypersonic Gliding Vehicles.* Science China Information Sciences, 2011. **54**(3): p. 542-550.
86.Darby, C.L. and A.V. Rao, *Minimum-Fuel Low-Earth Orbit Aeroassisted Orbital Transfer of Small Spacecraft.* Journal of Spacecraft and Rockets, 2011. **48**(4): p. 618-628.
87.Huan, Z., W. Zheng, and G. Tang, *Ballistic Error Propagation Algorithm for Glide Trajectory Based on Perturbation Theory.* Proceedings of the Institution of Mechanical Engineers Part G Journal of Aerospace Engineering, 2016. **231**(9): p. 1574-1592.
88.Kumar, A. and A.S. Rana, *Capturing Non-Equilibrium in Hypersonic Flows: Insights From a Two-Temperature Model in Polyatomic Rarefied Gases.* Physics of Fluids, 2024. **36**(10).
89.Li, G., et al., *Maneuver Modes Analysis for Hypersonic Glide Vehicles.* 2014.
90.Karabacak, M. and O. Turan, *Off-Design Analysis of the Inverted Brayton Cycle Engine.* Aircraft Engineering and Aerospace Technology, 2024. **96**(7): p. 954-963.
91.Wang, Z., *Optimal Trajectories and Normal Load Analysis of Hypersonic Glide Vehicles via Convex Optimization.* Aerospace Science and Technology, 2019. **87**: p. 357-368.
92.Cheng, Y., et al., *Aerodynamic Statistics-Based Trajectory Estimation of Hypersonic Maneuvering Target.* Ieee Access, 2020. **8**: p. 227642-227656.
93.Fan, Y.G., W.-H. Zhu, and G. Bai, *A Cost-Effective Tracking Algorithm for Hypersonic Glide Vehicle Maneuver Based on Modified Aerodynamic Model.* Applied Sciences, 2016. **6**(10): p. 312.
94.Li, H., et al., *A Two-Level Optimization Method for Hypersonic Periodic Cruise Trajectory.* International Journal of Aerospace Engineering, 2021. **2021**: p. 1-11.
95.Leishman, J.G., *Introduction to Aerospace Flight Vehicles.* 2024: Embry-Riddle Aeronautical University.
96.Zhao, Z., et al., *Effect of Terahertz Antenna Radiation in Hypersonic Plasma Sheaths With Different Vehicle Shapes.* Applied Sciences, 2022. **12**(4): p. 1811.

97.Kumar, A. and B. Thornber, *RANS Modelling of Shock-Wave Boundary Layer Interaction in a Mixed Compression Axisymmetric Hypersonic Intake.* Applied Mechanics and Materials, 2016. **846**: p. 61-66.
98.Jiang, H., et al., *Hypersonic Flow Control of Shock Wave/Turbulent Boundary Layer Interactions Using Magnetohydrodynamic Plasma Actuators.* Journal of Zhejiang University Science A, 2020. **21**(9): p. 745-760.
99.Wang, Z., et al., *Research on Drag Reduction and Heat Prevention Performance of Spike With the Channel Concept for Hypersonic Blunt Body.* Journal of Physics Conference Series, 2022. **2252**(1): p. 012004.
100.Spottswood, S.M., T.J. Beberniss, and T. Eason, *Structural Response Prediction: Full-Field, Dynamic Pressure and Displacement Measurements of a Panel Excited by Shock Boundary-Layer Interaction.* 2015.
101.Li, Y.H., X. Tang, and W. Zhou, *Aerodynamic and Numerical Study on the Influence of Spike Shapes at Mach 1.5.* Advanced Materials Research, 2014. **1046**: p. 177-181.
102.Vashishtha, A., Y. Watanabe, and K. Suzuki, *Bow-Shock Instability and Its Control in Front of Hemispherical Concave Shell at Hypersonic Mach Number 7.* Transactions of the Japan Society for Aeronautical and Space Sciences Aerospace Technology Japan, 2016. **14**(ists30): p. Pe_121-Pe_128.
103.Ghori, A., S. Narendar, and Y. Dubey, *Thermo-Structural Response Prediction of UHTCC Control Surface for Hypersonic Cruise Vehicle.* Applied Mechanics and Materials, 2023. **915**: p. 81-94.
104.Gnoffo, P.A., K.B. Thompson, and A.M. Korzun, *Tapping the Brake for Entry, Descent, and Landing.* 2016.
105.Meng, X., et al., *Fluid–Structure Interactions Between Oblique Shock Trains and Thin-Walled Structures in Isolators.* Aerospace, 2024. **11**(6): p. 482.
106.Wang, F., H. Ding, and M. Lei, *Aerodynamic Characteristics Research on Wide-Speed Range Waverider Configuration.* Science China Technological Sciences, 2009. **52**(10): p. 2903-2910.
107.McNamara, J.J., et al., *Approximate Modeling of Unsteady Aerodynamics for Hypersonic Aeroelasticity.* Journal of Aircraft, 2010. **47**(6): p. 1932-1945.
108.Abbas, L.K., X. Rui, and P. Marzocca, *Non-Linear Aerothermoelastic Modeling and Behavior of a Double-Wedge Lifting Surface.* Journal of Shanghai Jiaotong University (Science), 2009. **14**(5): p. 620-625.
109.Chen, H., et al., *Research on the Mechanism of Thermal-Mechanical Coupling Effects of Hypersonic Aircraft Rudder Structure.* Journal of Physics Conference Series, 2024. **2764**(1): p. 012050.
110.Wang, Z., et al., *Fluid-Thermal Interaction Simulation of a Hypersonic Aircraft Optical Dome.* Energies, 2022. **15**(22): p. 8619.

111.Riley, Z.B., J.J. McNamara, and H.B. Johnson, *Hypersonic Boundary Layer Stability in the Presence of Thermo-Mechanical Surface Compliance.* 2012.

112.Nakatani, H., et al., *Wind Tunnel Test of Mach 5 Class Hypersonic Airplane.* Transactions of the Japan Society for Aeronautical and Space Sciences Space Technology Japan, 2009. **7**(ists26): p. Pe_71-Pe_76.

113.Jiang, Z., H. Yu, and L. Zih-Chen, *Research Progress on High-Enthalpy and Hypersonic Flows.* 2007: p. 37-44.

114.Krishnamurthy, R., *Spiked Blunt Bodies for Hypersonic Flights.* 2021, GridPro Blog.

115.Jiang, Z., Y. Liu, and G. Han, *Conceptual Study on Non-Ablative TPS for Hypersonic Vehicles.* 2011.

116.Fan, D., et al., *Drag Reduction Investigation for Hypersonic Lifting-Body Vehicles With Aerospike and Long Penetration Mode Counterflowing Jet.* Aerospace Science and Technology, 2018. **76**: p. 361-373.

117.John, B., et al., *Drag and Heat Flux Reduction using Counterflow Jet and Spike-Analysis of their Equivalence for a Blunt Cone Geometry at Mach 8.* Journal of Applied Fluid Mechanics, 2020. **14**(2): p. 375-388.

118.Liu, Q., et al., *On the Drag Reduction Mechanism of Hypersonic Turbulent Boundary Layers Subject to Heated Wall Blowing.* Advances in Aerodynamics, 2023. **5**(1).

119.Wang, W., et al., *Hypersonic Vehicle Aerodynamic Optimization Using Field Metamodel-Enhanced Sequential Approximate Optimization.* International Journal of Aerospace Engineering, 2021. **2021**: p. 1-12.

120.Son, J.Y., C.-W. Son, and K. Yee, *A Novel Direct Optimization Framework for Hypersonic Waverider Inverse Design Methods.* Aerospace, 2022. **9**(7): p. 348.

121.Fan, B. and J. Huang, *Coupled Fluid-Thermal Investigation on Drag and Heat Reduction of a Hypersonic Spiked Blunt Body With an Aerodisk.* Aerospace, 2021. **9**(1): p. 19.

122.Li, W., et al., *Structure Design and Heat Transfer Performance Analysis of a Novel Composite Phase Change Active Cooling Channel Wall for Hypersonic Aircraft.* Micromachines, 2024. **15**(5): p. 623.

123.Le, V.T., N.S. Goo, and J.Y. Kim, *Thermomechanical Behavior of Superalloy Thermal Protection System Under Aerodynamic Heating.* Journal of Spacecraft and Rockets, 2019. **56**(5): p. 1432-1448.

124.Munk, D.J., G.A. Vio, and D. Verstraete, *A Hypersonic Aircraft Optimization Tool With Strong Aerothermoelastic Coupling.* Applied Mechanics and Materials, 2016. **846**: p. 494-499.

125.Steeves, C.A. and A.G. Evans, *Optimization of Thermal Protection Systems Utilizing Sandwich Structures With Low Coefficient of Thermal Expansion Lattice Hot Faces.* Journal of the American Ceramic Society, 2011. **94**(s1).

126.Le, V.T. and N.S. Goo, *Thermomechanical Performance of Bio-Inspired Corrugated-Core Sandwich Structure for a Thermal Protection System Panel.* Applied Sciences, 2019. **9**(24): p. 5541.

127.Wang, G.H., et al., *Optimized Design of Multilayer Thermal Insulations for Hypersonic Vehicles.* Key Engineering Materials, 2016. **697**: p. 449-452.

128.Zheng, H., J. Liu, and S. Duan, *Novel Nonstationarity Assessment Method for Hypersonic Flutter Flight Tests.* Mathematical Problems in Engineering, 2018. **2018**: p. 1-12.

129.Alkandry, H., K.M. Hanquist, and I.D. Boyd, *Conceptual Analysis of Electron Transpiration Cooling for the Leading Edges of Hypersonic Vehicles.* 2014.

130.Kimmel, R.L. and D.K. Prabhu, *HIFiRE-1 Turbulent Shock Boundary Layer Interaction -Flight Data and Computations.* 2015.

131.Ghanbari, M., S. Maddah, and J. Alinejad, *Influence of Coolant Multi-Jets on Heat Reduction of Nose Cone With Blunt Spike at Hypersonic Flow.* Scientific Reports, 2022. **12**(1).

132.Dai, P., et al., *Longitudinal Tracking Control for a Morphing Waverider Using Adaptive Super Twisting Control.* Ieee Access, 2021. **9**: p. 59692-59702.

133.Xu, B. and Z. Shi, *An Overview on Flight Dynamics and Control Approaches for Hypersonic Vehicles.* Science China Information Sciences, 2015. **58**(7): p. 1-19.

134.Viola, N., et al., *Aerodynamic Characterization of Hypersonic Transportation Systems and Its Impact on Mission Analysis.* Energies, 2021. **14**(12): p. 3580.

135.Ferretto, D., et al., *Integrated Flight Control System Characterization Approach for Civil High-Speed Vehicles in Conceptual Design.* Aerospace, 2023. **10**(6): p. 495.

136.Xu, B., et al., *Dynamic Surface Control of Constrained Hypersonic Flight Models With Parameter Estimation and Actuator Compensation.* Asian Journal of Control, 2013. **16**(1): p. 162-174.

137.Wang, R., et al., *Study on Ground Experimental Method of Stagnation Point Large Heat Flux of Typical Sharp Wedge Leading Edge Structure.* Aerospace, 2023. **10**(10): p. 871.

138.Su, L., J. Wang, and H. Chen, *A Real-Time and Optimal Hypersonic Entry Guidance Method Using Inverse Reinforcement Learning.* Aerospace, 2023. **10**(11): p. 948.

139.Nieto, A., et al., *Towards High Efficiency Hypersonic Flight-Hypersonic Bi-Directional Flying Wing.* 2012.

140.Kennell, C., et al., *Measurement of Vehicle Stability Coefficients in Hypersonic Wind Tunnels.* 2015.

141.Esposito, S., *Analysis of Plasma Formation During Hypersonic Flight in the Earth Atmosphere.* 2023.

142.Maxwell, J.R. and T.T. Hoang, *Two Phase Thermal Protection of the Hypersonic Leading Edge.* 2016.

143.Lv, C., Z. Cui, and Y. Han, *Light Propagation Characteristics of Turbulent Plasma Sheath Surrounding the Hypersonic Aerocraft*.* Chinese Physics B, 2019. **28**(7): p. 074203.

144.Kim, M. and A. Gülhan, *Plasma Manipulation Using a MHD-based Device for a Communication Blackout in Hypersonic Flights.* 2011.

145.Sun, H., et al., *Backward Scattering Characteristics of a Reentry Vehicle Enveloped by a Hypersonic Flow Field.* International Journal of Antennas and Propagation, 2018. **2018**: p. 1-14.

146.Shi, L., et al., *Characteristic of Plasma Sheath Channel and Its Effect on Communication.* Progress in Electromagnetics Research, 2012. **123**: p. 321-336.

147.Wang, C.-C. and L.-C. Hsu, *Numerical Investigation of Hypersonic Flow With Repetitive-Pulsed Plasma Actuators.* Proceedings of the Institution of Mechanical Engineers Part G Journal of Aerospace Engineering, 2017. **232**(9): p. 1715-1724.

148.Liu, Y., D. Xiao, and Y. Lu, *Research on Advanced Flight Control Methods Based on Actuator Constraints for Elastic Model of Hypersonic Vehicle.* Proceedings of the Institution of Mechanical Engineers Part G Journal of Aerospace Engineering, 2013. **228**(9): p. 1627-1637.

149.Cheng, J., et al., *Drag Reduction and Aero Heating Prevention of a Combined Configuration of Opposing Jet and Backward Jet for Hypersonic Vehicles at Different Angles of Attack.* 2024.

150.Fu, X., J. Wang, and Y. Ren, *Research on High-Temperature Resistant Resin Matrix Composites of Hypersonic Aircraft Structure.* Journal of Physics Conference Series, 2022. **2228**(1): p. 012014.

151.Shen, B.-x., H. Liu, and W. Liu, *Influence of Angle of Attack on a Combined Opposing Jet and Platelet Transpiration Cooling Blunt Nose in Hypersonic Vehicle.* Journal of Zhejiang University Science A, 2020. **21**(9): p. 761-769.

152.Estruch, D., et al., *Hypersonic Interference Heating in the Vicinity of Surface Protuberances.* Experiments in Fluids, 2010. **49**(3): p. 683-699.

153.Tantos, C., et al., *Effect of Vibrational Degrees of Freedom on the Heat Transfer in Polyatomic Gases Confined Between Parallel Plates.* International Journal of Heat and Mass Transfer, 2016. **102**: p. 162-173.

154.Истомин, В.А., Е.В. Кустова, and M. Mekhonoshina, *Eucken Correction in High-Temperature Gases With Electronic Excitation.* The Journal of Chemical Physics, 2014. **140**(18).

155.Wang, X. and X. Zhong, *Nonequilibrium and Reactive High-Speed Flow Simulations With a Fifth-Order WENO Scheme.* 2009.

156.Song, L., et al., *Analysis of Hypersonic Platform-Borne SAR Imaging: A Physical Perspective.* Remote Sensing, 2021. **13**(23): p. 4943.

157.Bertin, J.J. and R.M. Cummings, *Critical Hypersonic Aerothermodynamic Phenomena.* Annual Review of Fluid Mechanics, 2006. **38**(1): p. 129-157.

158.Wang, Z., Y.F. Han, and W. Cao, *Simplified Mixing Rules for Calculating Transport Coefficients of High-Temperature Air.* International Journal of Aerospace Engineering, 2023. **2023**: p. 1-15.

159.Mungiguerra, S., et al., *Arc-Jet Wind Tunnel Characterization of Ultra-High-Temperature Ceramic Matrix Composites.* Corrosion Science, 2019. **149**: p. 18-28.

160.Bruno, C. and A. Ingenito, *Some Key Issues in Hypersonic Propulsion.* Energies, 2021. **14**(12): p. 3690.

161.Lamorte, N. and P.P. Friedmann, *Hypersonic Aeroelastic and Aerothermoelastic Studies Using Computational Fluid Dynamics.* Aiaa Journal, 2014. **52**(9): p. 2062-2078.

162.McNamara, J.J. and P.P. Friedmann, *Aeroelastic and Aerothermoelastic Analysis of Hypersonic Vehicles: Current Status and Future Trends.* 2006.

163.Schlichting, H. and K. Gersten, *Boundary-Layer Theory*. 2000.

164.Olanrewaju, P.O., et al., *Effects of Internal Heat Generation, Thermal Radiation and Buoyancy Force on a Boundary Layer Over a Vertical Plate With a Convective Surface Boundary Condition.* South African Journal of Science, 2011. **107**(9/10).

165.Dong, M., et al., *Experimental Research on Effect of Zonal Heating Airfoil About Boundary Layer Flow and Heat Transfer.* Journal of Physics Conference Series, 2023. **2472**(1): p. 012049.

166.Makinde, O.D., *Second Law Analysis for Variable Viscosity Hydromagnetic Boundary Layer Flow With Thermal Radiation and Newtonian Heating.* Entropy, 2011. **13**(8): p. 1446-1464.

167.Pal, D., *Mixed Convection Heat Transfer in the Boundary Layers on an Exponentially Stretching Surface With Magnetic Field.* Applied Mathematics and Computation, 2010. **217**(6): p. 2356-2369.

168.Nagata, K., Y. Sakai, and S. Komori, *Effects of Small-Scale Freestream Turbulence on Turbulent Boundary Layers With and Without Thermal Convection.* Physics of Fluids, 2011. **23**(6).

169.Yang, B., et al., *Experimental Study on Hypersonic Double-Wedge Induced Flow Based on Plasma Active Actuation Array.* Aerospace, 2024. **11**(1): p. 60.

170.Zhang, Y., et al., *Influence of Expansion Waves on Cowl Shock/Boundary Layer Interaction in Hypersonic Inlets.* Journal of Propulsion and Power, 2014. **30**(5): p. 1183-1191.

171.Ligrani, P.M., et al., *Recent Investigations of Shock Wave Effects and Interactions.* Advances in Aerodynamics, 2020. **2**(1).

172.Ren, Y., H. Zheng, and L. Qiu, *Experimental Research on Online Monitoring of High Temperature and High Speed Airflow Ablation of Quartz Ceramic Thermal Protection Structure Using Guided Wave.* Structural Health Monitoring, 2023. **22**(5): p. 3356-3372.

173.Falkiewicz, N. and C.E.S. Cesnik, *Proper Orthogonal Decomposition for Reduced-Order Thermal Solution in Hypersonic Aerothermoelastic Simulations.* Aiaa Journal, 2011. **49**(5): p. 994-1009.

174.Chaudhuri, A., *On Shock Propagation Through Double-Bend Ducts by Entropy-Generation-Based Artificial Viscosity Method.* Entropy, 2019. **21**(9): p. 837.

175.Giepman, R., F.F.J. Schrijer, and B.W.v. Oudheusden, *High-Resolution PIV Measurements of a Transitional Shock Wave–boundary Layer Interaction.* Experiments in Fluids, 2015. **56**(6).

176.Qin, F. and X. Wu, *Response and Receptivity of the Hypersonic Boundary Layer Past a Wedge to Free-Stream Acoustic, Vortical and Entropy Disturbances.* Journal of Fluid Mechanics, 2016. **797**: p. 874-915.

177.Beketaeva, A.O., et al., *Simulation of the Shock Wave Boundary Layer Interaction in Flat Channel With Jet Injection.* Zamm Journal of Applied Mathematics and Mechanics / Zeitschrift Für Angewandte Mathematik Und Mechanik, 2023. **103**(8).

178.Tang, X., et al., *Receptivity Study of Hypersonic Boundary Layer Under Freestream Finite Amplitude Continuous Entropy Disturbance With Single Frequency.* Destech Transactions on Engineering and Technology Research, 2020(amee).

179.Giepman, R., et al., *Experimental Investigation of Boundary Layer Tripping Devices for Shock Wave - Boundary Layer Control.* 2015.

180.Souverein, L. and J.F. Debiève, *Effect of Air Jet Vortex Generators on a Shock Wave Boundary Layer Interaction.* Experiments in Fluids, 2010. **49**(5): p. 1053-1064.

181.Hu, B., et al., *Engineering Calculation and Analysis of Aerodynamic Heating for Hypersonic Rocket Sled.* Journal of Physics Conference Series, 2023. **2478**(8): p. 082009.

182.Lü, Y., et al., *Review of Research Progress in Nontraditional Machining of Ultrahigh-Temperature Ceramic Matrix Composites.* Coatings, 2023. **13**(1): p. 187.

183.Xie, Z., et al., *Synthesis and Characterization of Zirconium Diboride Precursor Based on Polycentric Bridge Bonds.* Materials Chemistry and Physics, 2015. **159**: p. 178-184.

184.Chang, X., et al., *Superelastic Carbon Aerogels: An Emerging Material for Advanced Thermal Protection in Extreme Environments.* Advanced Functional Materials, 2023. **33**(26).

185.Kim, Y.-N., et al., *Nonlinear Random Response Analyses of Panels Considering Transverse Shear Deformations Under Combined Thermal and Acoustic Loads.* Shock and Vibration, 2018. **2018**(1).

186.Sun, X., H. Yang, and T. Mi, *Heat Transfer and Ablation Prediction of Carbon/Carbon Composites in a Hypersonic Environment Using Fluid-Thermal-Ablation Multiphysical Coupling.* International Journal of Aerospace Engineering, 2020. **2020**: p. 1-13.

187.Saad, A.A., C. Martínez, and R.W. Trice, *Radiation Heat Transfer During Hypersonic Flight: A Review of Emissivity Measurement and Enhancement Approaches of Ultra high Temperature Ceramics*International Journal of Ceramic Engineering & Science, 2023. **5**(2).

188.Guo, L. and S. Fang, *Study on Aerodynamic Force and Aerothermal Effect of Two-Dimensional Wedge Wave Rider With Different Passivation Shape of the Head.* Journal of Physics Conference Series, 2022. **2403**(1): p. 012002.

189.Sziroczák, D. and H. Smith, *A Review of Design Issues Specific to Hypersonic Flight Vehicles.* Progress in Aerospace Sciences, 2016. **84**: p. 1-28.

190.Matsuno, Y., et al., *Multidisciplinary Design Optimization of Long or Short Range Hypersonic Aircraft.* Transactions of the Japan Society for Aeronautical and Space Sciences, 2014. **57**(3): p. 143-152.

191.Bian, X., et al., *Analysis of Infrared Radiation Characteristics and Detection Technology of Near-Space Hypersonic Vehicles.* 2024.

192.Fu, W., et al., *Ascent Trajectory Optimization for Hypersonic Vehicle Based on Improved Chicken Swarm Optimization.* Ieee Access, 2019. **7**: p. 151836-151850.

193.Zhong, Q. and W. Wu, *A Switching Based Interference Control for Booster Separation of Hypersonic Vehicle.* Wireless Communications and Mobile Computing, 2021. **2021**(1).

194.Li, S., et al., *Design and Aerodynamic Investigation of a Parallel Vehicle on a Wide-Speed Range.* Science China Information Sciences, 2014. **57**(12): p. 1-10.

195.Ding, F., et al., *An Airframe/Inlet Integrated Full-Waverider Vehicle Design Using as Upgraded Aerodynamic Method.* The Aeronautical Journal, 2019. **123**(1266): p. 1135-1169.

196.Huang, H., et al., *Numerical Research on the Performance of Solid Fuel Rocket Scramjet Combustor.* Proceedings of the Institution of Mechanical Engineers Part G Journal of Aerospace Engineering, 2021. **236**(7): p. 1450-1461.

197.Lv, Z., et al., *Experimental and Numerical Investigation of a Solid-Fuel Rocket Scramjet Combustor.* Journal of Propulsion and Power, 2016. **32**(2): p. 273-278.

198.Urzay, J., *Supersonic Combustion in Air-Breathing Propulsion Systems for Hypersonic Flight.* Annual Review of Fluid Mechanics, 2018. **50**(1): p. 593-627.

199.Hillion, L., J.-D. Parisse, and A. Mangeot, *Preliminary Sizing and Study of a Hybrid Rocket Based Combined Cycle.* Frontiers in Space Technologies, 2023. **4**.

200.Alkuam, E. and W.M. Alobaidi, *Experimental and Theoretical Research Review of Hybrid Rocket Motor Techniques and Applications.* Advances in Aerospace Science and Technology, 2016. **01**(03): p. 71-82.

201.Souza, A., et al., *On the Multidisciplinary Design of a Hybrid Rocket Launcher With a Composite Overwrapped Pressure Vessel.* Journal of Composites Science, 2024. **8**(3): p. 109.

202.Osemwengie, O. and A. Sa, *Parameters Influencing Regression Rate of Solid Rocket Fuels.* International Journal of Aeronautical Science & Aerospace Research, 2019: p. 164-170.

203.Gugulothu, S.K., *A Systematic Literature Review Based on Different Fuel Injection Strategies Used in Scramjet Combustors.* Heat Transfer-Asian Research, 2019. **48**(8): p. 3657-3681.

204.Goel, A., et al., *Retrospective Cost Adaptive Thrust Control of a 1D Scramjet With Mach Number Disturbance.* 2015.

205.Powell, O., et al., *Development of Hydrocarbon-Fueled Scramjet Engines: The Hypersonic Technology (HyTech) Program.* Journal of Propulsion and Power, 2001. **17**(6): p. 1170-1176.

206.Cecere, D., et al., *Hydrogen/Air Supersonic Combustion for Future Hypersonic Vehicles.* International Journal of Hydrogen Energy, 2011. **36**(18): p. 11969-11984.

207. Saccone, G., et al., *Hydrogen/Air Supersonic Combustion Modelling and Validation for Scramjet Applications.* Journal of Fluid Flow Heat and Mass Transfer, 2022.
208. Chandrasekhar, C., V. Ramanujachari, and T.K.K. Reddy, *Evaluation of Kerosene Fuelled Scramjet Combustor Using a Combination of Cooled and Uncooled Struts.* Defence Science Journal, 2014. **64**(1): p. 5-12.
209. Relangi, N., et al., *Design of Supersonic and Hybrid Engine Based Advanced Rocket (SHAR).* Iop Conference Series Materials Science and Engineering, 2022. **1226**(1): p. 012031.
210. Smart, M. and E. Ruf. *Free-jet testing of a REST scramjet at off-design conditions*. in *25th AIAA aerodynamic Measurement technology and ground testing Conference*. 2006.
211. Reardon, J.P., J.A. Schetz, and K.T. Lowe, *Computational analysis of unstart in variable-geometry inlet.* Journal of Propulsion and Power, 2021. **37**(4): p. 564-576.
212. Idris, A.C., M.R. Saad, and K. Kontis, *Potential of micro-vortex generators in enhancing the quality of flow in a hypersonic inlet-isolator.* Journal of Advanced Research in Fluid Mechanics and Thermal Sciences, 2021. **77**(1): p. 1-10.
213. Mogrekar, A. and R. Sivakumar, *CFD Analysis of Micro-Ramps for Hypersonic Flows.* Applied Mechanics and Materials, 2014. **592-594**: p. 1962-1966.
214. Veeran, S., A. Pesyridis, and L. Ganippa, *Ramjet Compression System for a Hypersonic Air Transportation Vehicle Combined Cycle Engine.* Energies, 2018. **11**(10): p. 2558.
215. Zahrolayali, N., et al., *Assessing the Performance of Hypersonic Inlets by Applying a Heat Source With the Throttling Effect.* Aerospace, 2022. **9**(8): p. 449.
216. Zheng, Y., et al., *Airframe-Propulsion Integration Design and Optimization.* 2019.
217. Sandeep, J. and A.V.S.S.K.S. Gupta, *Analytical Approach for Design of Double Ramp Hypersonic Intake for Scramjet Engine.* Advances in Science and Technology, 2023.
218. Laurence, S.J., et al., *Transient Fluid-Combustion Phenomena in a Model Scramjet.* Journal of Fluid Mechanics, 2013. **722**: p. 85-120.
219. Baccarella, D., et al., *Isolator-Combustor Interactions in a Circular Model Scramjet With Thermal and Non-Thermal Choking-Induced Unstart.* Journal of Fluid Mechanics, 2021. **917**.

220.Wei, J., et al., *Backstepping Control Based on Constrained Command Filter for Hypersonic Flight Vehicles With AOA and Actuator Constraints.* International Journal of Aerospace Engineering, 2021. **2021**: p. 1-16.
221.Rouzbar, R. and S. Eyi, *Reacting Flow Analysis of a Cavity-Based Scramjet Combustor Using a Jacobian-Free Newton–Krylov Method.* The Aeronautical Journal, 2018. **122**(1258): p. 1884-1915.
222.Pandey, K.M. and T. Sivasakthivel, *Recent Advances in Scramjet Fuel Injection - A Review.* International Journal of Chemical Engineering and Applications, 2010: p. 294-301.
223.Goodwin, G.B. and E.S. Oran, *Flame Instability and Transition to Detonation in Supersonic Reactive Flows.* 2018.
224.Guo, P., J. Xu, and M. Wang, *Optimization of Hydrogen Injection Flow Rate for Hydrogen Internal Combustion Engines Based on IQGA.* 2023.
225.Vergine, F., et al., *Supersonic Combustion and Flame-Holding Characteristics of Pylon Injected Hydrogen in a Mach 2.4 High Enthalpy Flow.* 2012.
226.Hamzehloo, A. and P.G. Aleiferis, *Computational Study of Hydrogen Direct Injection for Internal Combustion Engines.* 2013.
227.Liu, X., et al., *Comparison of Hydrogen Port Injection and Direct Injection (DI) in a Single-Cylinder Dual-Fuel Diesel Engine.* 2020.
228.Kamil, M., *Effect of Injection Hole Diameter on Operational Conditions of Common-Rail Fuel-Injection System for Port-Injection Hydrogen-Fueled Engine.* International Journal of Automotive and Mechanical Engineering, 2015. **11**: p. 2383-2395.
229.Darlami, K., et al., *Study and Analysis of Fuel Injection at Busemann Inlet at Mach 7.* American Journal of Mechanics and Applications, 2020. **8**(3): p. 33.
230.Qin, F., et al., *Large Eddy Simulation of Combustion Characteristics in a Hydrogen Fueled Model Scramjet Combustor.* 2015.
231.Nieman, R., et al., *Informing Air–carbon Ablation Modeling With Theoretical Calculations of Atomic Oxygen and Nitrogen Interacting With Carbon Surfaces.* Physical Chemistry Chemical Physics, 2023. **25**(22): p. 15479-15489.
232.Uyanna, O. and H. Najafi, *Thermal Protection Systems for Space Vehicles: A Review on Technology Development, Current Challenges and Future Prospects.* Acta Astronautica, 2020. **176**: p. 341-356.
233.Zhu, Y., et al., *Review on Active Thermal Protection and Its Heat Transfer for Airbreathing Hypersonic Vehicles.* Chinese Journal of Aeronautics, 2018. **31**(10): p. 1929-1953.
234.Park, J., S. Cho, and H. Kwon, *Aluminum-Ceramic Composites for Thermal Management in Energy-Conversion Systems.* Scientific Reports, 2018. **8**(1).

235.Lee, Y.-J., Y.-H. Park, and T. Hinoki, *Influence of Grain Size on Thermal Conductivity of SiC Ceramics.* Iop Conference Series Materials Science and Engineering, 2011. **18**(16): p. 162014.

236.Chen, F., S. Zhang, and H. Liu, *Modeling and Analysis of Fluid-Thermal-Structure Coupling Problems for Hypersonic Vehicles.* 2018.

237.Chen, J., et al., *Cellulose Nanofiber Supported 3D Interconnected BN Nanosheets for Epoxy Nanocomposites With Ultrahigh Thermal Management Capability.* Advanced Functional Materials, 2016. **27**(5).

238.McNamara, J.J. and P.P. Friedmann, *Aeroelastic and Aerothermoelastic Analysis in Hypersonic Flow: Past, Present, and Future.* Aiaa Journal, 2011. **49**(6): p. 1089-1122.

239.Wan, T. and L. Cm, *Drag Reduction Optimization for Hypersonic Blunt Body With Aerospikes.* Journal of Aeronautics & Aerospace Engineering, 2017. **06**(04).

240.Kulkarni, V., V. Menezes, and K.P.J. Reddy, *Effectiveness of Aerospike for Drag Reduction on a Blunt Cone in Hypersonic Flow.* Journal of Spacecraft and Rockets, 2010. **47**(3): p. 542-544.

241.Rao, D.S.R. and B. John, *A Numerical Study on the Single Pulsed Energy Addition Based Unsteady Wave Drag Reduction at Varied Hypersonic Flow Regimes.* Proceedings of the Institution of Mechanical Engineers Part G Journal of Aerospace Engineering, 2020. **235**(11): p. 1385-1401.

242.Zhu, Y., et al., *Transition in Hypersonic Boundary Layers: Role of Dilatational Waves.* Aiaa Journal, 2016. **54**(10): p. 3039-3049.

243.Maicke, B.A. and J. Majdalani, *Evaluation of CFD Codes for Hypersonic Flow Modeling.* 2010.

244.Ding, F., et al., *Novel Inlet–airframe Integration Methodology for Hypersonic Waverider Vehicles.* Acta Astronautica, 2015. **111**: p. 178-197.

245.Singh, N. and T.E. Schwartzentruber, *Nonequilibrium Internal Energy Distributions During Dissociation.* Proceedings of the National Academy of Sciences, 2017. **115**(1): p. 47-52.

246.Yang, F.H., et al., *Preparation and Performance of C/C-SiC Ceramic Matrix Composites.* Solid State Phenomena, 2018. **281**: p. 402-407.

247.Paul, A., et al., *UHTC–carbon Fibre Composites: Preparation, Oxyacetylene Torch Testing and Characterisation.* Journal of the European Ceramic Society, 2013. **33**(2): p. 423-432.

248.King, D.S., et al., *Novel Processing Approach to Polymer derived Ceramic Matrix Composites.* International Journal of Applied Ceramic Technology, 2017. **15**(2): p. 399-408.

249.Xue, R., et al., *A Survey on the Conceptual Design of Hypersonic Aircraft Powered by RBCC Engine.* Proceedings of the Institution of Mechanical Engineers Part C Journal of Mechanical Engineering Science, 2021. **237**(18): p. 4213-4245.

250.Panakarajupally, R.P., et al., *Fatigue Characterization of SiC/SiC Ceramic Matrix Composites in Combustion Environment.* Journal of Engineering for Gas Turbines and Power, 2020. **142**(12).

251.Kütemeyer, M., et al., *Fabrication of Ultra High Temperature Ceramic Matrix Composites Using a Reactive Melt Infiltration Process.* Journal of the European Ceramic Society, 2016. **36**(15): p. 3647-3655.

252.Hu, X., et al., *Model Predictive Control based Non linear Fault Tolerant Control for Air breathing Hypersonic Vehicles*Iet Control Theory and Applications, 2014. **8**(13): p. 1147-1153.

253.Wang, P., L.H. Liu, and J. Wu, *Nonlinear Hierarchy-Structured Predictive Control System Design for Hypersonic Flight Vehicle.* Applied Mechanics and Materials, 2012. **232**: p. 194-199.

254.Hu, Z.B., L. Deng, and B.B. Li, *Design of Air-Breathing Hypersonic Vehicle Control System.* Applied Mechanics and Materials, 2015. **719-720**: p. 324-329.

255.Xiao, D., et al., *A Design Method of Switching Control for Hypersonic Vehicles Based on Guardian Maps.* 2015.

256.Ji, Y., Q. Zong, and H. Zhou, *Command Filtered Back-Stepping Control of a Flexible Air-Breathing Hypersonic Flight Vehicle.* Proceedings of the Institution of Mechanical Engineers Part G Journal of Aerospace Engineering, 2013. **228**(9): p. 1617-1626.

257.Zhao, S., et al., *Prescribed Performance Tracking Control for Hypersonic Flight Vehicles With Model Uncertainties.* International Journal of Aerospace Engineering, 2019. **2019**: p. 1-11.

258.Luo, C., et al., *Adaptive Neural Control of Hypersonic Vehicles With Actuator Constraints.* International Journal of Aerospace Engineering, 2018. **2018**: p. 1-15.

259.Hao, L., et al., *Guardian Map Approach to Feasible Range of Static Stability Margin of Hypersonic Flight Vehicles With Input Saturation.* International Journal of Aerospace Engineering, 2020. **2020**: p. 1-11.

260.Wu, Q. and Y. Guo, *Neural Back-Stepping Control of Hypersonic Flight Vehicle With Actuator Fault.* Journal of Control Science and Engineering, 2018. **2018**: p. 1-5.

261.Ji, Y., H. Zhou, and Q. Zong, *Adaptive Active Fault-Tolerant Control of Generic Hypersonic Flight Vehicles.* Proceedings of the Institution of Mechanical Engineers Part I Journal of Systems and Control Engineering, 2014. **229**(2): p. 130-138.

262.Peng, S.-S., et al., *Effect of Kerosene Dual-Jet Spacing on the Mixing and Combustion Characteristics of a Scramjet Combustor.* Acs Omega, 2024. **9**(31): p. 33482-33493.

263.Li, H., et al., *Overall Heat Transfer Coefficient Evaluation Method for Uncracked Hydrocarbon Fuel in a Regeneratively-Cooled Heat Exchanger of a Scramjet.* Applied Sciences, 2022. **12**(13): p. 6590.

264.Dai, Y., et al., *A Rapid Method of Integrated Aeropropulsive Analysis for the Conceptual Design of Airbreathing Hypersonic Aircraft.* Aerospace, 2024. **11**(1): p. 89.

265. et al., *A Conceptual Design of Next Generation Hypersonic Air-Breathing Airplanes With Dual Waveriders Forebody.* 2015.

266.Laitón, S.N.P., et al., *Experimental Study of Single Expansion Ramp Nozzle Performance Using Pitot Pressure and Static Pressure Measurements.* International Journal of Aerospace Engineering, 2019. **2019**: p. 1-11.

267.Rosato, D.A., et al., *Stabilized Detonation for Hypersonic Propulsion.* Proceedings of the National Academy of Sciences, 2021. **118**(20).

268.Fleeman, E., *A Historical Overview of a Half Century of U.S. Missile Development.* 2019, Defense Systems Information Analysis Center.

269.Hertzberg, A., C.E. Wittliff, and J.G. Hall, *Summary of Shock Tunnel Development and Application to Hypersonic Research.* 1961.

270.Hunt, J.L., et al., *Performance Potential and Research Needs of a Hypersonic, Airbreathing, Lifting Missile Concept.* Journal of Aircraft, 1979. **16**(10): p. 666-673.

271.Nista, L. and B.H. Saracoglu, *Development of a Robust Solver to Model the Flow Inside the Engines for High-Speed Propulsion.* Matec Web of Conferences, 2019. **304**: p. 03013.

272.Hannemann, K., et al., *The High Enthalpy Shock Tunnel Göttingen of the German Aerospace Center (DLR).* Journal of Large-Scale Research Facilities JLSRF, 2018. **4**: p. A133.

273.Padbidiri, S.S., et al., *Design and Development of a New 0.5m Hypersonic Wind Tunnel Facility at Indian Institute of Science.* 2023: p. 67-74.

274.Chen, B., et al., *An Evaluation Model of Star Sensor Observation Capability Under Hypersonic Aerothermal Conditions.* Ieee Access, 2023. **11**: p. 646-654.

275.Zhenyu, C., et al., *Finite-Time Integrated Guidance and Control System for Hypersonic Vehicles.* Transactions of the Institute of Measurement and Control, 2020. **43**(4): p. 842-853.

276.Chen, J., et al., *Fractional Calculus Guidance Algorithm in a Hypersonic Pursuit-Evasion Game.* Defence Science Journal, 2017. **67**(6): p. 688.

277.Song, H. and T. Zhang, *Fast Robust Integrated Guidance and Control Design of Interceptors.* Ieee Transactions on Control Systems Technology, 2016. **24**(1): p. 349-356.

278.Chen, W., et al., *Research on Differential Game Guidance Law for Intercepting Hypersonic Vehicles.* 2022.

279.He, S., D. Lin, and J. Wang, *Integral Global Sliding Mode Guidance for Impact Angle Control.* Ieee Transactions on Aerospace and Electronic Systems, 2019. **55**(4): p. 1843-1849.

280.Zhu, C. and G. Zheng-yu, *Design of Head-Pursuit Guidance Law Based on Backstepping Sliding Mode Control.* International Journal of Aerospace Engineering, 2019. **2019**: p. 1-18.

281.Yan, L., et al., *Three-Dimensional United Biased Proportional Navigation Law for Interception of Maneuvering Targets With Angular Constraint.* Proceedings of the Institution of Mechanical Engineers Part G Journal of Aerospace Engineering, 2014. **229**(6): p. 1013-1024.

282.Liu, S., et al., *Bulletin of the Polish Academy of Sciences: Technical Sciences.* 2022.

283.Si, Y. and S. Song, *Continuous Reaching Law Based Three-Dimensional Finite-Time Guidance Law Against Maneuvering Targets.* Transactions of the Institute of Measurement and Control, 2018. **41**(2): p. 321-339.

284.Li, S., et al., *A New Hypersonic Wind Tunnel Force Measurement System to Reduce Additional Bending Moment and Avoid Time-Varying Stiffness.* Sensors, 2022. **22**(7): p. 2572.

285.Gao, H., et al., *A New Force-Measuring Balance for Large-Scale Model in Hypersonic Wind Tunnel.* Review of Scientific Instruments, 2021. **92**(11).

286.Hermann, T., I. Naved, and M. McGilvray, *Tool for Rapid Transient Transpiration-Cooled Reentry Simulation.* Aiaa Journal, 2020. **58**(2): p. 842-853.

287.Lee, C. and S. Chen, *Recent Progress in the Study of Transition in the Hypersonic Boundary Layer.* National Science Review, 2018. **6**(1): p. 155-170.

288.Hirschel, E.H. and C. Weiland, *Design of Hypersonic Flight Vehicles: Some Lessons From the Past and Future Challenges.* Ceas Space Journal, 2010. **1**(1-4): p. 3-22.

289.Purwar, A. and B. Basu, *Thermo structural Design of ZrB₂–SiC based Thermal Protection System for Hypersonic Space Vehicles.* Journal of the American Ceramic Society, 2017. **100**(4): p. 1618-1633.

290.Li, H., et al., *Thermal Insulation Characteristics of High-Temperature Heat Seals for Hypersonic Aero-Engines.* Journal of Physics Conference Series, 2023. **2566**(1): p. 012119.

291.Li, J., *An Integrated Algorithm for Hypersonic Fluid–thermal–structural Analysis of Aerodynamically Heated Cylindrical Leading Edge.* Proceedings of the Institution of Mechanical Engineers Part G Journal of Aerospace Engineering, 2019. **233**(14): p. 5177-5191.

292.Sha, X., et al., *Noise Level Effect on the Hypersonic Boundary Layer.* Advances in Mechanical Engineering, 2023. **15**(7).

293.Pelin, G., et al., *Oxy-Butane Ablation Testing of Thermal Protection Systems Based on Nanomodified Phenolic Resin Matrix Materials.* Polymers, 2023. **15**(19): p. 4016.

294.Grinstead, J.H., G.A. Raiche, and T. Gökçen, *Test Engineering for Arc Jet Testing of Thermal Protection Systems: Design, Analysis, and Validation Methodologies.* 2006.

295.S, P., et al., *RETALT_TPS Design and Manufacturing.* Ceas Space Journal, 2022. **14**(3): p. 595-604.

296.Bolender, M.A., et al., *HIFiRE 6: Overview and Status Update 2014.* 2015.

297.Morelli, E.A., *Flight-Test Experiment Design for Characterizing Stability and Control of Hypersonic Vehicles.* Journal of Guidance Control and Dynamics, 2009. **32**(3): p. 949-959.

298.Pehrson, J.C., et al., *Hypersonic FLEET Velocimetry and Uncertainty Characterization in a Tripped Boundary Layer.* Measurement Science and Technology, 2023. **35**(1): p. 015206.

299.Chen, Z. and Y. Zhao, *Aerothermoelastic Analysis of a Hypersonic Vehicle Based on Thermal Modal Reconstruction.* International Journal of Aerospace Engineering, 2019. **2019**: p. 1-13.

300.Seager, C. and R.K. Agarwal, *Hypersonic Blunt-Body Shape Optimization for Reducing Drag and Heat Transfer.* Journal of Thermophysics and Heat Transfer, 2017. **31**(1): p. 48-55.

301.Dharavath, M., P. Manna, and D. Chakraborty, *Computational Fluid Dynamics Simulation of Tip-to-Tail for Hypersonic Test Vehicle.* Journal of Propulsion and Power, 2015. **31**(5): p. 1370-1379.

302.McClinton, C.R., et al., *Preliminary X-43 Flight Test Results.* Acta Astronautica, 2005. **57**(2-8): p. 266-276.

303.Hank, J., J.M. Murphy, and R. Mutzman, *The X-51a Scramjet Engine Flight Demonstration Program.* 2008.

304.Rolim, T.C., et al., *Experimental Results of a Mach 10 Conical-Flow Derived Waverider to 14-X Hypersonic Airspace Vehicle.* Journal of Aerospace Technology and Management, 2011. **3**(2): p. 127-136.

305.Thoemel, J., et al., *In-Flight Testing of Critical Technologies and Experimentation of Aerothermodynamic Phenomena.* 2009.

306.Ding, F., et al., *Novel Approach for Design of a Waverider Vehicle Generated From Axisymmetric Supersonic Flows Past a Pointed Von Karman Ogive.* Aerospace Science and Technology, 2015. **42**: p. 297-308.

307.Xu, B., et al., *DOB-Based Neural Control of Flexible Hypersonic Flight Vehicle Considering Wind Effects.* Ieee Transactions on Industrial Electronics, 2017. **64**(11): p. 8676-8685.

308.Jiang, X., et al., *Numerical Study on Low-Temperature Region as Heat Sink and Its Heat Dissipation Capacity for Hypersonic Vehicle.* Aerospace, 2021. **8**(9): p. 238.

309.DeMange, J.J., et al., *An Evaluation of High Temperature Airframe Seals for Advanced Hypersonic Vehicles.* 2007.

310.Li, S., et al., *Research on the Drag Reduction Performance Induced by the Counterflowing Jet for Waverider With Variable Blunt Radii.* Acta Astronautica, 2016. **127**: p. 120-130.

311.Qu, Z., et al., *Numerical Study on Aerodynamic Performance of Waverider With a New Bluntness Method.* Proceedings of the Institution of Mechanical Engineers Part G Journal of Aerospace Engineering, 2020. **235**(10): p. 1225-1233.

312.Candon, M. and H. Ogawa, *Thrust Augmentation Optimization Through Supersonic After-Burning in Scramjet Engine Nozzles via Surrogate-Assisted Evolutionary Algorithms.* Acta Astronautica, 2015. **116**: p. 132-147.

313.Jing, Y. and Y.N. Jiao, *A Model for Gas Permeation in Three-Dimensional Braided Seal.* Journal of Industrial Textiles, 2017. **47**(7): p. 1664-1680.

314.Chiriac, C., *Countering Hypersonic Threat.* Bulletin of Carol I National Defence University, 2022. **11**(3): p. 38-45.

315.Li, Y., et al., *Cooperative Optimal Guidance of Hypersonic Glide Vehicles by Real-Time Optimization and Deep Learning.* Proceedings of the Institution of Mechanical Engineers Part G Journal of Aerospace Engineering, 2023. **237**(10): p. 2266-2283.

316.Zhang, T., et al., *Research on Multi-Domain Dimensionality Reduction Joint Adaptive Processing Method for Range Ambiguous Clutter of FDA-Phase-MIMO Space-Based Early Warning Radar.* Remote Sensing, 2022. **14**(21): p. 5536.

317.Pietrus, T., *IAMD (Integrated Air and Missile Defence) Strategic Aspects With SBAD Focus.* Safety & Defense, 2023. **9**(1): p. 69-73.
318.Gao, K. and X. Yan, *Study on the Optimal Strategy of Missile Interception.* Ieee Access, 2021. **9**: p. 22239-22252.
319.Lin, L., X. Zhang, and H. Yu, *Adaptive Real time Recursive Radial Distance time Plane Hough Transform Track before detect Algorithm for Hypersonic Target.* Iet Radar Sonar & Navigation, 2020. **14**(1): p. 138-146.
320.Peng, P., et al., *Dynamic RCS Feature of Ballistic Missile for Detection and Classification in the Boost Phase.* 2013.
321.Liu, B., et al., *Global Performance Analysis for Space-Based Missile Warning System.* Proceedings of the Institution of Mechanical Engineers Part G Journal of Aerospace Engineering, 2016. **230**(14): p. 2569-2580.
322.Gao, K., et al., *Optimal Interception Strategy of Air Defence Missile System Considering Multiple Targets and Phases.* Proceedings of the Institution of Mechanical Engineers Part O Journal of Risk and Reliability, 2021. **236**(1): p. 138-147.
323.Su, W., et al., *Cooperative Interception Strategy for Multiple Inferior Missiles Against One Highly Maneuvering Target.* Aerospace Science and Technology, 2018. **80**: p. 91-100.
324.Liu, S., et al., *Fractional-Order Sliding Mode Guidance Law for Intercepting Hypersonic Vehicles.* Aerospace, 2022. **9**(2): p. 53.
325.Chen, R.H., J.L. Speyer, and D. Lianos, *Optimal Intercept Missile Guidance Strategies With Autopilot Lag.* 2010.
326.Huang, Y. and Y. Li, *On the Anti-Missile Interception Technique of Unpowered Phase Based on Data-Driven Theory.* Mechanics & Industry, 2022. **23**: p. 26.
327.Gujar, S. and D. Vishwakarma, *Hypersonic Weapons Market - By Type (Hypersonic Glide Vehicles, Hypersonic Missiles), By Range (Short-Range (up to 1,000 kilometers), Medium-Range (1,000 - 3,000 kilometers), Long-Range (above 3,000 kilometers)), By Platform, By Technology, Forecast 2024 - 2032* 2024.
328.Ioanes, E., *America's hypersonic arms race with China, explained*. 2023, Vox.
329.Sayler, K.M., *Hypersonic Weapons: Background and Issues for Congress.* 2024.
330.Klare, M.T., *Assessing the Dangers: Emerging Military Technologies and Nuclear (In)Stability*. 2023.
331.Analytica, O., *Hypersonic arms alter US-Russian threat perceptions.* Emerald Expert Briefings, 2019(oxan-db).

332.Quackenbush, S.L., *Deterrence Theory: Where Do We Stand?* Review of International Studies, 2010. **37**(2): p. 741-762.
333.Johnson, J.C., B.A. Leeds, and A. Wu, *Capability, Credibility, and Extended General Deterrence.* International Interactions, 2015. **41**(2): p. 309-336.
334.Lichtenstein, W.L., *Conventional Military Deterrence - Its Rise to Dominance and Its Future.* 2002.
335.Danilovic, V., *The Sources of Threat Credibility in Extended Deterrence.* Journal of Conflict Resolution, 2001. **45**(3): p. 341-369.
336.Berejikian, J.D., *A Cognitive Theory of Deterrence.* Journal of Peace Research, 2002. **39**(2): p. 165-183.
337.He, H., A. Luo, and Q. Geng, *Deterrence and Security the Impact of Military Conflicts on Global Peace.* Pacific International Journal, 2023. **6**(3): p. 92-100.
338.Khan, R., G. Mustafa, and B.B. Liaqat, *Nuclear Deterrence and South Asia: Theory and its Pre-Requisites.* Annals of Human and Social Sciences, 2022. **3**(2): p. 284-291.
339.Elbahy, R., *Deterring Violent Non-State Actors: Dilemmas and Implications.* Journal of Humanities and Applied Social Sciences, 2019. **1**(1): p. 43-54.
340.Morgan, P.M., *Deterrence Now.* 2003.
341.Bugos, S. and K. Reif, *Understanding Hypersonic Weapons: Managing the Allure and the Risk.* 2024.
342.McLaughlin, P.J., *The Coming Hypersonic Era.* Journal of Global Strategic Studies, 2023. **3**(2): p. 47-66.
343.Bassey, S., et al., *Disruptive Technologies and Strategic Nuclear Stability: A Russian and Chinese Perspective.* 2024. **1**(2): p. 8-19.
344.Na'inna, A.M. and N.V. Anti-Dwanso, *Arms Race in the 21st Century: Consequences and Mitigating Measures.* Global Journal of Social Sciences, 2022. **21**(2): p. 45-76.
345.Dall'Agnol, A.C. and É.E. Duarte, *Military Power and Conventional Deterrence.* Revista De Relaciones Internacionales Estrategia Y Seguridad, 2022. **17**(1): p. 101-118.
346.Driver, D., *Deterrence in Eastern Europe in Theory and Practice.* Connections the Quarterly Journal, 2019. **18**(1-2): p. 11-24.
347.Zhao, S., Q. Sun, and J. Jia, *Parametric Influences of Geometric Dimensions on High Temperature Mechanical Behaviors and Damage Mechanisms of Ceramic Matrix Composite and Superalloy Double Bolted Joints.* International Journal of Aerospace Engineering, 2022. **2022**: p. 1-16.
348.Dassios, K.G., *A Geometry-Invariant Fracture Law for Ceramic Matrix Composites.* Journal of Composite Materials, 2013. **49**(1): p. 65-74.

349.Racle, E., et al., *Fatigue Lifetime of Ceramic Matrix Composites at Intermediate Temperature by Acoustic Emission.* Materials, 2017. **10**(6): p. 658.

350.Li, L., *Damage and Fracture of Fiber-Reinforced Ceramic-Matrix Composites Under Thermal Fatigue Loading in Oxidizing Atmosphere.* Journal of the Ceramic Society of Japan, 2019. **127**(2): p. 67-80.

351.Rashid, A.B., et al., *Breaking Boundaries with Ceramic Matrix Composites: A Comprehensive Overview of Materials, Manufacturing Techniques, Transformative Applications, Recent Advancements, and Future Prospects.* Advances in Materials Science and Engineering, 2024. **2024**(1): p. 2112358.

352.DiCarlo, J.A., *Advances in SiC/SiC Composites for Aero Propulsion*2014: p. 217-235.

353.Karadimas, G. and K. Salonitis, *Ceramic Matrix Composites for Aero Engine Applications—A Review.* Applied Sciences, 2023. **13**(5): p. 3017.

354.Heiß, C., N. Travitzky, and P. Greil, *Manufacturing of Silicon Carbide Knit Fabrics.* Advanced Engineering Materials, 2012. **14**(3): p. 162-165.

355.Wang, L., et al., *Effect of High pressure Vapor Pretreatment on the Microstructure Evolution and Tensile Strength of Zirconia Fibers.* Journal of the American Ceramic Society, 2019. **102**(8): p. 4450-4458.

356.Zou, X., et al., *Fabrication and Mechanical Behavior of 2d-C _f/Ta _{<i>x</I>}Hf _{1− <i>x</I>}C–SiC Composites by a Low-Temperature and Highly-Efficient Route.* Journal of Advanced Ceramics, 2023. **12**(10): p. 1961-1972.

357.Rashid, A., *Overview of Disc Brakes and Related Phenomena - A Review.* International Journal of Vehicle Noise and Vibration, 2014. **10**(4): p. 257.

358.Li, B., et al., *Ablation Behaviour of Carbon Fibre Reinforced BN–Si₃N₄matrix Composite.* Materials Technology, 2008. **23**(1): p. 45-47.

359.Zotti, A., et al., *Effect of the Mixing Technique of Graphene Nanoplatelets and Graphene Nanofibers on Fracture Toughness of Epoxy Based Nanocomposites and Composites.* Polymers, 2022. **14**(23): p. 5105.

360.Xu, N., et al., *Complete Degradation of a High-Performance Epoxy Thermoset Enabled by Γ-Ray-Sensitive Motifs Based on the Cascaded Synergetic Strategy.* Chemistry of Materials, 2022. **34**(10): p. 4732-4740.

361.Afzal, A., et al., *Aging Effects on the Mechanical Performance of Carbon Fiber-Reinforced Composites.* International Journal of Polymer Science, 2023. **2023**: p. 1-16.

362.Li, W., et al., *Readily Recyclable, High-Performance Catalyst-Free Tung Oil-Derived Vitrimer and Carbon Fiber Reinforced Composites.* 2023.

363.Wang, B., et al., *High-Performance, Biobased, Degradable Polyurethane Thermoset and Its Application in Readily Recyclable Carbon Fiber Composites.* Acs Sustainable Chemistry & Engineering, 2020. **8**(30): p. 11162-11170.

364.Campbell, F.C., *Structural Composite Materials.* 2010.

365.Zhang, X., et al., *Interfacial Microstructure and Properties of Carbon Fiber Composites Modified With Graphene Oxide.* Acs Applied Materials & Interfaces, 2012. **4**(3): p. 1543-1552.

366.Ma, L., et al., *Interfacial Properties and Impact Toughness of Dendritic Hexamethylenetetramine Functionalized Carbon Fiber With Varying Chain Lengths.* RSC Advances, 2014. **4**(74): p. 39156-39166.

367.Wu, Q., et al., *Effect of Sizing on Interfacial Adhesion of Commercial High Strength Carbon Fiber reinforced Resin Composites*Polymer Composites, 2014. **37**(1): p. 254-261.

368.Li, W., et al., *Effect of Plasma Modification on the Mechanical Properties of Carbon Fiber/Phenolphthalein Polyaryletherketone Composites.* Polymer Composites, 2013. **34**(3): p. 368-375.

369.Gao, B., et al., *Interfacial Microstructure and Enhanced Mechanical Properties of Carbon Fiber Composites Caused by Growing Generation 1–4 Dendritic Poly(amidoamine) on a Fiber Surface.* Langmuir, 2016. **32**(33): p. 8339-8349.

370.Han, J., et al., *Oxidation-Resistant ZrB2–SiC Composites at 2200°C.* Composites Science and Technology, 2008. **68**(3-4): p. 799-806.

371.Qu, S. and J. Li, *New Strategy for Preparation of Yttria Powders With Atypical Morphologies and Their Sintering Behavior.* Materials, 2023. **16**(7): p. 2854.

372.Li, J., X.S. Zhang, and P. Song, *Toughening Mechanism and Thermal Shock Resistance of ZrO₂(3Y)/Fe28at-%Al Composites Evaluated by Indentation Techniques.* Advances in Applied Ceramics Structural Functional and Bioceramics, 2014. **113**(7): p. 404-410.

373.Xu, C.H., C.Z. Huang, and X. Ai, *Compositional Design of Multiphase Composite Ceramic Tool Material Based on the Thermal Shock Resistance and Its Application.* Materials Science Forum, 2004. **471-472**: p. 21-25.

374.Florek, P., et al., *Carbon Fiber Reinforced Polymer Composites Doped With Graphene Oxide in Light of Spectroscopic Studies.* Materials, 2021. **14**(8): p. 1835.

375.Hao, Y., et al., *Recent Advances in Properties and Applications of Carbon Fiber-Reinforced Smart Cement-Based Composites.* Materials, 2023. **16**(7): p. 2552.

376.Li, Y.-F., et al., *Mechanical Behavior of Different Fiber Lengths Mix-Proportions Carbon Fiber Reinforced Concrete Subjected to Static, Impact, and Blast Loading.* International Journal of Protective Structures, 2022. **15**(1): p. 43-64.
377.Agarwal, N., et al., *An overview of carbon-carbon composite materials and their applications.* Frontiers in Materials, 2024. **11**: p. 1374034.
378.Hall, J.L., *Columbia and Challenger: Organizational Failure at NASA.* Space Policy, 2016. **37**: p. 127-133.
379.Fahrenthold, E.P. and Y.-K. Park, *Simulation of Foam Impact Effects on Components of the Space Shuttle Thermal Protection System.* 2004.
380.Cockrell, C.E., et al., *NASA Research Center Contributions to Space Shuttle Return to Flight (SSRTF).* 2005.
381.Bayazeid, S.M., et al., *Recovery of Impact-Damaged Carbon Fiber–reinforced Composites Using Induction Heating.* Journal of Composite Materials, 2021. **56**(4): p. 605-618.
382.Thomas, D.A., *Causes of Catastrophic Failure in Complex Systems.* 2010.
383.Li, Y.-F., et al., *Static and Dynamic Performances of Chopped Carbon-Fiber-Reinforced Mortar and Concrete Incorporated With Disparate Lengths.* Materials, 2021. **14**(4): p. 972.
384.Luo, Y., *Study on Three Major Carbon Fiber Composites.* Applied and Computational Engineering, 2023. **7**(1): p. 240-246.
385.Koncherry, V., P. Potluri, and A. Fernando, *Multifunctional Carbon Fibre Tapes for Automotive Composites.* Applied Composite Materials, 2016. **24**(2): p. 477-493.
386.Asp, L., et al., *A Structural Battery and Its Multifunctional Performance.* Advanced Energy and Sustainability Research, 2021. **2**(3).
387.Kumar, P. and J.T. Mahavar, *Emerging Trends in Composite and Advanced Composite Materials Technology for Oil and Gas Industry.* 2023: p. 73-79.
388.Rajak, D.K., et al., *Fiber-Reinforced Polymer Composites: Manufacturing, Properties, and Applications.* Polymers, 2019. **11**(10): p. 1667.
389.Li, Q., et al., Fabrication and properties of 3-D Cf/ZrB2–ZrC–SiC composites via polymer infiltration and pyrolysis. Journal of the American Ceramic Society, 2012. **95**(4): p. 1216-1219.
390.Nisar, A., et al., *Unveiling Enhanced Oxidation Resistance and Mechanical Integrity of Multicomponent Ultra high Temperature Carbides*Journal of the American Ceramic Society, 2021. **105**(4): p. 2500-2516.
391.Wang, J.P., et al., *Aqueous Gelcasting Synthesis of Zirconium Diboride.* Key Engineering Materials, 2014. **633**: p. 31-34.

392.Neuman, E.W. and G. Hilmas, *Strength of Zirconium Diboride to 2300°C.* Journal of the American Ceramic Society, 2012. **96**(1): p. 47-50.

393.Gangireddy, S., S.N. Karlsdóttir, and J.W. Halloran, *Liquid Oxide Flow During Oxidation of Zirconium Diboride-Silicon Carbide Ultra High Temperature Ceramics.* Key Engineering Materials, 2010. **434-435**: p. 144-148.

394.Williams, P.A., et al., *Oxidation of ZrB2–SiC Ultra-High Temperature Composites Over a Wide Range of SiC Content.* Journal of the European Ceramic Society, 2012. **32**(14): p. 3875-3883.

395.Walker, L.S., W.R. Pinc, and E.L. Corral, *Powder Processing Effects on the Rapid Low Temperature Densification of <scp><scp>ZrB₂–SiC</Scp></Scp> Ultra High Temperature Ceramic Composites Using Spark Plasma Sintering.* Journal of the American Ceramic Society, 2011. **95**(1): p. 194-203.

396.Sani, E., et al., *Optical Properties of Dense Zirconium and Tantalum Diborides for Solar Thermal Absorbers.* Renewable Energy, 2016. **91**: p. 340-346.

397.Zhang, S.C., G. Hilmas, and W.G. Fahrenholtz, *Mechanical Properties of Sintered ZrB2–SiC Ceramics.* Journal of the European Ceramic Society, 2011. **31**(5): p. 893-901.

398.Jalaly, M., F.J. Gotor, and M.J. Sayagués, *Self propagating Mechanosynthesis of HfB₂ Nanoparticles by a Magnesiothermic Reaction.* Journal of the American Ceramic Society, 2017. **101**(4): p. 1412-1419.

399.Gasch, M.J., S.M. Johnson, and J. Marschall, *Thermal Conductivity Characterization of Hafnium Diboride Based Ultra High Temperature Ceramics.* Journal of the American Ceramic Society, 2008. **91**(5): p. 1423-1432.

400.Verma, V., В.В. Чеверикин, and R.C. Cozza, *Review: Effect on Physical, Mechanical, and Wear Performance of ZrB₂ based Composites Processed With or Without Additives.* International Journal of Applied Ceramic Technology, 2020. **17**(6): p. 2509-2532.

401.Monteverde, F., et al., *Plasma Wind Tunnel Testing of Ultra-High Temperature ZrB2–SiC Composites Under Hypersonic Re-Entry Conditions.* Journal of the European Ceramic Society, 2010. **30**(11): p. 2313-2321.

402.Mukherjee, R. and B. Basu, *Opportunities and Challenges in Processing and Fabrication of Ultra High Temperature Ceramics for Hypersonic Space Vehicles: A Case Study With ZrB₂–SiC.* Advances in Applied Ceramics Structural Functional and Bioceramics, 2018. **117**(sup1): p. s2-s8.

403.Zhang, H., et al., *Properties of ZrB₂–SiC Ceramics by Pressureless Sintering.* Journal of the American Ceramic Society, 2009. **92**(7): p. 1599-1602.

404.Savino, R., et al., *Aero-Thermo-Chemical Characterization of Ultra-High-Temperature Ceramics for Aerospace Applications.* Journal of the European Ceramic Society, 2018. **38**(8): p. 2937-2953.

405.Karlsdóttir, S.N. and J.W. Halloran, *Rapid Oxidation Characterization of Ultra High Temperature Ceramics*Journal of the American Ceramic Society, 2007. **90**(10): p. 3233-3238.

406.Squire, T.H. and J. Marschall, *Material Property Requirements for Analysis and Design of UHTC Components in Hypersonic Applications.* Journal of the European Ceramic Society, 2010. **30**(11): p. 2239-2251.

407.Cheng, T., et al., *Heat Transfer and Failure Mode Analyses of Ultra-high-Temperature Ceramic Thermal Protection System of Hypersonic Vehicles.* Mathematical Problems in Engineering, 2014. **2014**: p. 1-11.

408.Ni, D.W., et al., *Advances in Ultra-High Temperature Ceramics, Composites, and Coatings.* Journal of Advanced Ceramics, 2021. **11**(1): p. 1-56.

409.Savino, R., S. Mungiguerra, and G.D.D. Martino, *Testing Ultra-High-Temperature Ceramics for Thermal Protection and Rocket Applications.* Advances in Applied Ceramics Structural Functional and Bioceramics, 2018. **117**(sup1): p. s9-s18.

410.Wang, H., et al., *Synthesis and Characterization of a Novel Precursor derived ZrC/ZrB2 Ultra high temperature Ceramic Composite* Applied Organometallic Chemistry, 2012. **27**(2): p. 79-84.

411.Larrimbe, L., et al., *High Heat Flux Laser Testing of HfB₂ Cylinders.* Journal of the American Ceramic Society, 2016. **100**(1): p. 293-303.

412.Dwivedi, A., M.M. Khan, and H.S. Pali, *Numerical Analysis of Microchannel Heat Sink Composed of SiC and CNT Reinforced ZrB2 Composites.* Journal of Engineering Research, 2022.

413.Rueschhoff, L.M., et al., *Processing of Fiber Reinforced Ultra High Temperature Ceramic Composites: A Review.* International Journal of Ceramic Engineering & Science, 2019. **2**(1): p. 22-37.

414.Bai, Y., et al., *Efficient Multiscale Strategy for Toughening HfB₂ Ceramics: A Heterogeneous Ceramic–metal Layered Architecture.* Journal of the American Ceramic Society, 2020. **104**(4): p. 1841-1851.

415.Kwak, T.S., *Trends of Removal Machining Technology of Silicon Nitride Ceramics.* Juniper Online Journal Material Science, 2023. **8**(1).

416.Ho, C.-Y., et al., *Study on Cutting of Laser-Preheated Mullite.* Applied Mechanics and Materials, 2011. **66-68**: p. 2142-2147.

417.Sciti, D., et al., *Properties of Large Scale Ultra-High Temperature Ceramic Matrix Composites Made by Filament Winding and Spark Plasma Sintering.* Composites Part B Engineering, 2021. **216**: p. 108839.

418.Zhang, Z., et al., *Fabrication and Mechanical Properties of Self-Toughening ZrB2–SiC Composites From in-Situ Reaction.* Journal of Advanced Ceramics, 2019. **8**(4): p. 527-536.

419.Zoli, L., et al., *On the Thermal Shock Resistance and Mechanical Properties of Novel Unidirectional UHTCMCs for Extreme Environments.* Scientific Reports, 2018. **8**(1).

420.Mohammad Bagheri, S., et al., *Numerical Optimization of Sample and Die Geometric Parameters to Increase the Attainable Temperature During Spark Plasma Sintering of TiC Ceramics.* Synthesis and Sintering, 2023. **3**(4).

421.Zhang, Y., et al., *Optimal Preparation of High-Entropy Boride-Silicon Carbide Ceramics.* Journal of Advanced Ceramics, 2020. **10**(1): p. 173-180.

422.Zapata-Solvas, E., et al., *Effect of Oxidation on Room Temperature Strength of ZrB₂- And HfB₂- Based Ultra-High Temperature Ceramics.* Advances in Applied Ceramics Structural Functional and Bioceramics, 2015. **114**(8): p. 407-417.

423.Nisar, A. and K. Balani, *Phase and Microstructural Correlation of Spark Plasma Sintered HfB2-ZrB2 Based Ultra-High Temperature Ceramic Composites.* Coatings, 2017. **7**(8): p. 110.

424.Wang, P., et al., *Polycrystalline ZrB₂coating Prepared on Graphite by Chemical Vapor Deposition.* Physica Status Solidi (B), 2016. **253**(8): p. 1590-1595.

425.Fahrenholtz, W.G., *Ultra-High Temperature Ceramics: Materials for Extreme Environments.* Scripta Materialia, 2017. **129**: p. 94-99.

426.Feng, B., et al., *High temperature Phase and Microstructure Evolution of Polymer derived SiZrCN and SiZrBCN Ceramic Nanocomposites*Journal of the American Ceramic Society, 2020. **103**(12): p. 7001-7013.

427.Gaydaychuk, A., et al., *Effect of Al Addition on the Oxidation Resistance of HfC Thin Films.* Coatings, 2021. **12**(1): p. 27.

428.Wang, Y., *Processing and Properties of High Entropy Carbides.* Advances in Applied Ceramics Structural Functional and Bioceramics, 2021. **121**(2): p. 57-78.

429.Garino, G., *Fracture Strength of Multi-Component Ultra-High Temperature Carbides.* 2022.

430.Peng, F., et al., *Simultaneous Hardening and Toughening of a High entropy (NbTaZrW)C Ceramic Carbide Using SiC Particle.* Journal of the American Ceramic Society, 2023. **106**(7): p. 4443-4454.

431.Yee, C., et al., *In Situ Ablation Recession Sensor for Ablative Materials Based on Ultraminiature Thermocouples.* Journal of Spacecraft and Rockets, 2014. **51**(6): p. 1789-1796.

432.Staggs, J.E.J., *Important Parameter Groups in Thermal Protection of Polymers.* Materials, 2015. **8**(8): p. 4679-4698.

433.Liu, S., B. Zhang, and W. Xie, *The Effects of Insulation Thickness on Heat Transfer for Thermal Protection System.* 2010.

434.Jackson, K., M. Gruber, and S. Buccellato, *Mach 6–8+ Hydrocarbon-Fueled Scramjet Flight Experiment: The HIFiRE Flight 2 Project.* Journal of Propulsion and Power, 2015. **31**(1): p. 36-53.

435.Gong, C., et al., *An Approach for Stress Analysis of Corrugated-Core Integrated Thermal Protection System Under Thermal and Mechanical Environment.* Composite Structures, 2018. **185**: p. 1-26.

436.Mansour, N.N., et al., *Flow Mechanics in Ablative Thermal Protection Systems.* Annual Review of Fluid Mechanics, 2024. **56**(1): p. 549-575.

437.Bárcena, J., et al., *Infra-Red and Vibration Tests of Hybrid Ablative/Ceramic Matrix Technological Breadboards for Earth Re-Entry Thermal Protection Systems.* Acta Astronautica, 2017. **134**: p. 85-97.

438.Rivier, M., J. Lachaud, and P.M. Congedo, *Ablative Thermal Protection System Under Uncertainties Including Pyrolysis Gas Composition.* Aerospace Science and Technology, 2019. **84**: p. 1059-1069.

439.Seraji, M.M. and A.L. Ahmad, *Thermal Ablation insulation Performance, Microstructural, and Mechanical Properties of Carbon Aerogel Based Lightweight Heat Shielding Composites.* Polymer Engineering & Science, 2021. **61**(5): p. 1338-1352.

440.Poloni, E., et al., *An Open Carbon–phenolic Ablator for Scientific Exploration.* Scientific Reports, 2023. **13**(1).

441.Yan, X., et al., *Assessment of a 3D Ablation Material Response Model for Lightweight Quartz Fiber Reinforced Phenolic Composite.* Polymer Composites, 2022. **43**(11): p. 8341-8355.

442.Li, W., et al., *Design of Ablation Resistant/Heat Insulation/Lightweight Integrated Thermal Protection Material for Extreme Aerothermodynamic Environment.* Polymer Composites, 2021. **42**(12): p. 6749-6763.

443.Sun, Y., et al., *Preparation and Properties of Phenolic Resin Impregnated Quartz Fiber Tile Ablative Composite.* Key Engineering Materials, 2016. **697**: p. 428-432.

444.Triantou, K., et al., *Novel Hybrid Ablative/Ceramic Layered Composite for Earth Re-Entry Thermal Protection: Microstructural and Mechanical Perfor-*

mance. Journal of Materials Engineering and Performance, 2015. **24**(4): p. 1452-1461.

445.Wang, H., et al., *Dual-Matrix Fiber-Reinforced Lightweight Ablative Thermal Protection Composites With Outstanding Micron-Scale Ablative Recession.* 2023.

446.Li, W., et al., *Evaluation Method and Key Factor Analysis for Thermal Protection Performance of Multifunctional Integrated Ablative Materials.* Polymer Composites, 2020. **41**(12): p. 5043-5058.

447.Ren, H., et al., *Experimental and Numerical Research on High temperature Thermal Insulation Performance of Lightweight Porous Ceramic/Nanomaterial Composite Structure.* Materialwissenschaft Und Werkstofftechnik, 2019. **50**(12): p. 1525-1536.

448.Li, W., et al., *Effects of Gradient Density on Effective Heat Capacity of Charring Ablative Material for Re-Entry Vehicles.* International Journal of Numerical Methods for Heat &Amp Fluid Flow, 2015. **25**(3): p. 472-483.

449.Zuo, H., et al., *Advances in Ablation or Oxidation Mechanisms and Behaviors of Carbon Fiber-Reinforced Si-Based Composites.* Molecules, 2023. **28**(16): p. 6022.

450.Tian, Y., H. Huang, and X. Xu, *Curing Kinetics and Curing Process of Phenolic Impregnated Carbon Ablator.* Journal of Applied Polymer Science, 2017. **134**(42).

451.Liu, X., et al., *Lightweight, Flexible, and Heat insulated Phenolic Impregnated Carbon Ablator (PICA) With Adjustable Flexibility and High Compressive Resilience Property.* Journal of Applied Polymer Science, 2021. **139**(9).

452.Mahzari, M., et al., *Preliminary Analysis of the Mars Science Laboratory's Entry Aerothermodynamic Environment and Thermal Protection System Performance.* 2013.

453.Mahzari, M., et al., *Inverse Estimation of the Mars Science Laboratory Entry Aeroheating and Heatshield Response.* Journal of Spacecraft and Rockets, 2015. **52**(4): p. 1203-1216.

454.Martin, A. and I.D. Boyd, *Assessment of Carbon-Phenolic-in-Air Chemistry Models for Atmospheric Reentry.* 2010.

455.Milos, F.S. and Y.-K. Chen, *Ablation and Thermal Response Property Model Validation for Phenolic Impregnated Carbon Ablator.* Journal of Spacecraft and Rockets, 2010. **47**(5): p. 786-805.

456.Kumar, C.V. and B. Kandasubramanian, *Advances in Ablative Composites of Carbon Based Materials: A Review.* Industrial & Engineering Chemistry Research, 2019. **58**(51): p. 22663-22701.

457.Dong, W., et al., *Fabrication of a Continuous Carbon Fiber-Reinforced Phenolic Resin Composites via in Situ-Curing 3D Printing Technology.* Composites Communications, 2023. **38**: p. 101497.
458.Kazano, S., et al., *Experimental and Analytical Investigation on Resin Impregnation Behavior in Continuous Carbon Fiber Reinforced Thermoplastic Polyimide Composites.* Mechanics of Advanced Materials and Modern Processes, 2018. **4**(1).
459.Wongsriraksa, P. and A. Nakai, *The Effect of Process Parameters on Impregnation Properties of Pultruded Carbon Fiber Reinforced Thermoplastic Epoxy Composites.* Ecs Transactions, 2018. **88**(1): p. 177-185.
460.Wang, R., et al., *Optimizing Processing Parameters of Pin-Assisted-Melt Impregnation of Fiber-Reinforced Plastics by Numerical Simulation.* Journal of Reinforced Plastics and Composites, 2012. **31**(11): p. 731-737.
461.Liu, B., A. Xu, and L. Bao, *Preparation of Carbon Fiber-Reinforced Thermoplastics With High Fiber Volume Fraction and High Heat-Resistant Properties.* Journal of Thermoplastic Composite Materials, 2015. **30**(5): p. 724-737.
462.Bogoeva Gaceva, G., et al.*Ultrasound Assisted Process for Enhanced Interlaminar Shear Strength of Carbon Fiber/Epoxy Resin Composites.* Macedonian Journal of Chemistry and Chemical Engineering, 2010. **29**(2): p. 149.
463.Dong, W., et al., *Interfacial Performance of Phenolic sized Continuous Carbon Fiber reinforced Phenolic Resin Composites With Different Impregnation Nozzle Diameters via <scp>3D</Scp> Printing.* Polymer Composites, 2023. **44**(12): p. 9063-9073.
464.Guadagno, L., et al., *Development of Multifunctional Carbon Fiber Reinforced Composites (CFRCs) - Manufacturing Process.* 2014.
465.Watanabe, K., et al., *Operational Characteristics of a Resistance Circuit Ablation Sensor for Ablative Heatshield Systems.* Transactions of the Japan Society for Aeronautical and Space Sciences Aerospace Technology Japan, 2014. **12**(ists29): p. Po_2_7-Po_2_12.
466.White, T.R., et al., *Proposed Analysis Process for Mars Science Laboratory Heat Shield Sensor Plug Flight Data.* 2011.
467.Karthiselva, N.S. and S.R. Bakshi, *Reactive Spark Plasma Sintering and Mechanical Properties of Zirconium Diboride–Titanium Diboride Ultrahigh Temperature Ceramic Solid Solutions.* Technologies, 2016. **4**(3): p. 30.
468.Orrù, R. and G. Cao, *Comparison of Reactive and Non-Reactive Spark Plasma Sintering Routes for the Fabrication of Monolithic and Composite Ultra High Temperature Ceramics (UHTC) Materials.* Materials, 2013. **6**(5): p. 1566-1583.

469.Purwar, A., et al., *Development of ZrB₂–SiC–Ti by Multi Stage Spark Plasma Sintering at 1600°C.* Journal of the Ceramic Society of Japan, 2016. **124**(4): p. 393-402.

470.Sun, S.K., et al., *Thickness Effects on the Sinterability, Microstructure, and Nanohardness of SiC based Ceramics Consolidated by Spark Plasma Sintering.* Journal of the American Ceramic Society, 2023. **107**(2): p. 777-784.

471.Harimkar, S.P., T. Borkar, and A.K. Singh, *Spark Plasma Sintering of Amorphous-Crystalline Laminated Composites.* Materials Science and Engineering A, 2011. **528**(3): p. 1901-1905.

472.Laszkiewicz-Łukasik, J., et al., *Spark Plasma Sintering/Field Assisted Sintering Technique as a Universal Method for the Synthesis, Densification and Bonding Processes for Metal, Ceramic and Composite Materials.* Journal of Applied Materials Engineering, 2021. **60**(2): p. 53-69.

473.Zavjalov, A.P., Н.П. Шапкин, and Е.К. Папынов, *Spark Plasma Sintering of SiC Ceramics Based on Natural Renewable Raw Materials.* Materials Science Forum, 2020. **992**: p. 759-763.

474.Stuer, M., P. Bowen, and Z. Zhao, *Spark Plasma Sintering of Ceramics: From Modeling to Practice.* Ceramics, 2020. **3**(4): p. 476-493.

475.Huang, J., et al., *Components of Thermal Protection Mechanism of Carbon/Phenolic Composites Exposed to Severe Aerodynamic Heat.* Polymer Composites, 2022. **44**(1): p. 136-147.

476.Purwar, A., T. Venkateswaran, and B. Basu, *Experimental and Computational Analysis of Thermo oxidative structural Stability of ZrB₂–SiC–Ti During Arc jet Testing*Journal of the American Ceramic Society, 2017. **100**(10): p. 4860-4873.

477.Wagner, A., et al., *Pressure-Assisted Sintering and Characterization of Nd:YAG Ceramic Lasers.* Scientific Reports, 2021. **11**(1).

478.Zhang, Z.H., et al., *A New Rapid Route for <i>In Situ</I> Synthesizing Monolithic <scp>TiB</Scp> Ceramic.* Journal of the American Ceramic Society, 2011. **94**(9): p. 2754-2756.

479.Paydayesh, A., M. Kokabi, and A.R. Bahramian, *High Temperature Ablation of Highly Filled Polymer layered Silicate Nanocomposites*Journal of Applied Polymer Science, 2012. **127**(4): p. 2776-2785.

480.Butler, B.D., et al., *Characterization of Candidate Materials for Remote Recession Measurements of Ablative Heat Shield Materials.* 2016.

481.Kong, D., et al., *Preparation and Characteristic of the Novel Multiple-Layer Thermal Insulation Nanocomposite Materials.* Acs Sustainable Chemistry & Engineering, 2023. **11**(30): p. 11067-11076.

482.Lu, S., et al., *A Numerical Study of Crack Shielding/Anti-Shielding in Layered Architectures.* International Journal of Fatigue, 2019. **124**: p. 503-519.
483.Burke-Veliz, A., S. Syngellakis, and P.A.S. Reed, *Fatigue Crack Shielding and Deflection in Plain Bearings Under Large-Scale Yielding.* Engineering Failure Analysis, 2010. **17**(3): p. 648-657.
484.Zhang, X., et al., *Effects of Particle Diameter and Plate Thickness on Transpiration Cooling for Double-Layer Porous Plates.* Frontiers in Energy Research, 2022. **10**.
485.Kajal, K., *China's new cooling device increases hypersonic missile flight time to reach US.* 2024, Interesting Engineering.
486.Liu, Y. and Z. Jiang, *Concept of Non-Ablative Thermal Protection System for Hypersonic Vehicles.* Aiaa Journal, 2013. **51**(3): p. 584-590.
487.Liu, L., Q. Guo, and T. He, *Thermal-Acoustic Fatigue of a Multilayer Thermal Protection System in Combined Extreme Environments.* Advances in Mechanical Engineering, 2014. **6**: p. 176891.
488.Wu, D., et al., *Thermal Protection Performance of Metallic Honeycomb Core Panel Structures in Non-Steady Thermal Environments.* Experimental Heat Transfer, 2014. **29**(1): p. 53-77.
489.Rong, Y., *Heat-Balance Thermal Protection With High Thermal Conductivity Materials for Hypersonic Vehicle.* Destech Transactions on Engineering and Technology Research, 2017(apetc).
490.Li, W., et al., *Nonlinear Pyrolysis Layer Model for Thermal Behavior of Nonhomogeneous Charring Materials.* Journal of Applied Polymer Science, 2015. **132**(31).
491.East, R., J. Edwards, and U. DERA, *Hypersonic Missiles–Some Aerothermodynamic Problem Areas.* 2000.
492.Du, Z., Z. Wan, and C. Yang, *Robust Aeroelastic Design Optimization of Hypersonic Wings Considering Uncertainty in Heat Flux.* Transactions of the Japan Society for Aeronautical and Space Sciences, 2017. **60**(3): p. 152-163.
493.Weaver, A., et al., *Flowfield Uncertainty Analysis for Hypersonic CFD Simulations.* 2010.
494.Rêgo, I., et al., *Calculation of the Vehicle Drag and Heating Reduction at Hypervelocites With Laser-Induced Air Spike.* Journal of Aerospace Technology and Management, 2013. **5**(1).
495.ÖZel, C., C.K. MacİT, and M. ÖZel, *Investigation of Flight Performance of Notched Delta Wing Rockets on Different Types of Nose Cones.* Turkish Journal of Science and Technology, 2023. **18**(2): p. 435-447.

496.Reimer, T., et al., *Transpiration Cooling Tests of Porous CMC in Hypersonic Flow.* 2011.

497.Narayan, A., S. Narayanan, and R. Kumar, *Hypersonic Flow Past Nose Cones of Different Geometries: A Comparative Study.* Simulation, 2017. **94**(8): p. 665-680.

498.Watkins, T., et al., *Hypersonic Flow Over Closed and Open Nose Missile Bodies: Raw and SVD-Enhanced Schlieren Imaging, Numerical Modeling, and Physical Analysis.* 2023.

499.Iranmanesh, R., et al., *Numerical Investigation of Compressible Flow Around Nose Cone With Multi-Row Disk and Multi Coolant Jets.* Scientific Reports, 2023. **13**(1).

500.Lin, J., L.J. Zhao, and J. Wang, *Numerical Investigation on Aero-Thermodynamic Characteristics of Wedge Shaped Nose Cone of Near Space Supersonic Flight.* Applied Mechanics and Materials, 2014. **541-542**: p. 608-612.

501.Kim, B.K. and A.S.M. Al-Obaidi, *Investigation of the Effect of Nose Shape and Geometry at Supersonic Speeds for Missile Performance Optimisation.* Journal of Physics Conference Series, 2023. **2523**(1): p. 012010.

502.Tahsini, A.M., *Parametric Study of the Effect of Exothermic Coating Properties on Blunt Nose Drag Reduction in Hypersonic Flights.* Proceedings of the Institution of Mechanical Engineers Part G Journal of Aerospace Engineering, 2013. **228**(9): p. 1461-1465.

503.Fan, D., et al., *Spike Effects on Drag Reduction for Hypersonic Lifting Body.* Journal of Spacecraft and Rockets, 2017. **54**(6): p. 1185-1195.

504.Ma, P., T. Chao, and M. Yang, *Modeling and Simulation Methodology of Multifield Coupling for Hypersonic Vehicle.* 2012: p. 177-184.

505.Qin, D., Q. Zhu, and C. Wang, *Dynamic Characteristic and Control of a Hypersonic Flight Vehicle.* 2013.

506.Xiao, Y., et al., *Preliminary Study of Aerodynamic Performance for Waverider-Based Hypersonic Vehicles With Dorsal Mounted Engines.* 2017.

507.Bao, C., P. Wang, and G. Tang, *Integrated Guidance and Control for Hypersonic Morphing Missile Based on Variable Span Auxiliary Control.* International Journal of Aerospace Engineering, 2019. **2019**: p. 1-20.

508.Lewis, M.J., I.D. Boyd, and C.E. Cockrell, *Aerodynamics for Optimal Engine-Integrated Airbreathing Launcher Configurations.* 2004.

509.Sura, J., et al., *Estimation of Skin Friction Drag on a Model in Hypersonic Shock Tunnel.* Applied Mechanics and Materials, 2012. **232**: p. 234-239.

510.Rea, J.R., *Orion Entry Performance-Based Center-of-Gravity Box.* 2010.

511.Ma, Y., et al., *Hypersonic Lifting Body Aerodynamic Shape Optimization Based on the Multiobjective Evolutionary Algorithm Based on Decompo-*

sition. Proceedings of the Institution of Mechanical Engineers Part G Journal of Aerospace Engineering, 2014. **229**(7): p. 1246-1266.
512.Cui, K., D. Zhao, and G. Yang, *Waverider Configurations Derived From General Conical Flowfields.* Acta Mechanica Sinica, 2007. **23**(3): p. 247-255.
513.Żaba, K., et al., *Comparative Analysis of Properties and Microstructure of the Plastically Deformed Alloy Inconel®718, Manufactured by Plastic Working and Direct Metal Laser Sintering.* Archives of Metallurgy and Materials, 2016. **61**(1): p. 143-148.
514.Joseph, C., C. Persson, and M.H. Colliander, *Influence of Heat Treatment on the Microstructure and Tensile Properties of Ni-Base Superalloy Haynes 282.* Materials Science and Engineering A, 2017. **679**: p. 520-530.
515.Sugahara, T., et al., *Creep Behavior of the Inconel 718 Superalloy.* Defect and Diffusion Forum, 2012. **326-328**: p. 509-514.
516.Huang, K., et al., *High Temperature Oxidation and Thermal Shock Properties of La2Zr2O7 Thermal Barrier Coatings Deposited on Nickel-Based Superalloy by Laser-Cladding.* Coatings, 2020. **10**(4): p. 370.
517.Mudgal, D., S. Singh, and S. Prakash, *High Temperature Cyclic Oxidation Behavior of Ni and Co Based Superalloys.* Journal of Minerals and Materials Characterization and Engineering, 2012. **11**(03): p. 211-219.
518.Tomczak, A., *Influence of Casting Parameters on Fatigue Properties of the Żs6u-Wi Alloy.* Scientific Letters of Rzeszow University of Technology - Mechanics, 2021: p. 53-64.
519.Hilal, H., et al., *High Temperature Mechanical Deformation of an Additive Manufactured Nickel Based Superalloy Using Small Scale Test Methods.* Ubiquity Proceedings, 2018. **1**(S1): p. 21.
520.Xuan, L., et al., *Multi-Scale Mechanical-Thermal Coupling Analysis of Multi-Layer Thermal Protection Structure of Ceramic Matrix Composites.* Journal of Physics Conference Series, 2024. **2707**(1): p. 012109.
521.Liu, J., et al., *Heat Transfer and Flow Structures of Supercritical N-Decane in a Regenerative Cooling Channel Loaded With Non-Uniform Heat Flux.* Frontiers in Energy Research, 2022. **10**.
522.Qin, J., et al., *Irreversible Cycle Analysis of a Re-Cooled Cycle for a Scramjet.* Proceedings of the Institution of Mechanical Engineers Part G Journal of Aerospace Engineering, 2010. **224**(8): p. 919-926.
523.Bao, W., J. Qin, and W. Zhou, *Heat Transfer Characteristic Modelling and the Effect of Operating Conditions on Re-Cooled Cycle for a Scramjet.* The Aeronautical Journal, 2011. **115**(1164): p. 83-90.

524.Wang, Y., S. Li, and M. Dong, *Numerical Study on Heat Transfer Deterioration of Supercritical N-Decane in Horizontal Circular Tubes.* Energies, 2014. **7**(11): p. 7535-7554.

525.Tao, Z., et al., *Numerical Study of Flow and Heat Transfer of <i>n</I>-Decane With Pyrolysis and Pyrolytic Coking Under Supercritical Pressures.* Energy & Fuels, 2017. **31**(8): p. 8698-8707.

526.Luo, S., et al., *A Review of Regenerative Cooling Technologies for Scramjets.* Applied Thermal Engineering, 2021. **190**: p. 116754.

527.Qin, J., et al., *Performance Cycle Analysis of an Open Cooling Cycle for a Scramjet.* Proceedings of the Institution of Mechanical Engineers Part G Journal of Aerospace Engineering, 2009. **223**(6): p. 599-607.

528.Zhong, F., et al., *Heat Transfer of Aviation Kerosene at Supercritical Conditions.* Journal of Thermophysics and Heat Transfer, 2009. **23**(3): p. 543-550.

529.Jiang, J., et al., *Regeneratively Cooled Scramjet Heat Transfer Calculation and Comparison With Experimental Data.* Proceedings of the Institution of Mechanical Engineers Part G Journal of Aerospace Engineering, 2013. **228**(8): p. 1227-1234.

530.Pan, Y., et al., *Supercritical Pyrolysis and Coking of JP-10 in Regenerative Cooling Channels.* Energy & Fuels, 2020. **34**(2): p. 1627-1638.

531.Carpenter, C., S.B. Verma, and J. Kapat, *Numerical Study of Enhancement of Regenerative Cooling Using Ribs.* 2013.

532.Ning, W., Y. Pan, and J. Zhang, *Research Status of Active Cooling of Endothermic Hydrocarbon Fueled Scramjet Engine.* Proceedings of the Institution of Mechanical Engineers Part G Journal of Aerospace Engineering, 2012. **227**(11): p. 1780-1794.

533.Zhao, J., et al., *Numerical Analysis of Flow and Heat Transfer Characteristics of Supercritical Endothermic Hydrocarbon Fuel in Taper Tube.* Journal of Physics Conference Series, 2023. **2633**(1): p. 012011.

534.Li, L., et al., *Effect of Dimple Depth-Diameter Ratio on the Flow and Heat Transfer Characteristics of Supercritical Hydrocarbon Fuel in Regenerative Cooling Channel.* International Journal of Aerospace Engineering, 2021. **2021**: p. 1-9.

535.Sun, A., et al., *Morphological and Compositional Analysis on Thermal Deposition of Supercritical Aviation Kerosene in Micro Channels.* Molecules, 2023. **28**(11): p. 4508.

536.Rao, P.N. and D. Kunzru, *Thermal Cracking of JP-10: Kinetics and Product Distribution.* Journal of Analytical and Applied Pyrolysis, 2006. **76**(1-2): p. 154-160.

537.Zhang, D., et al., *Quasi-one-dimensional model of scramjet combustor coupled with regenerative cooling.* Journal of Propulsion and Power, 2016. **32**(3): p. 687-697.
538.Hu, J., et al., *A Parametrical Study on Convective Heat Transfer Between High-Temperature Gas and Regenerative Cooling Panel.* Energies, 2021. **14**(6): p. 1784.
539.Trelewicz, J.R., et al., *Heat Flux Measurements in a Scramjet Combustor Using Embedded Direct-Write Sensors.* Journal of Propulsion and Power, 2015. **31**(4): p. 1003-1013.
540.Bao, W., et al., *Hydrogen-Fueled Scramjet Cooling System Investigation Using Combustor and Regenerative Cooling Coupled Model.* Proceedings of the Institution of Mechanical Engineers Part G Journal of Aerospace Engineering, 2013. **228**(6): p. 820-830.
541.Li, X. and Z. Wang, *Study the Performance of an Organic Rankine Cycle for Thermal Absorption and Recovery From the Scramjet Cooling Heat.* International Journal of Aeronautical Science & Aerospace Research, 2019: p. 192-199.
542.Bao, W., et al., *Power Generation and Heat Sink Improvement Characteristics of Recooling Cycle for Thermal Cracked Hydrocarbon Fueled Scramjet.* Science China Technological Sciences, 2011. **54**(4): p. 955-963.
543.Zhou, W., et al., *Deterioration in Heat Transfer of Endothermal Hydrocarbon Fuel.* Journal of Thermal Science, 2011. **20**(2): p. 173-180.
544.Liu, Z.H., et al., *Hydraulic and Thermal Effects of Coke Deposition During Pyrolysis of Hydrocarbon Fuel in a Mini-Channel.* Energy & Fuels, 2012. **26**(6): p. 3672-3679.
545.Gao, M., et al., *Coke Deposition Inhibition for Endothermic Hydrocarbon Fuels in a Reforming Catalyst-Coated Reactor.* Energy & Fuels, 2019. **33**(7): p. 6126-6133.
546.Zhu, Y., et al., *Experimental Investigation of Flow Coking and Coke Deposition of Supercritical Hydrocarbon Fuels in Porous Media.* Energy & Fuels, 2018. **32**(3): p. 2941-2948.
547.Jiang, R., G. Liu, and X. Zhang, *Thermal Cracking of Hydrocarbon Aviation Fuels in Regenerative Cooling Microchannels.* Energy & Fuels, 2013. **27**(5): p. 2563-2577.
548.Xiong, Y., et al., *Quasi-One-Dimensional Model of Hydrocarbon-Fueled Scramjet Combustor Coupled With Regenerative Cooling.* International Journal of Aerospace Engineering, 2022. **2022**: p. 1-14.
549.He, N., et al., *Progress of Coupled Heat Transfer Mechanisms of Regenerative Cooling System in a Scramjet.* Energies, 2023. **16**(3): p. 1025.

550.Liu, J., et al., *Heat Transfer and Flow Structure Characteristics of Regenerative Cooling in a Rectangular Channel Using Supercritical CO2.* Aerospace, 2023. **10**(6): p. 564.
551.Lee, S.-H., et al., *Heat Transfer Characteristics of Kerosene Phase Change Based on Fuel Flow Rates in the Regenerative Cooling Heat Exchanger of Scramjet Engines.* Journal of Mechanical Science and Technology, 2021. **35**(6): p. 2733-2741.
552.Elefteriu, G. and W. Freer, *Hypersonic weapon systems: High expectations.* 2023, Council on Geostrategy.
553.Tracy, C., *The Accuracy of Hypersonic Weapons: Media Claims Miss the Mark.* 2020, Union of Concerned Scientists.
554.Chen, K., et al., *SINS/BDS Integrated Navigation for Hypersonic Boost-Glide Vehicles in the Launch-Centered Inertial Frame.* Mathematical Problems in Engineering, 2020. **2020**: p. 1-16.
555.Jain, S.K., et al., *Rapid and Accurate INS Transfer Alignment for Air Launched Tactical Missile Using Kalman Filter.* Defence Science Journal, 2023. **73**(4): p. 496-502.
556.Gao, P., et al., *A Self-Calibration Method for Tri-Axis Rotational Inertial Navigation System.* Measurement Science and Technology, 2016. **27**(11): p. 115009.
557.Zhang, Q., et al., *Singular Value Decomposition-Based Robust Cubature Kalman Filtering for an Integrated GPS/SINS Navigation System.* Journal of Navigation, 2014. **68**(3): p. 549-562.
558.Reid, Z. and J.N. Ash, *Leveraging 3D Models for SAR-based Navigation in GPS-denied Environments.* 2018.
559.Jurevičius, R., V. Marcinkevičius, and J. Šeibokas, *Robust GNSS-denied Localization for UAV Using Particle Filter and Visual Odometry.* Machine Vision and Applications, 2019. **30**(7-8): p. 1181-1190.
560.Turek, P., S. Grzywiński, and W. Bużantowicz, *Selected Issues and Constraints of Image Matching in Terrain-Aided Navigation: A Comparative Study.* 2021.
561.Chen, R., Q. Zhang, and L. Zhao, *Optimal Selection and Adaptability Analysis of Matching Area for Terrain Aided Navigation.* Iet Radar Sonar & Navigation, 2021. **15**(12): p. 1702-1714.
562.Zhao, L., et al., *A Novel Terrain-Aided Navigation Algorithm Combined With the TERCOM Algorithm and Particle Filter.* Ieee Sensors Journal, 2015. **15**(2): p. 1124-1131.
563.Zhang, T., L. Yao, and J. Tong, *An Autonomous Underwater Vehicle Positioning Matching Method Based on Iterative Closest Contour Point Algorithm*

and Affine Transformation. Proceedings of the Institution of Mechanical Engineers Part M Journal of Engineering for the Maritime Environment, 2016. **231**(3): p. 711-722.

564.Vadlamani, A. and M.U.d. Haag, *Flight Test and Simulation Results of an Integrated Dual Airborne Laser Scanner (DALS)/INS Navigator.* 2010.

565.Zhou, L., X. Cheng, and Y. Zhu, *Terrain Aided Navigation for Autonomous Underwater Vehicles With Coarse Maps.* Measurement Science and Technology, 2016. **27**(9): p. 095002.

566.Wu, L., et al., *Research on the Relative Positions-Constrained Pattern Matching Method for Underwater Gravity-Aided Inertial Navigation.* Journal of Navigation, 2015. **68**(5): p. 937-950.

567.Qingtang, F. and G. Wang, *A Solution to Terrain Elevation Matching Problem Based on HMM.* 2014.

568.Jang, S. and H. Bang, *Acquisition Method for Terrain Referenced Navigation Using Terrain Contour Intersections.* Iet Radar Sonar & Navigation, 2017. **11**(9): p. 1444-1450.

569.Yan, B., et al., *Terrain Matching Based on Adaptive Digital Elevation Map.* 2018.

570.Zheng, H., et al., *Simulation Research on Gravity-Geomagnetism Combined Aided Underwater Navigation.* Journal of Navigation, 2012. **66**(1): p. 83-98.

571.Chen, R., Q. Zhang, and L. Zhao, *Real Time Airborne Terrain Aided Inertial Navigation System in Multi-Resolution Terrain.* 2021: p. 423-433.

572.Zhao, L., *An Algorithm for Terrain-Aided Inertial Navigation Based on Nonlinear Optimal Filtering.* Science China Physics Mechanics and Astronomy, 2011. **54**(6): p. 1083-1088.

573.Lu, S.L., L. Zhao, and C.Y. Zhang, *Improved Tercom Based on Fading Factor.* Applied Mechanics and Materials, 2011. **143-144**: p. 770-774.

574.Wen, C.B., et al., *Gravity Matching Aided Navigation Using a New Mixed Filter Algorithm.* Applied Mechanics and Materials, 2014. **644-650**: p. 2067-2070.

575.Zhao, S., et al., *Improving Matching Accuracy of Underwater Gravity Matching Navigation Based on Iterative Optimal Annulus Point Method With a Novel Grid Topology.* Remote Sensing, 2021. **13**(22): p. 4616.

576.Huh, Y., K. Yu, and J. Heo, *Detecting Conjugate-Point Pairs for Map Alignment Between Two Polygon Datasets.* Computers Environment and Urban Systems, 2011. **35**(3): p. 250-262.

577.Wang, H., et al., *Simulation Research on a Minimum Root-Mean-Square Error Rotation-Fitting Algorithm for Gravity Matching Navigation.* Science China Earth Sciences, 2011. **55**(1): p. 90-97.
578.Wang, D., et al., *Underwater Terrain-Matching Algorithm Based on Improved Iterative Closest Contour Point Algorithm.* Measurement and Control, 2024. **57**(7): p. 893-902.
579.Chen, P., et al., *Underwater Terrain Positioning Method Based on Markov Random Field for Unmanned Underwater Vehicles.* Frontiers in Marine Science, 2023. **10**.
580.Ashrafifar, A. and M.F. Jegarkandi, *Adaptive Fin Failures Tolerant Integrated Guidance and Control Based on Backstepping Sliding Mode.* Transactions of the Institute of Measurement and Control, 2020. **42**(10): p. 1823-1833.
581.Yu, P., Y.B. Shtessel, and C. Edwards, *Continuous Higher Order Sliding Mode Control With Adaptation of Air Breathing Hypersonic Missile.* International Journal of Adaptive Control and Signal Processing, 2016. **30**(8-10): p. 1099-1117.
582.Niu, K., et al., *A New Sliding Mode Control Algorithm of IGC System for Intercepting Great Maneuvering Target Based on EDO.* Sensors, 2022. **22**(19): p. 7618.
583.Nivison, S. and P.P. Khargonekar, *A Sparse Neural Network Approach to Model Reference Adaptive Control With Hypersonic Flight Applications.* 2018.
584.Peters, A.B., et al., *Materials design for hypersonics.* Nature communications, 2024. **15**(1): p. 3328.
585.Li, L., *Optimal Index Shooting Policy for Layered Missile Defense System.* Journal of Systems Engineering and Electronics, 2020: p. 118-129.
586.Lee, K. and K. Oh, *Prediction of Possible Intercept Time by Considering Flight Trajectory of Nodong Missile.* International Journal of Aerospace System Engineering, 2016. **3**(2): p. 14-21.
587.Lin, C.C., et al., *Hybrid Guidance Law for Interception of Ballistic Targets.* Proceedings of the Institution of Mechanical Engineers Part G Journal of Aerospace Engineering, 2006. **220**(5): p. 525-536.
588.Shtessel, Y.B., I.A. Shkolnikov, and A. Levant, *Missile Interceptor Guidance and Control Using Second Order Sliding Modes.* Ifac Proceedings Volumes, 2005. **38**(1): p. 854-859.
589.Garrett, R.K., et al., *Managing the Interstitials, a System of Systems Framework Suited for the Ballistic Missile Defense System.* Systems Engineering, 2010. **14**(1): p. 87-109.

590.Yao, J., Q. Huang, and W. Wang, *Analyzing Ballistic Missile Defense System Effectiveness Based on Functional Dependency Network Analysis.* The Open Cybernetics & Systemics Journal, 2015. **9**(1).
591.Bradley, C., et al., *Missile Defense in the 21st Century Acquisition Environment: Exploring a BMD-Capable LCS Mission Package.* 2013.
592.Liu, S., et al., *Cooperative Guidance Law for Intercepting a Hypersonic Target With Impact Angle Constraint.* The Aeronautical Journal, 2022. **126**(1300): p. 1026-1044.
593.Cheng, Z., F. Liu, and Y. Zhang, *Multi-Agent Decision Support System for Missile Defense Based on Improved PSO Algorithm.* Journal of Systems Engineering and Electronics, 2017. **28**(3): p. 514.
594.Sankaran, J. and S. Fetter, *Defending the United States: Revisiting National Missile Defense Against North Korea.* International Security, 2022. **46**(3): p. 51-86.
595.Fang, F. and Y. Cai, *Optimal Cooperative Guidance With Guaranteed Miss Distance in Three-Body Engagement.* Proceedings of the Institution of Mechanical Engineers Part G Journal of Aerospace Engineering, 2016. **232**(3): p. 492-504.
596.Dou, J., et al., *Evaluation Method of Formation Ship-to-Air Missile Air Defense Area Capability.* Scientific Journal of Technology, 2023. **4**(12): p. 7-12.
597.Na, H. and J.-W. Ahn, *Framework to Operate Multiple Ship Defense Missiles Under Uncertain Evasive Maneuvers of Target.* Proceedings of the Institution of Mechanical Engineers Part G Journal of Aerospace Engineering, 2017. **232**(11): p. 2001-2018.
598.Kim, I., et al., *A Study on the Interception Using Fine Fragments and Particles to Hypersonic Vehicles.* Journal of the Korea Institute of Military Science and Technology, 2024. **27**(2): p. 285-293.
599.Zhang, B., et al., *RBFNN-Based Anti-Input Saturation Control for Hypersonic Vehicles.* Aerospace, 2024. **11**(2): p. 108.
600.Mackey, L.E. and I.D. Boyd, *Analysis of Hypersonic Flow Effects on Sensor Performance.* 2016.
601.Bao, Z., J. Li, and L. Guo, *Simulation and Feature Extraction of the Dynamic Electromagnetic Scattering of a Hypersonic Vehicle Covered With Plasma Sheath.* Remote Sensing, 2020. **12**(17): p. 2740.
602.Cernat, R., *Impactul Tehnologiilor Emergente Și Disruptive Asupra Domeniului Militar.* Gândirea Militară Românească, 2021. **2021**(4): p. 230-245.

603.Nan, Y., et al., *Influencing Factor Analysis of Interception Probability and Classification-Regression Neural Network Based Estimation.* Journal of Systems Engineering and Electronics, 2023. **34**(4): p. 992-1006.

604.Чеков, А.Д., *Transformation of U.S. Missile Defense Policy in the Context of Countering Hypersonic Weapons.* World Economy and International Relations, 2023. **67**(4): p. 40-52.

605.Li, F., et al., *Optimal Penetration Guidance Law for High-Speed Vehicles Against an Interceptor With Modified Proportional Navigation Guidance.* Symmetry, 2023. **15**(7): p. 1337.

606.Fu, B., H. Guo, and J. Yan, *Interception of the Near Space Hypersonic Weapons Based on the Partial Integrated Guidance and Control Method.* 2016.

607.Freeland, S., *Applying the Jus in Bello to Military Uses of Outer Space: A Square Peg in a Round Hole?* 2016: p. 109-122.

608.lakshmi, E.V., N.N. Sastry, and B.P. Rao, *Effect of Errors on Miss Distance of Missile Trackers in Active Decoy Environment.* International Journal of Electrical and Computer Engineering (Ijece), 2019. **9**(6): p. 4696.

609.Wang, Z., et al., *Collaborative Innovation Efficiency and Influencing Factors of Civil-Military Integration Enterprises.* Mathematical Problems in Engineering, 2021. **2021**: p. 1-11.

610.Tortonesi, M., et al., *Leveraging Internet of Things Within the Military Network Environment — Challenges and Solutions.* 2016.

611.Manso, M., et al., *Connecting the Battlespace: C2 and IoT Technical Interoperability in Tactical Federated Environments.* 2022.

612.Jiang, T., Z. Cheng-jie, and H. Zhang, *Research on the Architecture and Governance Mechanisms of a Civil-Military Integration Technology Innovation Network.* 2018.

613.Liu, R., Z. Xianping, and G. Xiang, *Civil-Military Integration and Technical Innovation of Private Listed Companies.* 2019.

614.Alim, E., *Hipersonik Füzeler Ve Stratejik İstikrar: Rusya-Abd İlişkileri Üzerine Bir İnceleme.* Savunma Bilimleri Dergisi, 2021(39): p. 137-177.

615.Singh, S.K., *India-Australia Ties and Emergence of the Quad: The Strategic Cooperation in the Indo-Pacific Region.* Towards Excellence, 2023: p. 344-354.

616.Lee, C.A., *Technology Acquisition and Arms Control: Thinking Through the Hypersonic Weapons Debate.* Texas National Security Review, 2022. **5**(4): p. 29-48.

617.Cooper, C.G., *Ensuring Lawful Use of Autonomous Weapons.* Journal of International Humanitarian Legal Studies, 2024: p. 1-30.

618.Asaro, P., *On Banning Autonomous Weapon Systems: Human Rights, Automation, and the Dehumanization of Lethal Decision-Making.* International Review of the Red Cross, 2012. **94**(886): p. 687-709.

619.Maqbool, A. and A. Anwar, *Warfare and Machines: An in-Depth Study of Autonomous Weapons in the Context of International Humanitarian Law.* 2023. **2**(1): p. 01-14.

620.Rosert, E. and F. Sauer, *Prohibiting Autonomous Weapons: Put Human Dignity First.* Global Policy, 2019. **10**(3): p. 370-375.

621.Banasik, M., *Global Challenges and Threats of Hypersonic Weapons – The Russian Context.* Safety & Defense, 2021. **7**(2): p. 51-64.

622.Walravens, S., et al., *Characteristics of Medical Evacuation by Train in Ukraine, 2022.* Jama Network Open, 2023. **6**(6): p. e2319726.

623.Beier, J.M., *"This Changes Things": Children, Targeting, and the Making of Precision.* Cooperation and Conflict, 2021. **57**(2): p. 210-225.

624.Slade, R., R. Tickner, and P. Wynn-Pope, *Protecting Humanity From the Catastrophic Humanitarian Consequences of Nuclear Weapons: Reframing the Debate Towards the Humanitarian Impact.* International Review of the Red Cross, 2015. **97**(899): p. 731-752.

625.Gazit, N. and R.J. Brym, *State-Directed Political Assassination in Israel: A Political Hypothesis.* International Sociology, 2011. **26**(6): p. 862-877.

626.Chin, J.J., et al., *Reshaping the Threat Environment: Personalism, Coups, and Assassinations.* Comparative Political Studies, 2022. **55**(4): p. 657-687.

627.Molloy, T., *Qassem Soleimani, Targeted Killing of State Actors, and Executive Order 12,333.* Victoria University of Wellington Law Review, 2021. **52**(1): p. 163-196.

628.Faruque, J.A. and F. Durrat, *The Dynamics of the US-Iran Relationship After the Assassination of Qassem Suleimani: Explanations and Predictions.* Social Science Review, 2023. **40**(1): p. 185-201.

629.Pratt, S., *Crossing Off Names: The Logic of Military Assassination.* Small Wars and Insurgencies, 2014. **26**(1): p. 3-24.

630.Gill, T.D., *The <i>Jus Ad Bellum Anno</I> 2040: An Essay on Possible Trends and Challenges.* 2022: p. 25-40.

631.Tannenwald, N., *Life Beyond Arms Control: Moving Toward a Global Regime of Nuclear Restraint &Amp; Responsibility.* Daedalus, 2020. **149**(2): p. 205-221.

632.Evans, C.P., *Creating an Environment for Nuclear Disarmament (CEND): A Good Faith, Effective Measure Pursuant to Article VI NPT or Empty Gesturing?* Cambridge International Law Journal, 2020. **9**(2): p. 201-224.

633.Wilkening, D.A., *Hypersonic Weapons and Strategic Stability.* Survival, 2019. **61**(5): p. 129-148.

634.Czajkowski, M., *Hypersonic Weapons – Selected Political and Strategic Issues.* Politeja, 2022. **19**(4 (79)).

635.Hu, Y., et al., *Novel Trajectory Prediction Algorithms for Hypersonic Gliding Vehicles Based on Maneuver Mode on-Line Identification and Intent Inference.* Measurement Science and Technology, 2021. **32**(11): p. 115012.

636.Guo, J., S.-H. Yang, and Z. Guo, *Robust Tracking for Hypersonic Vehicles Subjected to Mismatched Uncertainties via Fixed-Time Sliding Mode Control.* Proceedings of the Institution of Mechanical Engineers Part G Journal of Aerospace Engineering, 2021. **235**(14): p. 2145-2153.

637.George, K., et al., *Recent Developments in Elastomeric Heat Shielding Materials for Solid Rocket Motor Casing Application for Future Perspective.* Polymers for Advanced Technologies, 2017. **29**(1): p. 8-21.

638.Rigatos, G., et al., *Nonlinear Optimal Control for Autonomous Hypersonic Vehicles.* Aerospace Systems, 2019. **2**(2): p. 197-213.

639.Lee, G.S. and T. Lee, *Fluid and Combustion Dynamics in Dual-Mode Scramjets.* Frontiers in Aerospace Engineering, 2023. **1**.

640.Isono, T., et al., *Arrangement of Rocket Engine Location in a Combined Cycle Engine.* Transactions of the Japan Society for Aeronautical and Space Sciences Aerospace Technology Japan, 2019. **17**(2): p. 165-174.

641.Zhao, D., et al., *Effect of Different Particle Sizes on Combustion Characteristics of DCR Engines.* Journal of Physics Conference Series, 2024. **2746**(1): p. 012012.

642.Chen, B., et al., *Engineering C_f/ZrB₂ SiC Y₂O₃ for Thermal Structures of Hypersonic Vehicles With Excellent Long Term Ultra-high Temperature Ablation Resistance.* Advanced Science, 2023. **10**(34).

643.Hall, S.E., et al., *Mechanical Properties of High-Temperature Fiber-Reinforced Thermoset Composites With Plain Weave and Unidirectional Carbon Fiber Fillers.* Journal of Composites Science, 2022. **6**(7): p. 213.

644.Zeren, A., *Effect of the Graphite Content on the Tribological Properties of Hybrid Al/SiC/Gr Composites Processed by Powder Metallurgy.* Industrial Lubrication and Tribology, 2015. **67**(3): p. 262-268.

645.Kausar, A., et al., *Self-Healing Nanocomposites—Advancements and Aerospace Applications.* Journal of Composites Science, 2023. **7**(4): p. 148.

646.Blaiszik, B.J., et al., *Self-Healing Polymers and Composites.* Annual Review of Materials Science, 2010. **40**(1): p. 179-211.

647.Chiang, Y., T.W. Chiu, and S.W. Chang, *ImageMech: From Image to Particle Spring Network for Mechanical Characterization.* Frontiers in Materials, 2022. **8**.

648.Sun, H., et al., *A Knowledge Transfer Framework for General Alloy Materials Properties Prediction.* Materials, 2022. **15**(21): p. 7442.

649.Zhao, Y.P., et al., *Numerical Study on Heat and Drag Reduction by Transpiration in Hypersonic Flow.* International Journal of Numerical Methods for Heat &Amp Fluid Flow, 2023. **33**(8): p. 2815-2852.

650.Karagöz, S., *Integrated Multi-Scale Modelling-Simulation (MMS) and Machine Learning (ML) Based Design and Development of Novel Systems/Technologies.* 2023.

651.Parupelli, S.K. and S. Desai, *A Comprehensive Review of Additive Manufacturing (3D Printing): Processes, Applications and Future Potential.* American Journal of Applied Sciences, 2019. **16**(8): p. 244-272.

652.Adeniyi Kehinde Adeleke, N., et al., *3D Printing in Aerospace and Defense: A Review of Technological Breakthroughs and Applications.* World Journal of Advanced Research and Reviews, 2024. **21**(2): p. 1149-1160.

653.Oliveira, J.M.B., E. Stefani, and I. Costa, *Silent Skirmishes: Cyber Warfare's Erosion of Westphalian Principles in the Prelude to the 2022 Russian-Ukrainian War.* Revista Contemporânea, 2023. **3**(9): p. 13420-13440.

654.Scott, K.M., *'Out Beyond Jointery': Developing a Model for Gaming Multi-Domain Warfare.* International Conference on Cyber Warfare and Security, 2022. **17**(1): p. 599-608.

655.Komalasari, R. and C. Mustafa, *Combating International Cyber Conflict: A Healthy Just War and International Law Analysis of NATO and Indonesian Policies.* Jurnal Pertahanan Media Informasi TTG Kajian & Strategi Pertahanan Yang Mengedepankan Identity Nasionalism & Integrity, 2023. **9**(3): p. 559-570.

656.Colarik, A.M. and L.J. Janczewski, *Establishing Cyber Warfare Doctrine.* Journal of Strategic Security, 2012. **5**(1): p. 31-48.

657.Igakuboon, A.N., *An Appraisal of the Legal Framework for the Protection of Civilians in Cyber-Warfare Under International Humanitarian Law.* International Journal of Research and Scientific Innovation, 2022. **09**(07): p. 14-26.

658.Ibrahim, A., et al., *Cyber Warfare Impact to National Security - Malaysia Experiences.* Kne Social Sciences, 2019.

659.Robinson, M., et al., *An Introduction to Cyber Peacekeeping.* Journal of Network and Computer Applications, 2018. **114**: p. 70-87.

660.Choi, S.-H., et al., *Cyber-Resilience Evaluation Methods Focusing on Response Time to Cyber Infringement.* Sustainability, 2023. **15**(18): p. 13404.

661.Johnson, J., *Artificial Intelligence &Amp; Future Warfare: Implications for International Security.* Defense and Security Analysis, 2019. **35**(2): p. 147-169.

662.Goswami, S.S., et al., *Exploring the Impact of Artificial Intelligence Integration on Cybersecurity: A Comprehensive Analysis.* Journal of Industrial Intelligence, 2024. **2**(2): p. 73-93.

663.Dhir, N., et al., *Prospective Artificial Intelligence Approaches for Active Cyber Defence.* 2021.

664.Delaney, W.P., *Perspectives on Defense Systems Analysis.* 2015.

665.Smeets, M., *U.S. Cyber Strategy of Persistent Engagement &Amp; Defend Forward: Implications for the Alliance and Intelligence Collection.* Intelligence & National Security, 2020. **35**(3): p. 444-453.

666.Huskaj, G., *Offensive Cyberspace Operations for Cyber Security.* International Conference on Cyber Warfare and Security, 2023. **18**(1): p. 476-479.

667.Liu, J., et al., *Space Infrared Tracking of a Hypersonic Cruise Vehicle Using an Adaptive Scaling UKF.* Aerospace Systems, 2020. **3**(4): p. 287-296.

668.Buchanan, B., *A National Security Research Agenda for Cybersecurity and Artificial Intelligence.* 2020.

669.Healey, J., *The Implications of Persistent (And Permanent) Engagement in Cyberspace.* Journal of Cybersecurity, 2019. **5**(1).

670.Raju, N. and T. Erästö, *The Role of Space Systems in Nuclear Deterrence.* 2023.

671.Masood, M. and M.A. Baig, *Potential Impact of Lethal Autonomous Weapon Systems on Strategic Stability and Nuclear Deterrence in South Asia.* Margalla Papers, 2023. **27**(2): p. 27-43.

672.Lavers, C.R. and F. Moustakis, *Space-Based Assets, Applications, User Importance, and Maritime Vulnerabilities.* 2020.

673.Pavur, J. and I. Martinovic, *The Cyber-Asat: On the Impact of Cyber Weapons in Outer Space.* 2019.

674.Acton, J.M., *Escalation Through Entanglement: How the Vulnerability of Command-and-Control Systems Raises the Risks of an Inadvertent Nuclear War.* International Security, 2018. **43**(1): p. 56-99.

675.Ferreira-Snyman, A., *Selected Legal Challenges Relating to the Military Use of Outer Space, With Specific Reference to Article IV of the Outer Space Treaty.* Potchefstroom Electronic Law Journal/Potchefstroomse Elektroniese Regsblad, 2015. **18**(3): p. 487-529.

676.Anantatmula, V., *U.S. Initiative to Place Weapons in Space: The Catalyst for a Space-Based Arms Race With China and Russia.* Astropolitics, 2013. **11**(3): p. 132-155.
677.Unachukwu Ugochukwu Vitus, N., *Outer Space Weaponisation: International Security and Nigeria in the Outer Space.* World Journal of Advanced Research and Reviews, 2023. **20**(3): p. 314-323.
678.U.S. Department of Defense, *MDA, SDA Announce Upcoming Launch of the Hypersonic and Ballistic Tracking Space Sensor and Tranche 0 Satellites* 2024, U.S. Department of Defense.
679.New Sapace Economy, *What is the Proliferated Warfighter Space Architecture and Why is it Important?* 2024, New Sapace Economy,.
680.Sankaran, J., *Missile Defenses and Strategic Stability in Asia: Evidence From Simulations.* Journal of East Asian Studies, 2020. **20**(3): p. 485-508.
681.Korda, M. and H.M. Kristensen, *US Ballistic Missile Defenses, 2019.* Bulletin of the Atomic Scientists, 2019. **75**(6): p. 295-306.
682.Erwin, S., *L3Harris gets green light to produce 16 space-based hypersonic missile trackers* 2023, SpaceNews.
683.Peters, A.B., et al., *Materials Design for Hypersonics.* Nature Communications, 2024. **15**(1).
684.Wu, F., et al., *Research on Demonstration and Validation of Space Rapid Response Launch System Based on Data Link.* Journal of Physics Conference Series, 2022. **2252**(1): p. 012055.
685.Havlick, D.G. and T.H.D. Dao, *Revealing Vertical Geopolitics: Quantifying the Volume of Militarised Restricted Airspaces in the <scp>USA</Scp> Using <scp>GIS</Scp>.* Transactions of the Institute of British Geographers, 2023. **48**(4): p. 797-810.
686.Chen, H., H. Gao, and H. Zhang, *Integrated Navigation Approaches of Vehicle Aided by the Strapdown Celestial Angles.* International Journal of Advanced Robotic Systems, 2020. **17**(3).
687.Xuejun, D., et al., *A Disaster Warning Model for a Spacecraft Launch Based on Information Fusion and Network Inference.* International Journal of Risk Assessment and Management, 2012. **16**(1/2/3): p. 142.
688.Pan, X.Q., et al., *Influence of External Factors on Airborne Missile's Horizontal Backward Launching.* International Journal of Aerospace Engineering, 2021. **2021**: p. 1-12.
689.Cheng, X., et al., *A Liquid Launch Vehicle Safety Assessment Model Based on Semi-Quantitative Interval Belief Rule Base.* Mathematics, 2022. **10**(24): p. 4772.

690.OĞUz, Ş. and M.S. Erol, *End of the INF Treaty: Are We Entering a New Cold War Era?* Gazi Akademik Bakış, 2021. **14**(28): p. 1-20.
691.Brockmann, K. and N. Raju, *NewSpace and the Commercialization of the Space Industry: Challenges for the Missile Technology Regime.* 2022.
692.Grego, L., *Anti-Satellite and Space Weapons.* 2024: p. 557-574.
693.Kopeć, R., *Broń Antysatelitarna. U Progu Drugiego Etapu Militaryzacji Kosmosu.* Politeja, 2018. **15**(53): p. 45-71.
694.Byers, M. and A.C. Boley, *Who Owns Outer Space?* 2023.
695.Petrova, I., *Overview of the Evolution Process of the Soviet and American ASAT Weapons in the Second Half of the XX Century.* Russia and America in the 21st Century, 2023(5): p. 0.
696.Luo, X., et al., *Longitudinal Control of Hypersonic Vehicles Based on Direct Heuristic Dynamic Programming Using ANFIS.* 2014.
697.Ruhm, B.C., *Finding the Middle Ground: The U.S. Air Force, Space Weaponization, and Arms Control.* 2003.
698.Strobeyko, A., *Space for Change: The ASAT Tests in Outer Space in Light of the UN Liability Convention.* Polish Review of International and European Law, 2020. **8**(1): p. 91-102.
699.Padden, H., *Does Space Law Prevent Patterns of Antarctic Imperialism in Outer Space?* Global Policy, 2022. **13**(3): p. 346-357.
700.Berthoud, L. and S. Agass, *Identifying Space Threats for SpaceAware Resilience- A Spacecraft and Satellite Service Resilience Model.* 2022.
701.Wardana, A., G. Gunaryo, and Y.H. Yogaswara, *Development of Cyber Weapons to Improve Indonesia's Cyber Security.* Journal of Social Science, 2022. **3**(3): p. 453-459.
702.Chen, K., et al., *SINS/BDS Tightly-Coupled Integrated Navigation Algorithm for Hypersonic Vehicle.* 2021.
703.Lei, H., et al., *Variable-Time-Domain Online Neighboring Optimal Trajectory Modification for Hypersonic Interceptors.* International Journal of Aerospace Engineering, 2017. **2017**: p. 1-12.
704.Yang, B., et al., *A New Compensation Method of Dynamic Lever-Arm Effect Error for Hypersonic Vehicles.* Ieee Access, 2022. **10**: p. 59878-59888.
705.Scarabosio, A., et al., *Radiation and Scattering of EM Waves in Large Plasmas Around Objects in Hypersonic Flight.* 2021.
706.Luo, X., *Competition Between Great Powers and a Looming Strategic Arms Race in the Asia–Pacific.* China International Strategy Review, 2021. **3**(1): p. 123-136.

707.Zeng, J., et al., *Research on Anti-Saturation Attack Model of Ship Formation for Anti-Ship Missile Targets.* Applied Mechanics and Materials, 2014. **615**: p. 276-281.

708.Huang, H., et al., *Defense Strategy of Aircraft Confronted With IR Guided Missile.* Mathematical Problems in Engineering, 2017. **2017**(1).

709.Faqih, K., et al., *Smart Guided Missile Using Accelerometer and Gyroscope Based on Backpropagation Neural Network Method for Optimal Control Output Feedback.* Mechatronics Electrical Power and Vehicular Technology, 2020. **11**(2): p. 55-63.

710.Kim, K., *Engagement-Scenario-Based Decoy-Effect Simulation Against an Anti-Ship Missile Considering Radar Cross Section and Evasive Maneuvers of Naval Ships.* Journal of Ocean Engineering and Technology, 2021. **35**(3): p. 238-246.

711.Li, W., et al., *Study on General Technology of Smart Floating Munition.* Journal of Physics Conference Series, 2023. **2460**(1): p. 012053.

712.Guo, R. and B.-y. Wang, *On Structure Complexity and Behavior Complexity of Naval Warfare.* 2011.

713.Zhongyao, M., K. Wu, and Z. Liu, *Multi-Ship Cooperative Air Defense Model Based on Queuing Theory.* 2022.

714.Gao, W., Y. Sun, and Y. Wei, *Guidance Flight With the Unmanned Airship Steered by a Homing Seeker.* 2014.

715.Na'inna, A.M. and N.V. Anti-Dwanso, *Arms Race in the 21st Century: Consequences and Mitigating Measures.* Global Journal of Social Sciences, 2022. **21**(1): p. 45-76.

716.Torres, G.L. and L.D. López, *Space, Speed, and Sovereignty: Hypersonic Tensions in the Southern Hemisphere.* 2024, Center for Strategic & International Studies.

717.Berg, G.C., *Does Latin America Comprise Transnational 'Subregions'?* The World Economy, 2011. **34**(2): p. 298-312.

718.Brahmana, D.P., et al., *Indonesia's Defense Policy In Addressing Space Threats In Perspective Of Defense Management.* International Journal Of Humanities Education and Social Sciences, 2024. **3**(4).

719.Zhou, W., H. Jiang, and Z. Chen, *Trade vs. Security: Recent Developments of Global Trade Rules and China's Policy and Regulatory Responses from Defensive to Proactive.* World Trade Review, 2023. **22**(2): p. 193-211.

720.Huang, C., *China's take on national security and its implications for the evolution of international economic law.* Legal Issues of Economic Integration, 2021. **48**(2).

721.Andornino, G.B., *Economic–security nexus in the AIIB: China's quest for security through Eurasian connectivity.* Global Policy, 2019. **10**(4): p. 604-613.
722.Liu, K., *The economics of China's Holistic View of National Security: A preliminary assessment.* Economic Affairs, 2024. **44**(2): p. 218-244.
723.Beavers, L., *National space policy: International comparison of policy and the 'Gray'area.* Space Education & Strategic Applications, 2021. **2**(1).
724.Sheng, X., *Research on the differences of national security education between the united states and china.* EDU, 2023. **9**(9).
725.Chen, C.-K., *China in Latin America then and now: A systemic Constructivist analysis of China's Foreign policy.* Journal of Current Chinese Affairs, 2021. **50**(2): p. 111-136.
726.Jenne, N., *The value of security cooperation in Sino–South American relations.* Asian Education and Development Studies, 2021. **10**(3): p. 433-444.
727.Chen, J., et al., *Shape Optimization of the Cross Section for Noncircular Hypersonic Missile Forebody.* International Journal of Aerospace Engineering, 2020. **2020**(1): p. 8885494.
728.Permal, S., *China's Military Capability and Anti-access Area-denial Operations, Maritime Affairs: Journal of the National Maritime Foundation of India, Vol. 10, Issue 2.* 2014.
729.Matsushima, F.R.R., *One Belt One Road Initiative and Its Implications for Latin America.*
730.Wright, D. and C. Tracy, *Hypersonic weapons are mediocre. It's time to stop wasting money on them.* 2024, Bulletin of the Atomic Scientists.
731.Kramer, C., *U.S. Hypersonic Weapons and Alternatives.* 2023.
732.Team, C.P., *What Does China Really Spend on its Military?* 2024.
733.RE: Russia, *The Militarisation Slide: having the world's third largest military budget does not make Russia stronger.* 2023.
734.Arthur, G., *Despite security threats, India's defense budget remains static.* 2024, Defense News.

Index

Made in the USA
Monee, IL
08 November 2026

67812141R00286